NEW FRONTIERS IN POLYMER RESEARCH

NEW FRONTIERS IN POLYMER RESEARCH

ROBERT K. BREGG
EDITOR

Nova Science Publishers, Inc.
New York

For permission to use material from this book please contact us:
Telephone 631-231-7269; Fax 631-231-8175
Web Site: http://www.novapublishers.com

LIBRARY OF CONGRESS CATALOGING-IN-PUBLICATION DATA:

New frontiers in polymer research / Robert K. Bregg (editor).
p. cm.

Includes bibliographical references and index.
ISBN 1-59454-956-7
1. Polymers--Research. 2. Plastics--Additives. 3. Polymeric composites. I. Bregg, Robert K.
TP1087.N49 2006
668.9--dc22 2005037268

Published by Nova Science Publishers, Inc. ✦ New York

CONTENTS

PREFACE

Polymers are substances containing a large number of structural units joined by the same type of linkage. These substances often form into a chain-like structure. Starch, cellulose, and rubber all possess polymeric properties. Today, the polymer industry has grown to be larger than the aluminum, copper and steel industries combined. Polymers already have a range of applications that far exceeds that of any other class of material available to man. Current applications extend from adhesives, coatings, foams, and packaging materials to textile and industrial fibers, elastomers, and structural plastics. Polymers are also used for most composites, electronic devices, biomedical devices, optical devices, and precursors for many newly developed high-tech ceramics. This new book presents leading-edge research in this rapidly-changing and evolving field.

Adsorption of protein onto functional microspheres shows considerable usability on understanding complex bio-phenomena, however, the imperfection of related theories demands further study. Chapter 1 provides some important information, especially the intrinsical information involved by protein adsorption process. With sulfonated microspheres (MS) as matrix and bovine serum albumin (BSA) as model protein, the adsorption process was investigated as the functions of time, protein concentration, pH and ionic strength. The adsorption appears to be a complex process and the presence of sulfonate groups play important roles, so as to increase the adsorption rate and the amount of protein-adsorbing. Close to the isoelectric point of BSA (pI 4.7), the amount of protein adsorbed reaches a maximum. A higher or lower pH results in a significant decrease of adsorption amount. Analogously, the ionic strength can also affect the adsorption process. To help learning the probable essence, the classical theories such as Langmuir and Freundlich theories are used to investigate this process. The result shows that the applicability of the classical theories has their limitations and cannot well describe the practical process. The deviation from classical theory becomes more evident with the increase in the concentration of sulfonate groups of functional microspheres. This means the practical process is much complex than the ideal situation; more influential factors may be involved. As a development, a new model, in $\frac{C^m}{Q} = \frac{1}{KQ_m} + \frac{1}{Q_m}C^m$ form, derived from classical theory has been established and its efficiency has been proven in present study. To learn the effect of pH, the related theories are inherited and developed. A considerably interesting result is that the adsorption amounts at different pH conditions are actually correlative; also the role of electrostatic interaction can be characterized quantitatively by kinetic method. Based on salt-effect, a systematic theory that describes the role of salt in the adsorption process of protein was established. The result shows that the effect of ionic strength in protein adsorption can be determined by the relative magnitude of inhibitory force and driving force. At pH 2.2, the driving force for protein adsorption is larger than the inhibitory force, and while the inhibitory force is beyond the driving force at pH 4.3. The change in the driving force may be responsible for the change of protein-absorbed amount.

Polyurethanes are becoming increasingly important as engineering materials because they have excellent abrasion resistance and the properties of both elastomers and plastics. The growth of science and technology of polyurethanes leads to the development of new materials with more desirable properties. Such kinds of materials include metal-containing polyurethanes, poly(urethane-urea)s and poly(urethane-ether)s with isocyanate structural units combining the properties of enhanced thermal stability, fire retardancy, flexibility and solubility. Ionic diols containing metal salts are used as important starting materials for the synthesis of metal-containing polyurethanes in which the metal is firmly incorporated in the backbone of the polymer chain. Incorporation of metal into polyurethanes has led to wide application as aqueous thickeners, impregnates, textile sizers, adhesives, additives, resins and catalysts. The aim of chapter 2 is to present an overview of the various methods of preparation and properties of metal-containing polyurethanes and their copolymers.

Fiberboard has been used widely in the fields of construction, insulation, soundproofing, decoration, and packaging. The most popular adhesives for the fiberboard industry are petroleum-based and starch-based adhesives. The petroleum-based adhesives impact the environment with toxicity issues and problems during manufacturing, transportation, and utilization. On the other hand, the starch-based adhesives are environmentally friendly but usually have low adhesion strength and water resistance. Therefore, the development of an environmentally friendly adhesive with high performance is greatly needed. Modified soy protein has great potential as an adhesive because of its natural structure. Chapter 3 summarizes the research results on soy protein adhesives for fiberboard packaging applications. Soy protein adhesives can be obtained by modifying soy material with at least 50% protein. Variables affecting soy protein adhesion performance include modifier and its concentration, soy protein concentration, pressing conditions, and assembly time. Under optimum processing conditions, soy protein-based adhesives gave appropriate working viscosity, adhesion strength, and water resistance as a fiberboard packaging sealant. After modification, soy protein-based adhesives can provide dry shear strength of about 1.9 MPa and wet shear strength of about 0.4 MPa for solid fiberboard. For corrugated fiberboard, the dry shear strength was about 1.0 MPa and wet shear strength was about 0.2 MPa. Both the dry and wet solid and/or corrugated fiberboard exhibited 100% cohesive failure within fiberboard. The percentage of water-soluble-mass of the modified soy protein adhesive had a value as low as 1.7%, which is low compared to the requirement. Both the solid and corrugated fiberboard packaging boxes sealed with the soy protein adhesives could be easily transported with no damage or alteration in the box after subjecting the boxes to the standard tests of exposure to rain and impact tests. The adhesives developed are geared toward packaging applications, however, they have the potential to be used for many other applications.

Study of the adhesion phenomena at a molecular scale is fundamental to obtain a better control of the properties and the performances of the materials for numerous modern applications in various domains (as biotechnology, microelectronics...). In chapter4, the Atomic Force Microscope (AFM) was used to perform force measurements in quasi-static mode to investigate surface properties of model systems at the nanoscale. In particular, in the experimental part of the paper, the authors describe the systematic calibration procedure of AFM measurements that was performed to produce quantitative data. The influence of capillary condensation of atmospheric water vapor at the tip-sample interface on the measured forces is also discussed. The main part of the paper focused on two types of model systems. The first one was composed of a rigid substrate (silicon plates), which was chemically modified by molecular self-assembling (SAMs) to display different surface properties (methyl and amine functional groups). The second system consists of model polymer networks (cross-linked polydimethylsiloxane or PDMS) of variable mechanical properties, whose surfaces were chemically modified with the same groups as before with silicon substrates (methyl and amine functional groups). Quantitative assessment values were obtained on the model polymer networks from the AFM force curve measurements in considering that viscoelasticity and adhesion can influence the experimental data. The comparison of the force curves obtained from the two model systems shows that the viscoelastic or mechanical contribution dominates in the adhesion on polymer substrates. Finally, a relationship, which expresses the separation energy at a local scale as a function of the energy dissipated within the contact zone, on one hand and the surface properties of the polymer on the other hand, is proposed.

Poly(3-hydroxybutyrate) (PHB) is a naturally occurring biodegradable polyester produced by many bacteria. In contrast to traditional polymers, the main remarkable characteristics of PHB lie in its biodegradability and biocompatibility. PHB also shows many special properties, such as piezoelectricity, optical activity and so on. However, PHB is a highly crystalline polymer due to its exceptional stereochemical regularity. This gives rise to its low impact strength, narrow processing window and poor hydrophilicity. Poly(ethylene glycol) (PEG) is a synthetic soluble polymer known to be highly hydrophilic, biocompatible and flexible, and frequently used for modifications of PHB. In the second part of chapter 5, four methods were discussed for the preparations of PHB/PEG derivatives. They include: physical blending, graft copolymerization, multiblock copolymerization and interpenetrating network technique. The characterizations of their properties, such as thermal properties, mechanical properties and hydrophilicity were also reviewed. The degradation mechanism and factors influencing the degradation behavior were described in the final part. PHB is known to be degraded via three kinds of mechanisms: enzymatic degradation, hydrolytic degradation and thermal degradation. Many factors have effects on the process of PHB's degradation and these factors can de divided into two catalogues: exterior factors and material characteristics. The common exterior factors comprise pH, temperature, environmental conditions. Crystallinity, hydrophilicity, tacticity are main material characteristics. In the final section, the degradation behavior of PHB/PEG blend and PHB/PEG multiblock copolymer were discussed in detail. The degradation of PHB/PEG derivatives proceeded at a higher rate than that of homopolymer PHB. This acceleration is caused by two reasons: the improvement of hydrophilicity within the matrix and the decrease of the crystallinity in the PHB phase.

With a focus directed toward noncovalent interactions, the supramolecular approach is allowing scientists to create quite remarkable molecules and supermolecules. Enhanced knowledge of molecular recognition is being used in an ever increasingly sophisticated manner to produce different classes of compounds, such as crown ethers, cryptands, spherands, cyclophanes, cyclodextrins, cleft-like structures, and other cage-like compounds mainly for complexation of inorganic and organic ions and organic molecules as described in chapter 6. The design of improved ion-sensing materials has attracted intense attention from chemists in recent years. Base-catalyzed condensation of *tert*-butylphenol and formaldehyde leads to cyclic oligomers called "calixarenes" or "calix[n]arenes", with n being the number of aromatic rings joined by methylene bridges. These molecules can undergo chemical modifications either by reaction at the hydroxy groups (lower rim) or by replacement of the *tert*-butyl groups (upper rim) by other substituents. Modified calixarenes have been found to be extremely useful in host-guest chemistry due to their tendency to form complexes with cations, anions, and neutral molecules in adsorption, filtration and separation processes, as ionophores for ISFETs (Ion Selective Field Effect Transistors), as catalysts in two-phase reactions, as ion carriers in membrane technology, in nanosize self-assembling systems, which are prospective candidates as materials for microelectronics, mesoscopic devices, *etc.* Inspired by the increasing use of calixarenes, interest in the design of polymer -appended calixarenes increased as well to make them suitable as new recognition agents with higher effiency. In recent years, various attempts have been undertaken to incorporate calixarenes into different polymers, including self-assembled systems. There are two general strategies for preparing covalently linked, calixarene-based polymers. One is attachment of a calixarene compounds by reaction with a suitably functionalized polymer or oligomer. The other is preparation of a calixarene monomer and its polymerization or copolycondensation with typical vinyl monomers. This survey is intended to provide an overview of calixarene derivatives linked with polymers by conventional covalent bonds.

In chapter 7, a set of new organo-inorganic composites stabilized by hydrogen bonds have been obtained by the reaction of silicic acid and tetraethyl orthosilicate with organic polymers: poly(1-vinylpyrrolidone-2), poly(acrylamide), poly(ethylene glycol) and poly(4-vinylpyridine). Considerable influence of the structure of primary silica particles on the composition and morphology of the resulting products have been found: stoichiometry from 1: 2 to 2: 1 and morphology from friable substance of 65±16 nm particles to spherical particles 2.9±1.1 μm in diameter can be obtained in aqueous medium with the same polymer [poly(1-vinylimidazole)]. The use of poly(vinyl butyral) and oxirane-containing polymers as a matrixes in hydrolysis of tetraethyl orthosilicate gives rise to graft copolymer with side polysiloxane chains. Membranes, highly functionalized surfaces for DNA and immuno-analysis, sorbents of heavy metals have been designed on the bases of new nanocomposites.

The parameters describing Dynamic Light Scattering (intensity (I), correlation radius (ζ) and diffusion coefficient (D)) are reported for some (polystyrene + solvent) systems. Measurements were performed for polystyrenes of molecular weight (Mw) 7.5k to 400k amu. dissolved in protiated or deuterated methylcyclohexane, cyclohexane and acetone. DLS measurements were performed under isothermal and/or isobaric conditions for semidilute polymer solutions at high enough concentrations to ensure precipitation (liquid-liquid demixing). The results describe measurements over broad ranges of temperature, pressure, solution concentration, and solvent isotope label (H/D). Chapter 8 concentrates on description of the diffusion of polymer chains in three different small molecule solvents. Two, cyclohexane and methylcyclohexane, are classified as θ solvents. The third, acetone, is a poor solvent. The investigations of the demixing of polystyrene-solvent systems beginning in the homogenous region show that DLS correlograms, initially monomodal, split during temperature and/or pressure quenching, and became much more complicated as precipitation is approached. There is an appreciable concentration dependence on the shape of the correlogram. As concentration increases a bimodal pattern develops. Changes in intensity and correlation radius of the diffusive mode as functions of temperature and pressure are discussed.

In: New Frontiers in Polymer Research
Editor: Robert K. Bregg, pp. 1-30
ISBN: 1-59454-956-7

Chapter 1

FUNCTIONAL POLYMER MICROSPHERES: PROTEIN ADSORPTION AND THEORY

Songjun Li[1*] and Bailing Liu[2]
[1] College of Chemistry, Central China Normal University, Wuhan, 430079, People's R. China
[2] Chengdu Institute of Organic Chemistry, Chinese Academy of Sciences, Chengdu 610041, People's R. China

ABSTRACT

Adsorption of protein onto functional microspheres shows considerable usability on understanding complex bio-phenomena, however, the imperfection of related theories demands further study. The present research provided some important information, especially the intrinsical information involved by protein adsorption process. With sulfonated microspheres (MS) as matrix and bovine serum albumin (BSA) as model protein, the adsorption process was investigated as the functions of time, protein concentration, pH and ionic strength. The adsorption appears to be a complex process and the presence of sulfonate groups play important roles, so as to increase the adsorption rate and the amount of protein-adsorbing. Close to the isoelectric point of BSA (pI 4.7), the amount of protein adsorbed reaches a maximum. A higher or lower pH results in a significant decrease of adsorption amount. Analogously, the ionic strength can also affect the adsorption process. To help learning the probable essence, the classical theories such as Langmuir and Freundlich theories are used to investigate this process. The result shows that the applicability of the classical theories has their limitations and cannot well describe the practical process. The deviation from classical theory becomes more evident with the increase in the concentration of sulfonate groups of functional microspheres. This means the practical process is much complex than the ideal situation; more influential factors may be involved. As a development, a new model, in $\frac{C^m}{Q}=\frac{1}{KQ_m}+\frac{1}{Q_m}C^m$ form, derived from classical theory has been established and its efficiency has been proven in present study. To learn the effect of pH, the related

* E-mail address: Lsjchem@hotmail.com; Tel: +86-28-85260436

theories are inherited and developed. A considerably interesting result is that the adsorption amounts at different pH conditions are actually correlative; also the role of electrostatic interaction can be characterized quantitatively by kinetic method. Based on salt-effect, a systematic theory that describes the role of salt in the adsorption process of protein was established. The result shows that the effect of ionic strength in protein adsorption can be determined by the relative magnitude of inhibitory force and driving force. At pH 2.2, the driving force for protein adsorption is larger than the inhibitory force, and while the inhibitory force is beyond the driving force at pH 4.3. The change in the driving force may be responsible for the change of protein-absorbed amount.

Keywords: protein adsorption; sulfonated microspheres; adsorption theory; model establishment.

1. INTRODUCTION

There is considerable interest in adsorption of proteins (especially blood proteins) onto microspheres because of its importance in a wide range of biomedical applications, such as artificial tissues and organs [1], drug delivery systems [2], biosensors [3], solid-phase immunoassays [4], immunomagnetic cell separation [5] and immobilized enzymes or catalyst [6]. Such applications need absorbing or coupling proteins onto the surface of microspheres. The earliest example of coupling proteins to microspheres can be derived as early as 1970s, applying microsphere-coating antibodies to capture antigens [7]. It has been well known that the interactions between proteins and microspheres or proteins themselves play important roles in the absorption process. Main interaction involved in such process can be classified as electrostatic, hydrophobic and hydrogen-bonding etc. [8]. Many researchers have reported about absorbing proteins onto functional microspheres and shown the specific roles of various interactions in the absorption process. However, the main viewpoints, so far, on such process are inconsistent with one another. Some researchers have concluded that hydrophobic interaction is the main aspect of protein adsorption, and an increase in the hydrophilicity will result in a lower level of adsorption [9]. Others have claimed that hydrogen-bonding is more important than hydrophobic interaction [10,11]. More recently, Haynes et al [12,13] indicated that the electrostatic contribution is the key point among interactions. Indeed, the specific behaviors of various interactions, especially their intrinsic mechanisms, are not yet clear to us; more works remain to be presented.

Adsorption of proteins onto microspheres is actually a considerably complex process and often involves many dynamic steps, from the initial attachment of proteins onto the surface of microspheres to the final equilibrium reached; also in many examples [14,15], absorption of protein molecules to a solid-surface is not in mono-molecular form but in an associated form. Moreover, the conformation of proteins probably occurs to a certain degree of deformation or structural change due to a large area of contact [16]. Therefore, it would be far from enough to tell the mechanism simply from a qualitative or quantitative relationship between equilibrium absorption-amount of proteins and hydrophilic-hydrophobic interaction, and the more influential factors should be taken into account. It would be especially necessary to investigate the adsorption process with dynamic view. Also the probability of intermolecular association of proteins in the adsorption process, as well as protein molecules interaction, should be considered.

Recently, much interest has been shown in sulfonated microspheres [17-19], as they can present an environment of electric charge and hydrogen bond. Based on such process, it would be helpful to understand certain bio-phenomena with analogous molecular interactions. Yoon et al [20,21] have investigated the interaction involved by adsorbing bovine serum albumin (BSA) onto sulfonated polystyrene microspheres. A considerably interesting result is that a transition point, the exact balance point between electrostatic interaction and hydrophobic interaction, can be observed. Below the transition point, hydrophobic interaction would govern the adsorption, while above the transition point, electrostatic interaction became the dominator during the process. Unfortunately, the further information about the intrinsical behavior has not been shown in their works. Although we have also noted that many researchers conducted similar works and attempted to give logical interpretation on the interaction mechanism, scanty investigations, so far, have been performed in the full range, from not modified to highly modified surfaces. Specially, for the probable intermolecular association of proteins, there is almost not any description available in literature; a quantitative model that can describe such association behavior in the adsorption process is quite deficient.

In present study, BSA (bovine serum albumin) was chosen as the model protein and sulfonated PMMA [poly (methyl methacrylate)] microspheres as the matrix to study the adsorption process. Because of its highly purified form, easily soluble in water and homogeneous, BSA is convenient to use in a series of repeated experiments for comparison of the adsorption behaviors of BSA under different conditions. The adsorption isotherms and the kinetics, as well as the effects of protein concentration, pH conditions and ionic strength on the process have been studied. To help learning the intrinsic behavior, the related theories have established and used to investigate this process. The purpose is to show some important information, especially the intrinsic mechanism, involved by the adsorption process.

2. EXPERIMENTAL

2.1. Materials

Methyl methacrylate (MMA), coming from Bodi Chemicals Co. (China), was purified by pressure-reduced distillation prior to using (0.7-0.8 atm, 40-50℃). Polyvinyl pyrrolidone stabilizer (PVP) (MW, 10000-70000) was obtained from Silong Chemical Co. (China). Benzoyl peroxide (BPO) was purchased from the Chemical Factory of Hubei University (China). 2-acrylamido-2-methylpropane sulfoacid (AMPS) and BSA (catalog number A-7906, pI 4.7) were purchased from Sigma Chemical Co. (USA).

2.2. Preparation of Functional Microspheres

Sulfonated PMMA microspheres (MS, 3.4 ± 0.2 μm in average diameter) were prepared by dispersion polymerization [22]. In methanol medium (80g), basing on 8 gram of MMA, 2 % (w/w) of BPO as initiator and 1% (w/w) of PVP as stabilizer were used in the polymerization. At one time these reagents were put into a 100 ml of three-necked flask fitted with mechanical stirrer, nitrogen inlet and reflex condenser. After deaerating with nitrogen, the polymerization was performed at 70℃ for 10 hours with 250 rpm of agitation speed, and then a certain amount of AMPS (2, 4 or 8 %-w/w) as functional monomer was added dropwise into the system. After the addition, the mixture was kept in the same condition for another 10 hours. The resulted particles were separated by centrifugation and profusely washed with double-distilled water. To clearly elucidate, the particles obtained in the presence of specific amount of AMPS (0, 2, 4 and 8 %-w/w) are named respectively as MS-0, MS-2, MS-4 and MS-8. Conductometric titration (titrant, 0.01M NaOH solution) shows the concentration of sulfonate groups on the surface of these microspheres are respectively 4.258×10^{-2} $mmol\cdot g^{-1}$ MS (MS-2), 5.362×10^{-2} $mmol\cdot g^{-1}$ MS (MS-4) and 6.315×10^{-2} $mmol\cdot g^{-1}$ MS (MS-8).

2.3. Evaluation of Adsorption Isotherms

According to the description of literatures [20,21,23], the adsorption isotherms have been determined carefully. The typical batch experiments were carried out in water bath (37℃). BSA was dissolved in acetic buffer (pH 4.3 except for additional illustration). The solid content of MS was about 40mg/ml in each operation (ionic strength, 0.01 except for additional illustration). The sample solutions (in triplicate) were stirred for desired time and were then centrifuged to remove particles. The residual protein concentrations were determined by spectrophotometric method with U-2010 spectrophotometer (Hitachi, Japan). For each experiment, control was set up at certain BSA concentration (without MS), which allows for correction to be made for protein loss in this system. The adsorbed amount per gram MS was determined by the mass balance of the protein, and the average value of triplicate samples was presented.

3. THEORIES

3.1. Langmuir Theory and Origin of Freundlich Model

Models are the basis of developing related theories. Historically, too many models or profiles have been proposed, such as Langmuir adsorption-theory [24], Freundlich model [25], Temkin model [26] and multi-layers adsorption-theory of BET (Brunauer-Emmett-Teller) model [27] etc. Indeed, among these models, Langmuir model, in manner, holds the dominative place, similar to the state equation of ideal gas in gas research, and is generally accepted as the fundamentality, while other models, to more or less extent, are the derivatives of Langmuir model. It has been well known that Langmuir model, an ideal model, describes a dynamic adsorption process with the presences of adsorption of molecules in bulk-phase and desorption of molecules in the solid surface. Before equilibrium-reached, the rate of adsorption is larger than the rate of desorption; when the rate of adsorption becomes equal to the rate of desorption, the equilibrium is obtained. Although Langmuir model has been proven its efficiency in many adsorption processes, an obvious drawback of this model is that the interaction between protein molecules or protein and surface of microspheres cannot be summarized. Abandoning such drawback from Langmuir model, Freundlich model has validated on many adsorption processes. However, the origin of Freundlich was only empirical, presenting no distinct source or deriving mathematically from outline or trendline of data-points. Temkin model was originally used to study the behavior of absorption of ammonia onto the surface of Fe-based catalyst, and has been confirmed as only model applicable to chemical adsorption. BET multi-layers theory can well summarize physical adsorption in multi-layers form, especially in infinite multi-layers form. However, it would be difficult to form infinite multi-layers adsorption under a specific bio-environment. In present study, BSA is absorbed onto electricity-charged microspheres, which involves almost no chemical adsorption or infinite multi-layers adsorption. Hence Temkin model and BET theory would not be discussed detailedly here.

As is apparent, based on the present theories, it would be difficult to discuss such an adsorption process with the possible intermolecular association of proteins. So deriving or developing the classical theory would be highly necessary. To deepen the background of classical Langmuir theory, a physical profile for BSA adsorption is shown in Fig.1. As afore-elucidated, adsorption is a dynamic process with the presence of adsorption of BSA molecules in bulk-phase and desorption of BSA molecules in the surface of MS. According to Langmuir theory, the rates of BSA adsorption and desorption would be [24]:

$$v_a = k_a C_{\text{surf}}(1-\theta)C$$
$$v_d = k_d C_{\text{surf}}\theta \tag{1}$$

Here v_a and v_d are the rates of BSA adsorption and desorption respectively, k_a and k_d, the rate constants of adsorption and desorption, θ, the coverage degree of BSA over the MS surface, C, the protein concentration in the bulk-phase, and C_{surf}, the concentration of surface of microspheres. Apparently, when reaching equilibrium, the rate of adsorption equals to the rate of desorption:

$$k_a C_{\text{surf}}(1-\theta)C = k_d C_{\text{surf}}\theta \tag{2a}$$

or

$$\theta = \frac{KC}{1+KC} \tag{2b}$$

Here K is a thermodynamic constant, obtained from $K = \frac{k_a}{k_d}$. Now, substituting $\theta = \frac{Q}{Q_m}$ into Eq.(2b), we can seek Eq.(2b) into following form:

$$\frac{Q}{Q_m} = \frac{KC}{1+KC} \tag{3}$$

Rearrangement of above relationship can show:

$$\frac{C}{Q} = \frac{1}{KQ_m} + \frac{1}{Q_m}C \tag{4}$$

Here Q and Q_m are the actual and maximal absorbed-amount of protein. As in Eq.(4), if the adsorption of protein onto MS is a Langmuir-typed adsorption, the plot of C/Q versus C would be normally expected to be a straight-line. From the intercept and slope, the thermodynamic constant *K*, characterizing the affinity of proteins to the surface of MS can be obtained. Actually, the practical adsorption process is much complex than the ideal situation; the interaction, more or less, can be involved in the adsorption process. These may be as the results of the interaction between protein molecules or protein and surface of MS. Also as a fact, with the adsorption going on, the more BSA molecules are adsorbed, and the earlier adsorbed molecules will hinder the later adsorption. Therefore, the specific examples of Langmuir-typed bio-macromolecules adsorption, such as BSA adsorption are considerably limited [28].

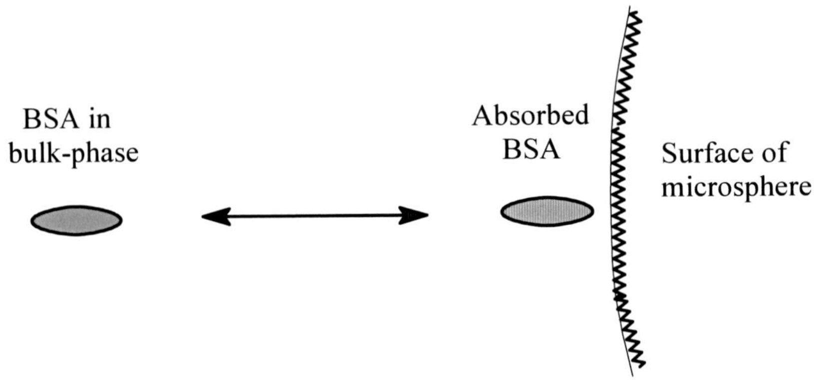

Figure1. Physical profile of Langmuir adsorption

As one of derivatives of Langmuir model, Freundlich model can overcome such drawback, though empirical in origin. Historically, much endeavor has been made to reveal the probable essence of Freundlich model. Many researches [29-31] have indicated that the activation energy for adsorption cannot be thought simply as fixed-value in the adsorption process but related to the change of coverage degree of proteins; also the activation energy for desorption is virtually associated with the change of coverage degree of proteins. This is mainly due to the presence of interactions between protein molecules or protein and surface of MS. Fig.2 presents a physical profile of Freundlich adsorption applied to BSA adsorption. As is apparent, such profile considers more than the Langmuir profile, and specially, the interaction can be depicted in the adsorption process. Expectedly, the applicable scope of Freundlich model would be much broader than that of Langmuir model, since the probable interaction has been taken into account. Now, it would be easy to know the rates of protein adsorption and desorption [32]:

$$\begin{aligned} v_a &= k_a(\theta)C_{\text{surf}}(1-\theta)C \\ &= A\exp[-\frac{E_a(\theta)}{RT}]C_{\text{surf}}(1-\theta)C \\ v_d &= k_d(\theta)C_{\text{surf}}\theta \\ &= D\exp[-\frac{E_d(\theta)}{RT}]C_{\text{surf}}\theta \end{aligned} \tag{5}$$

Here A and D are the pre-exponential factors of adsorption and desorption in Ahrrenius equation. Analogously, when reaching equilibrium, the rate of protein adsorption equals the rate of desorption:

$$A\exp[-\frac{E_a(\theta)}{RT}]C_{\text{surf}}(1-\theta)C = D\exp[-\frac{E_d(\theta)}{RT}]C_{\text{surf}}\theta \tag{6a}$$

or

$$\begin{aligned} \frac{\theta}{1-\theta} &= \frac{A}{D}\exp[-\frac{E_a(\theta)-E_d(\theta)}{RT}]C \\ &= L\exp[\frac{q_m(\theta)}{RT}]C \end{aligned} \tag{6b}$$

Here L is a constant, obtained from $L=\frac{A}{D}$, $q_m(\theta)$, the adsorption heat of BSA adsorption, obtained from $q_m(\theta)=E_d(\theta)-E_a(\theta)$. Many investigations have revealed that the change in adsorption heat of practical process with coverage degree can be often classified as logarithm- or linearity-typed [32]:

$$q_m(\theta) = q_m^0 - \alpha\ln\theta \tag{7a}$$

$$q_m(\theta) = q_m^0 - \beta\theta \tag{7b}$$

As a fact, the linearity-typed change of adsorption heat leads ultimately to the discovery of mathematic relationship in Temkin model, while the logarithm-typed one gestates partly the essence of Freundlich model. As afore-elucidated, the details of Temkin model would not be discussed here, since the present study involves almost no chemical process. Now substituting Eq.(7a) into (6b), one can obtain:

$$\theta^{\alpha/RT}(\frac{\theta}{1-\theta}) = L\exp(\frac{q_m^0}{RT})C \tag{8}$$

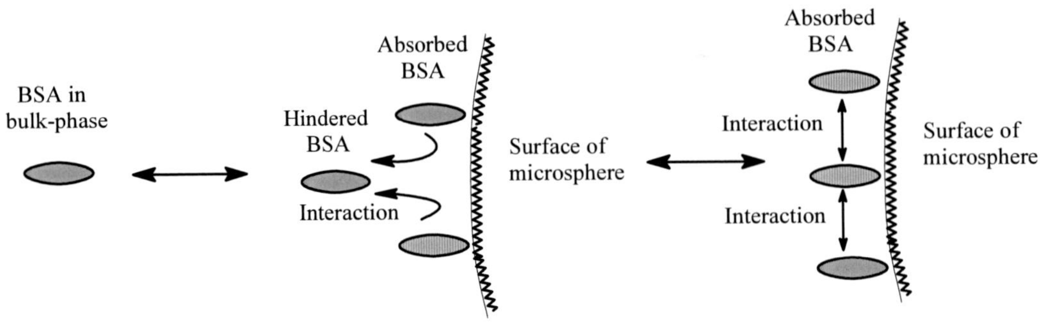

Figure 2. Physical profile of Freundlich adsorption

On the left of Eq.(8), the change of θ in exponential form is much larger than that in fraction form. Thus, the effect of fraction moiety in Eq.(8) is considerably limited, especially in a middle coverage degree. As a reasonable approximation, Eq.(8) can be rewritten into a form as:

$$\begin{aligned}\theta &= [L\exp(\frac{q_m^0}{RT})]^{RT/\alpha} C^{RT/\alpha} \\ &= KC^{1/n}\end{aligned} \tag{9a}$$

or

$$\frac{Q}{Q_m} = KC^{1/n} \tag{9b}$$

Here K and n are constants under a given condition, obtained respectively from:

$$K = L\exp(\frac{q_m^0}{RT})$$

and

$$n = \frac{\alpha}{RT};$$

To conveniently use, Eq.(9) is generally sought to such form in literature:

$$\ln Q = \left(\frac{1}{n}\right)\ln C + \ln KQ_m \tag{10}$$

As in Eq.(10), if the adsorption of protein onto MS is a Freundlich-typed adsorption, the diagram (lnQ versus lnC) would be a straight-line. Much example reported in literature has shown that the usability of Eq.(10) is actually much higher than that of Eq.(4). This is mainly due to the presence of intermolecular interaction.

3.2. Specific Derivation of Classical Theory

It has been well known that many proteins in solution are in the association form due to intermolecular interaction. Unfortunately, almost no model has involved such consideration in adsorption; more works remain to be presented. Here, we will take carefully consideration on such process. As shown in Fig.3, suppose the presence of BSA in bulk-phase is in a probable association state with the association constant K_A. Like both processes described above, such profile can involve the interaction of protein molecules or not. Thus, the final state of absorbed BSA can be classified correspondingly as two different categories, i.e. no (or weak) and strong interaction systems. Now, we can reduce their quantitative relationship when basing on both systems.

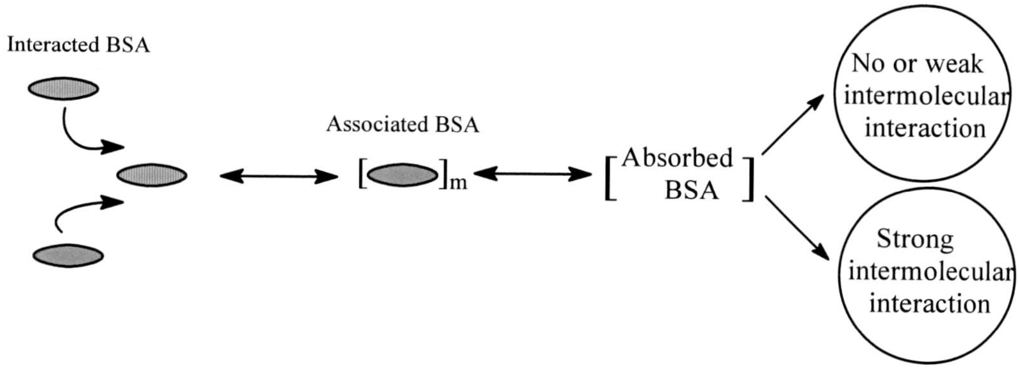

Figure 3. Physical profile for BSA adsorption in association form

3.2.1. No (or Weak) Interaction System

As shown in Fig.3, the presence of protein in bulk-phase is in a probable association state with the association constant K_A. Here suppose the average association degree of protein molecules is *m*, (i.e. the presence of BSA molecules is in *m*-aggregator form) and the concentration of BSA is *C*, the concentration of associated BSA then would be:

$$C_{[\mathrm{BSA}]_m} = K_A C^m \tag{11}$$

This available is in accordance to the thermodynamic relationship:

$$\frac{C_{[\mathrm{BSA}]_m}}{C^m} = K_A$$

Here, it is necessary to point out that if molar concentration of BSA is used, the association would affect the actual concentration of BSA. To eliminate such effect, it is necessary to employ the mass concentration in origin, from the determination of standard curves to the evaluation of isotherms. Now, the rates of BSA adsorption and desorption in present system would be:

$$\begin{aligned} v_a &= k_a C_{\text{surf}} (1-\theta) C_{[\text{BSA}]_\text{m}} \\ &= K_A k_a C_{\text{surf}} (1-\theta) C^m \end{aligned} \tag{12}$$

$$v_d = k_d C_{\text{surf}} \theta$$

Basing on the above discussion, one would easily obtain following relationship with the similar processing:

$$\frac{Q}{Q_m} = \frac{KC^m}{1+KC^m} \tag{13}$$

Here K is a constant, obtained from $K = K_A k_a / k_d$, and other parameters are analogous to those of Langmuir theory. Some works [20,21,23] have indicated that Eq.(13) can be more efficient than Langmuir model on describing the adsorption of BSA over carboxylated and sulfonated polystyrene microspheres. Although these works discussed also the probable arrangement of absorbed molecules over the surface of microspheres, the intrinsic reason has not been presented. In present research, this will be discussed (in the results and discussion). Now, for the purpose to convenient application, one can rearrange Eq.(13) as following form:

$$\frac{C^m}{Q} = \frac{1}{KQ_m} + \frac{1}{Q_m} C^m \tag{14}$$

As in Eq.(14), if the adsorption of protein onto MS is in association form, the diagram (C^m/Q versus C^m) would be a straight-line. From the intercept and slope, the constant *K*, characterizing the affinity of proteins to the surface of MS can be obtained.

3.2.2. Interaction System

Like the discussion in section 3.2, interaction can exist between protein molecules or absorbed protein and the surface of MS in the process of adsorption. Thus the development of such model, to some extent, is similar to the origin of Freundlich model. Now, connecting Eq.(11) with Eq.(5), one can obtain the rates of protein adsorption and desorption in such system:

$$\begin{aligned} v_a &= K_A k_a(\theta) C_{\text{surf}} (1-\theta) C^m \\ &= K_A A \exp[-\frac{E_a(\theta)}{RT}] C_{\text{surf}} (1-\theta) C^m \end{aligned}$$

$$\begin{aligned} v_d &= k_d(\theta) C_{\text{surf}} \theta \\ &= D \exp[-\frac{E_d(\theta)}{RT}] C_{\text{surf}} \theta \end{aligned} \tag{15}$$

By the similar treatment as in Eq.(6) and Eq.(7), following relationship can be obtained:

$$\theta^{\alpha/RT} (\frac{\theta}{1-\theta}) = L \exp(\frac{q_m^0}{RT}) C^m \tag{16}$$

Here L is a constant, obtained from $L = \frac{K_A A}{D}$, and other parameters are similar to those in Freundlich model. Analogously, the effect of fraction moiety in Eq.(16) can be neglected. As the reasonable approximation, Eq.(16) can be sought into such form:

$$\begin{aligned} \theta &= [L \exp(\frac{q_m^0}{RT})]^{RT/\alpha} C^{mRT/\alpha} \\ &= KC^{\frac{m}{n}} \end{aligned} \tag{17a}$$

or

$$\frac{Q}{Q_m} = KC^{\frac{m}{n}} \tag{17b}$$

Here K is also constant, obtained from:

$$K = L \exp(\frac{q_m^0}{RT})$$

The physical meaning of n is the same as in Freundlich model. Now, linearization of Eq.(17b) can show:

$$\ln Q = \left(\frac{m}{n}\right) \ln C + \ln KQ_m \tag{18}$$

Interestingly, Eq.(18), in form, is as the same as Eq.(10), though different in some constant items. This indicates that Freundlich model can describe virtually the adsorption in a protein-associated form. Now, it is not difficult for us to learn the probable reason as the fittings reported from Freundlich relationship are always better than those from Langmuir fitting, whether what environment the specific adsorption has been conducted.

4. Results and Discussion

4.1. Time Profile of BSA Adsorption

Fig.4 presents the time profile of BSA adsorption onto various microspheres. As is apparent, the absorption amount increases with the progress of time. There is an initial period of rapid adsorption and then a slower approach to a limiting value; the adsorption amount of protein at about 120 mins shows no visible difference from that at 240 mins. This indicates that the adsorption can actually reach the equilibrium within 2 hours. Thus in present study, the isotherms were determined with 2 hours' time. As can also be seen from the profile, the increase in protein-absorbed amount is evidenced by the increase in the amount of sulfonate groups of MS. This indicates that electrostatic interaction plays an important role in aiding increasing the adsorption amount of protein. The similar results are also shown in Yoon et al's works [20,21]. As they have explained, the increase in protein-absorbed amount with the increase of sulfonate groups is a characteristic of electrostatic interaction.

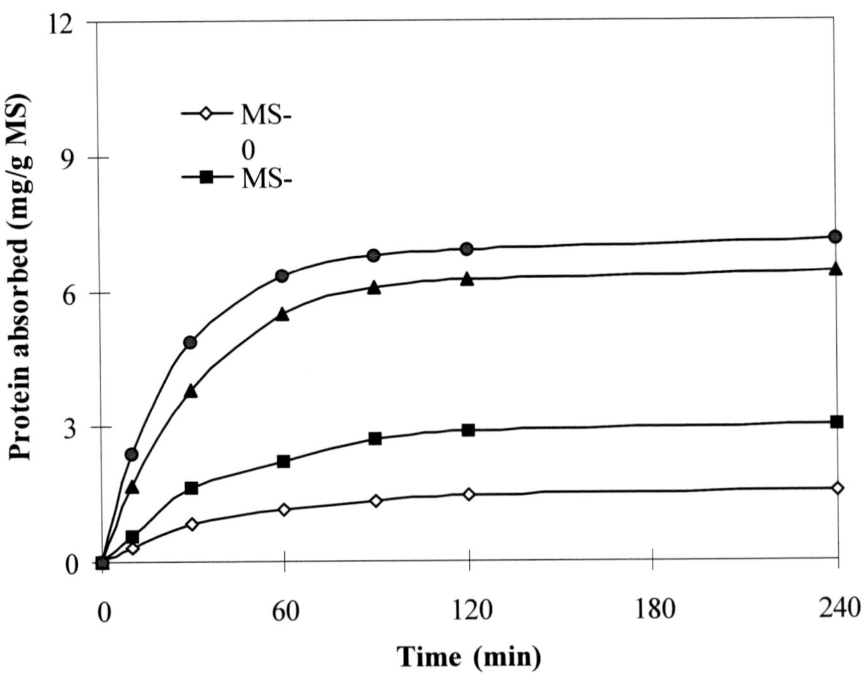

Fig.4 Time profile of BSA adsorption

As afore-mentioned, the time profile shows initially a rapid increase in the protein-absorbed amount and then a slower approach to a limiting value. This indicates that this process can be described as apparent first-order kinetic process [33]:

$$-\frac{dC}{dt} = kC \tag{19}$$

Integration of Eq.(19) can show:

$$C = C_0 \exp(-kt) \tag{20}$$

Here C_0 and C are the original and actual concentrations of BSA, and k, the apparent rate constant of protein adsorption. Clearly, the amount of protein-absorbed with time would be:

$$Q = Q_m[1 - \exp(-kt)] \tag{21}$$

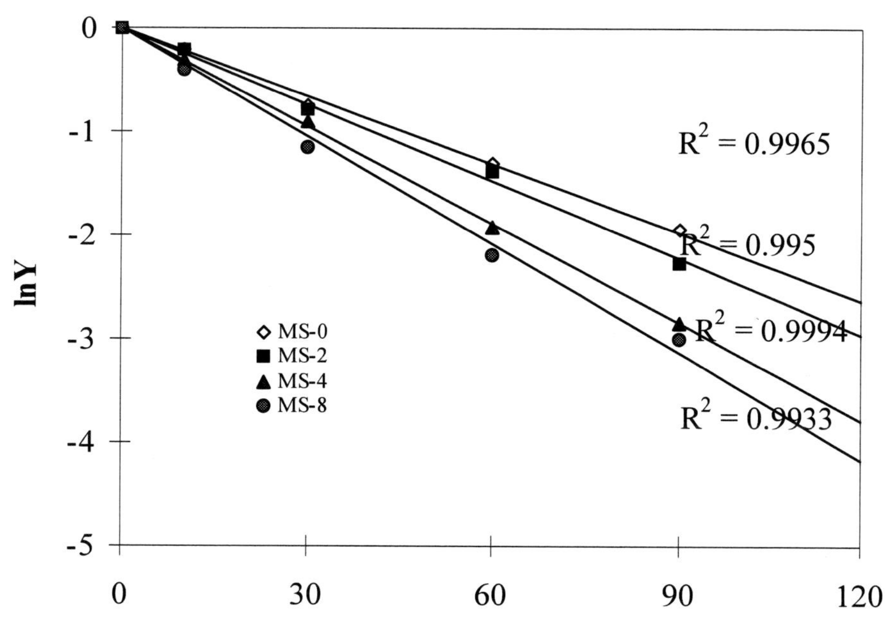

Fig.5 Kinetic curve of BSA adsorption

To use conveniently, one can rearrange Eq.(21) into such form:

$$\begin{cases} \ln Y = -kt \\ Y = \dfrac{Q_m - Q}{Q_m} \end{cases} \tag{22}$$

Here Q_m and Q are the maximal and actual adsorption amount of protein. According to Eq.(22), the diagram (lnY versus t) would be a straight-line. As shown in Fig.5, the correlativity of such lines is actually quite good. From the slopes of these straight-lines, the apparent rate constants, characterizing the speedy extent of protein adsorption, can be obtained and whose results are 2.19, 2.47, 3.16 and 3.47 $\times 10^{-2}$ min^{-1}, responding respectively to the adsorption of BSA onto MS-0, MS-2, MS-4 and MS-8. Clearly, this can be related to the roles of electrostatic interaction in the adsorption process.

4.2. Adsorption Isotherms of BSA

Fig.6 presents the isotherms of BSA adsorption onto various microspheres. The protein-absorbed amount increases with increasing protein concentration in equilibrium. Basing on the discussion as Section 3, here three models, i.e. Langmuir, Freundlich and derivative models, are used to fit the experimental results. As afore-elucidated, Langmuir model, an ideal model, describes a dynamic adsorption process with the reciprocation between absorbed and unabsorbed proteins with while no interaction present between protein molecules or absorbed molecules and surface. Like the work carried out by Wassell et al [28], this model, to some extent, can describe the adsorption behavior of BSA over the surface of weakly electricity-charged microspheres, such as MS-0 and MS-2 (Fig.7). However, the fittings from higher electricity-charged MS-4 and MS-8 are considerably poor. This indicates that the process of BSA adsorption onto electricity-charged microspheres is more complex than the ideal situation. The presence of electrostatic interaction apparently plays an important role in such adsorption process. In the case of low electricity-charged microspheres such as MS-0 and MS-2, the electrostatic interaction is weak, so that the surface of microspheres is relatively uniform. As a result, the effect of electrostatic interaction in such systems is comparably limited, and thus the adsorption of BSA over such microsphere surfaces shows a certain degree of Langmuir correlativity. However, in the higher electricity-charged cases, the effect of electrostatic interaction cannot be neglected; accordingly the fittings of BSA adsorption, to some extent, deviates the origin of Langmuir theory. Expectedly, if these interactions were taken into account, the fittings from Freundlich model would be in better correlativity than those from Langmuir model. As shown in Fig.8, such fittings can improve the linear correlativity. Specifically, it is noted that the correlative coefficient of MS-4 is increased from R^2=0.178 to R^2=0.9783. As also noted, the correlativity in the case of MS-8 is only enhanced from R^2=0.5796 to R^2=0.7987. Theoretically, the Freundlich model should show more usability in the case of MS-8 than in MS-4, since the origin of Freundlich model has considered the probable interaction. This indicates that some additional information, especially the intrinsic information in the present process has not been revealed by Freundlich's empirical theory. Fig.9 presents the fittings of derivative model (here m=2 is selected as a probable value). As is apparent, such fittings are virtually much better than Langmuir or Freundlich. The linear correlativity in each electricity-charged MS is larger than 0.995 (R^2). Specially, the adsorption of BSA over highly electricity-charged microspheres can be well summarized, such as MS-4 and MS-8. This indicates that derivative model can reveal the more intrinsic information in present process than Langmuir or Freundlich model. Basing on the discussion as Section 3, it would appear that BSA adsorption over such functional microspheres is probably in an association form. Virtually, it is difficult for BSA to exist completely as monomolecular form in common condition. Indeed, even if the commercially available crystal BSA contains still about 10% dimmer and a small amount of oligomer [23]. Thus, the present example is easily learnable.

To ascertain the probable specific, electric charge may be created within the protein itself. Clearly, if the protein were electricity-charged, electrostatic repulsive interaction would decrease the association probability. Accordingly, the probability of such adsorption in monomolecular form would increase. Principally, Langmuir model would show more usability in such example than non- or weak electricity-charged case; also the fittings of derivative model in such example would be worse than those from non- or weak electricity-charged case due to the decrease of association probability. Here the condition of pH 2.2 is additionally selected. As is apparent, pH 2.2 is much lower than the pH 4.3 (approximately the isoelectric point of BSA, pI 4.7) and the protein would be obviously positively electricity-charged. As expected, the Langmuir fittings (Fig.11) from the adsorption isotherms of electricity-charged proteins (Fig.10) are virtually much better than those in Fig.7. Also the fittings of derivative model in such example (Fig.12) are actually worse than those in Fig.9. Clearly, intermolecular association plays a role in present process. The more information about the effect of pH conditions on the adsorption process will be shown in Section 4.3. It is also necessary to point out that the present discussion is somewhat tentative; further work is necessary and is currently underway. Other aspects that, have not, as yet, been addressed here, are also expected to be significant.

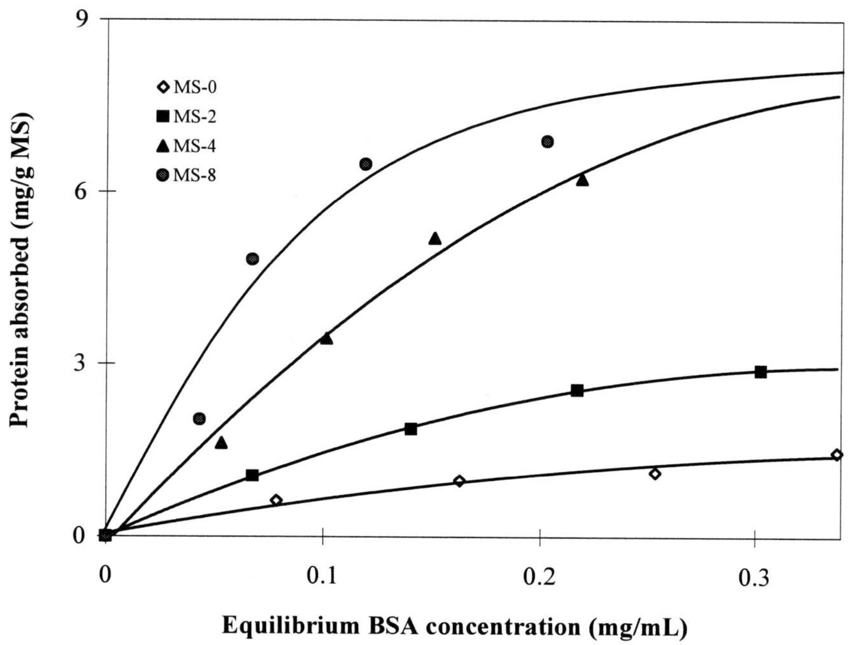

Fig.6 Adsorption isotherms of BSA

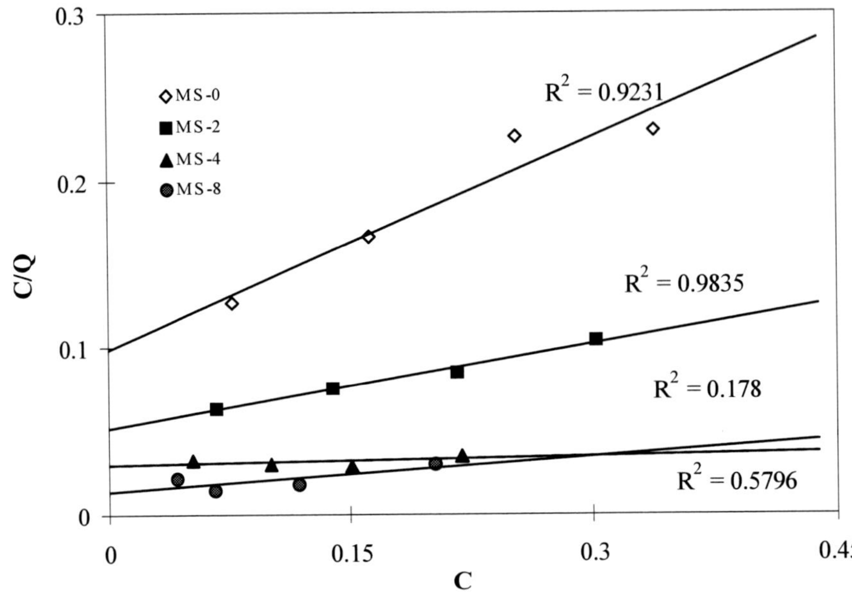

Fig.7 Langmuir fitting curves

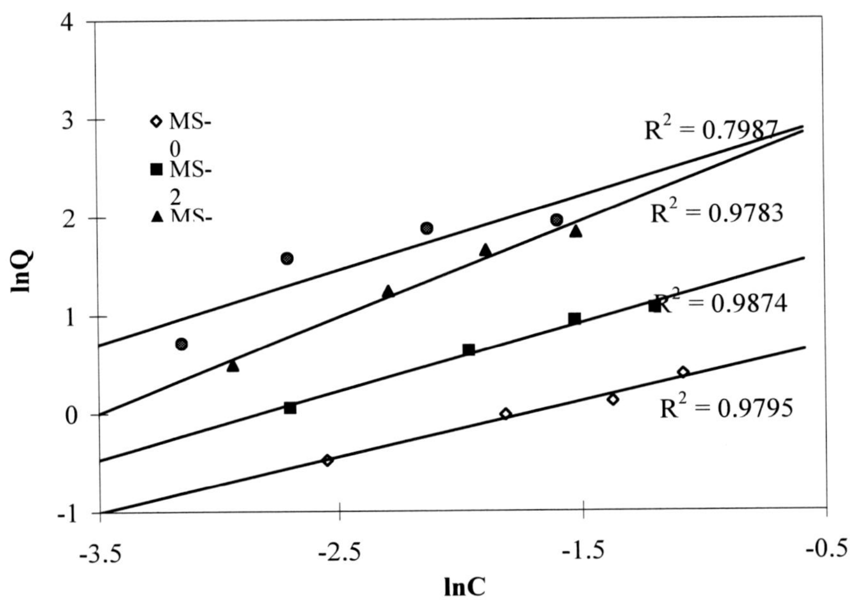

Fig.8 Freundlich fitting curves

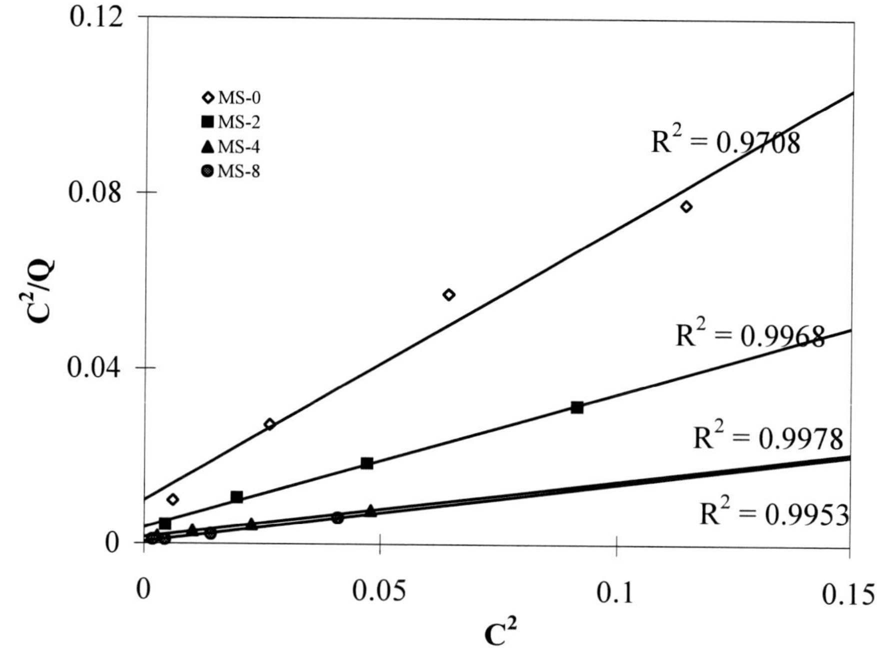

Fig.9 Fitting curves of derivative model

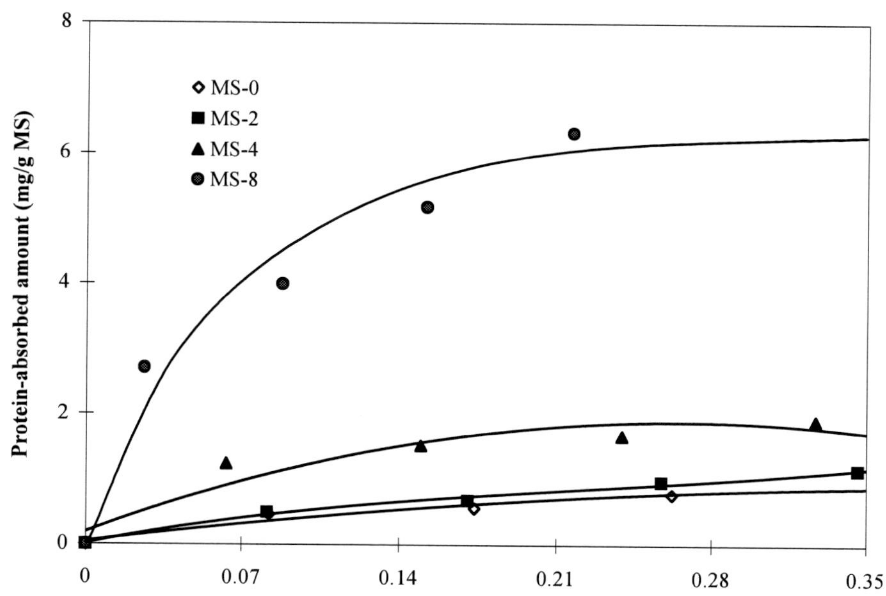

Fig.10 Adsorption isotherms (pH 2.2)

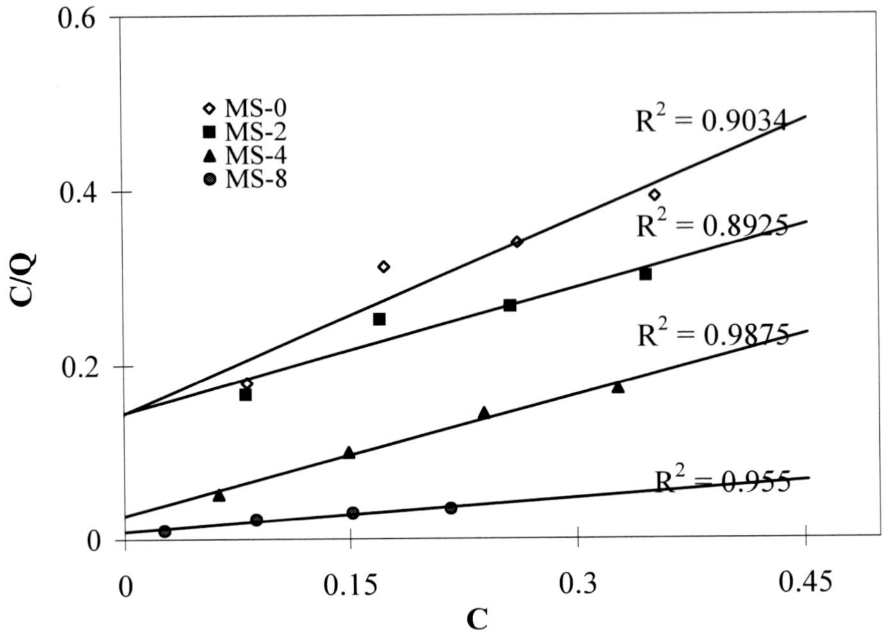

Fig.11 Langmuir fitting curves (pH 2.2)

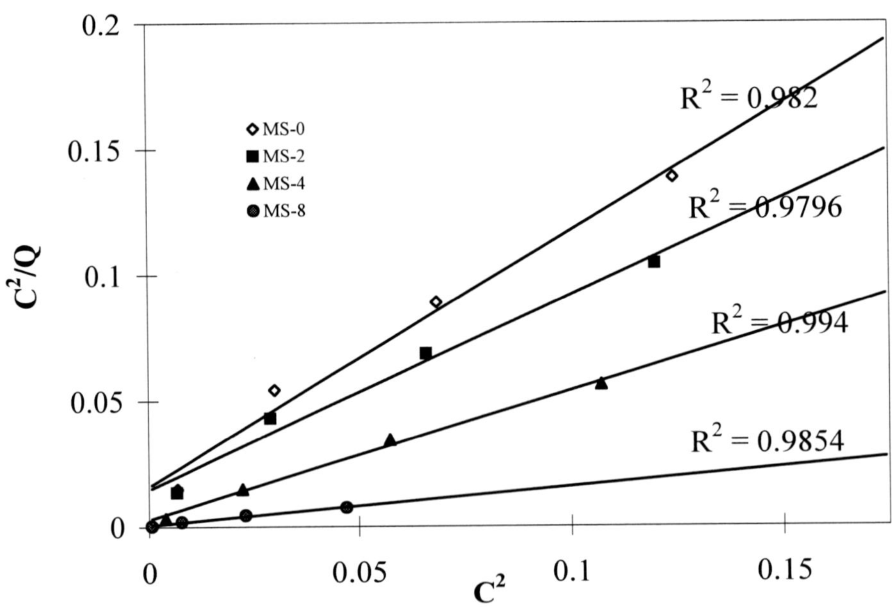

Fig.12 Fitting curves of derivative model (pH 2.2)

4.3. Ph Dependence

Effect of pH on adsorption amount of protein is presented in Fig.13a and b (the date in-follow-to BSA are the original concentration of BSA). Three typical pH values (2.2, 4.3, 7.4) are selected in present study. The pH of 2.2 is more acidic than the isoelectric point of BSA (pI 4.7). The pH 4.3 is near the pI of BSA, and while the pH 7.4 is more basic than the pI. As is apparent, close to the pI of BSA, the adsorption amount of protein exhibits a maximal value. A higher or lower pH condition would result in a lower level of adsorption. The adsorption amount of protein, to a high extent, depends on the pH conditions. Near the pI, the maximum adsorption-amount can be related to BSA conformation, as already explained [34,35], which is more folded at the pI than at other pH regions. The quick decrease in adsorption amount of protein under a higher or lower pH condition can be attributed to the bilateral electrostatic repulsion between BSA molecules, since they are with same electric charge, positive at pH 2.2 and negative at 7.4. Table 1 presents the affinity constants (derivative model). The effect of pH on the K-value in the case of MS-0 appears to be comparably limited, and while it is evident in the case of MS-4. This indicates that anionic end groups in MS-4 can affect the affinity. As also noted, the amounts of protein-absorbed from the sulfonate-charged MS-4 are actually higher than corresponding that from MS-0. This can related to the electrostatic attractions between anionic end groups and positive charge of protein molecules [36].

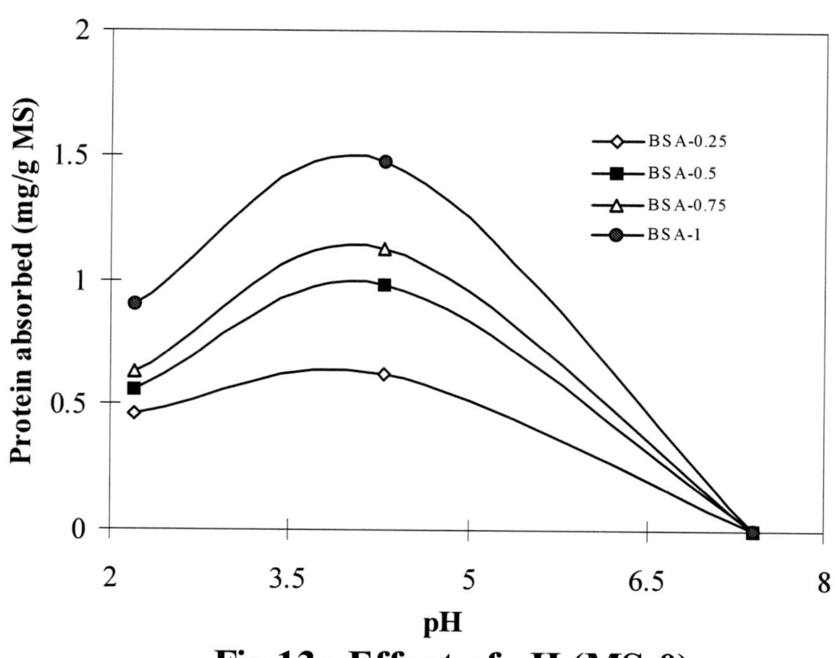

Fig.13a Effect of pH (MS-0)

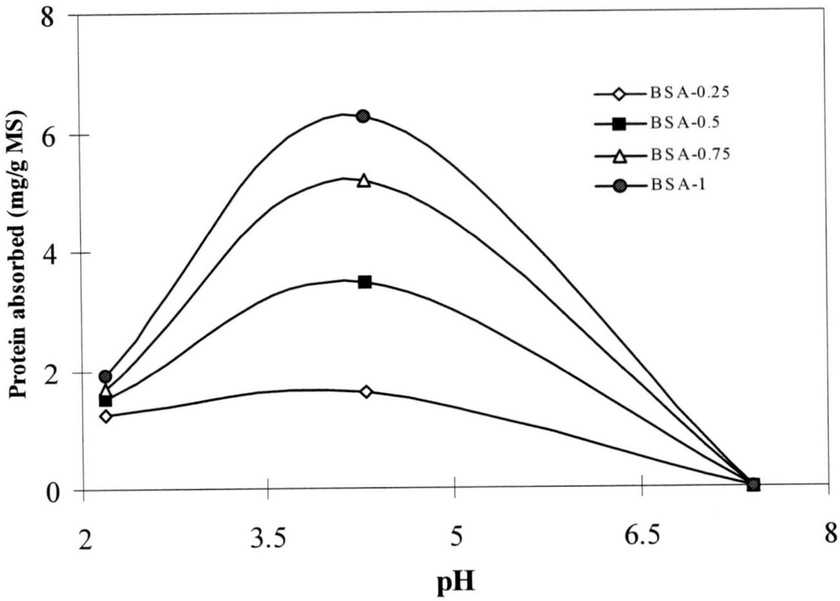

Fig.13b Effect of pH (MS-4)

Table 1 Affinity constants of BSA adsorption onto microspheres

MS-0		MS-4	
pH 2.2	pH 4.3	pH 2.2	pH 4.3
64.1	64.02	182.32	86.46

As in many literatures [34-36], the viewpoints as above, to some extent, interpret the effect of pH on protein adsorption-process. Virtually, so far, the interpretation is only qualitative and quantitative aspect still needs more investigation. Also such interpretation cannot summarize satisfactorily all phenomena concerning protein adsorption. The clear examples in literature [28,37] are profuse; these like that under a comparable condition, the K-value (Langmuir model), characterizing the affinity of protein to microspheres, do not seemingly correspond completely to the adsorption amount of protein. Also the quantitative relationship about the protein-absorbed amounts at different pH conditions remains to be presented. It is thus clear that taking additional considerations on this process is necessary. As afore-shown, we suppose the present process can be considered as apparent first-order kinetic process:

$$-\frac{dC}{dt} = kC$$

For different pH conditions, each has a similar expression. Apparently, connecting two similar relationships, one may obtain:

$$\frac{dC_1}{k_1C_1} = \frac{dC_2}{k_2C_2} \tag{23}$$

Here the subscripts 1 and 2 show different pH conditions. Integral of Eq.(23) from the beginning to the equilibrium can show:

$$\ln\frac{C_1^e}{C_1^0} = \frac{k_1}{k_2}\ln\frac{C_2^e}{C_2^0} \tag{24}$$

Here the subscripts 0 and e mean the original and equilibrium concentration of protein in corresponding system. To correlate the experimental results, Eq.(24) can be changed into such form when related to Eq.(20) and (21):

$$\ln\left(1-\frac{Q_1}{Q_{m1}}\right) = \frac{k_1}{k_2}\ln\left(1-\frac{Q_2}{Q_{m2}}\right) \tag{25}$$

Defining a new symbol S_r as follows:

$$S_r \equiv \frac{k_1}{k_2} \tag{26}$$

Based on general theory, we know S_r is a fixed-value under a given condition. Obviously, under comparable conditions, S_r may reflect the relatively speedy extent of protein adsorption. Now, Eq.(25) can be sought into such form:

$$\ln\left(1-\frac{Q_1}{Q_{m1}}\right) = S_r \ln\left(1-\frac{Q_2}{Q_{m2}}\right) \tag{27}$$

As in Eq.(27), the adsorption amounts of protein at different pH conditions are actually correlative; also the effect of electrostatic interaction (EI) at different pH conditions can be compared directly. Now, with kinetic method in use, the effect of pH condition on protein adsorption, to a certain degree, can be quantitatively characterized. S_r-value may be obtained from the diagram [$\ln\left(1-\frac{Q_1}{Q_{m1}}\right)$ versus $\ln\left(1-\frac{Q_2}{Q_{m2}}\right)$] (Fig.13) and whose result is shown in Table 2. At pH 2.2, the presence of electrostatic interaction increases the adsorption extent by 20.41%, and while increases only 1.92% at pH 4.3. This may be explained as that near the isoelectric point of BSA (pI 4.7) such as pH 4.3, the protein is almost non-charged, so the effect of electrostatic interaction is considerably limited; at pH 2.2, the electrostatic attractions between anionic end groups of MS-4 and positive charge of protein molecules can play an important role in promoting the adsorption. As can also be seen from Table 2, the adsorption at pH 2.2 is quicker than that at pH 4.3, especially in the case of MS-4. Many analogous results are also shown in literature [35,36]. This, as already explained, can be related to the decreases of the degree of protein hydration and charge near the isoelectric point, which decreases the accessibility of protein to microspheres. As a result, the rate of protein adsorption would decrease.

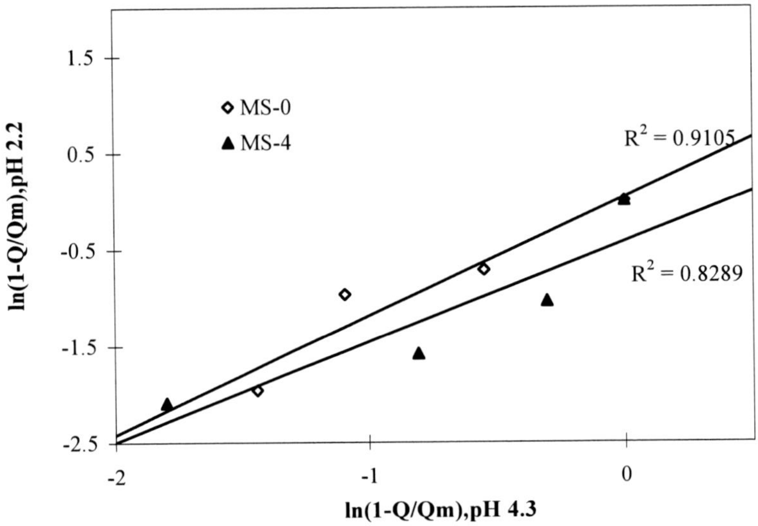

Fig.14a Relation of ln(1-Q/Qm) at pH 2.2 and 4.3

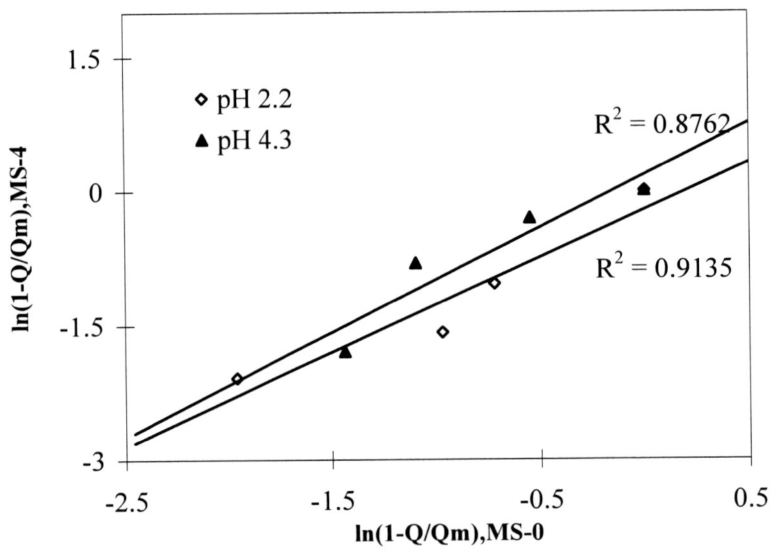

Fig.14b Relation of ln(1-Q/Qm) of MS-0 and MS-4

Table 2 Effects of pH and electrostatic interaction on S_r-value

	S_r-value		Effect of EI	Effect of pH
MS-4 / MS-0	pH 2.2	1.2041	20.41%	/
	pH 4.3	1.0192	1.92%	/
pH 2.2 / pH 4.3	MS-0	1.1981	/	19.81%
	MS-4	1.3466	/	36.66%

4.4. Effect of Ionic Strength

Effects of ionic strength on BSA adsorption are shown in Fig.15a and b. At pH 2.2, the increase of NaCl concentration results in a slight increase of protein-absorbed amount. However, at pH 4.3, the tendency is conversed; the increase in the NaCl concentration would produce a less adsorption. This indicates that the effect of ionic strength on protein adsorption actually relies on pH conditions. Many similar results are also shown in the earlier literatures [38-40]. The increase in adsorption amount of protein at pH 2.2 may be due to the suppression of NaCl on the electric double-layer of BSA [38], which can decrease the bilateral electrostatic repulsion between protein molecules. As also explained [39,40], the decrease near pI can be attributed to the shift of the isoelectric point of protein when increasing the ionic strength. As a result, the increase in NaCl concentration would increase the electrostatic repulsion between BSA molecules, thereby decreasing the adsorption amount.

It is necessary to point out that some theories concerning salt-effect have been widely described. However, so far, few workers in biochemical field have employed it to investigate a specific adsorption process. Virtually, salt-effect theory can give much additional information involved by the adsorption process. As afore-elucidated, Freundlich model has been proven as efficient in many adsorption processed though empirical in origin, sustaining a more or less interaction to exist. In present case, the interaction between NaCl and protein obviously plays a role in the adsorption process. Thus here Freundlich model is used to study the effect of NaCl on the adsorption process. As in Eq.(9) and Eq.(17), following equation can describe the adsorption process of protein:

$$\frac{Q}{Q_m} = KC^{\xi} \tag{28}$$

Here it is necessary to point out that the concentration should be corrected to activity when existing salt in the system:

$$\frac{Q_S}{Q_{mS}} = K_S a^{\xi} = K_S (\gamma C)^{\xi} \tag{29}$$

As in Eq.(29), the probable interaction involved by the adsorption process has been summarized; also the effect of salt on protein activity has been taken into account. So, principally Eq.(29) can be applied to describe the adsorption process. Now, connecting Eq.(29) with (28) under the comparable conditions, one can obtain:

$$\frac{Q_S}{Q} = \frac{K_S Q_{mS}}{K Q_m} \gamma^{\xi} = A\gamma^{\xi} \tag{30}$$

Here A is a constant under the specific condition, obtained from $A = K_S Q_{mS} / KQ_m$, and γ, the activity coefficient, can be obtained from [41]:

$$\gamma = \sqrt{\gamma_{Na^+}\gamma_{Cl^-}} \tag{31}$$

According Debye-Huckel law, the activity coefficient of specific ion can be expressed as [42]:

$$\lg \gamma_i = \frac{-\lambda Z_i^2 \sqrt{I}}{1 + Ba\sqrt{I}} \tag{32}$$

Here λ and B are constants, a, also a constant, relating to volume parameter of specific ion, Z, the ionic charge, and I, the ionic strength, obtained from:

$$I = \frac{1}{2}\sum c_i Z_i^2 \tag{33}$$

Actually, Eq.(32) can be simplified into Eq.(34) when the concentration of salt is considerably low [42]:

$$\lg \gamma_i = -0.5 Z_i^2 \sqrt{I} \tag{34}$$

Now, substitution of Eq.(34) into Eq.(31) can show:

$$\lg \gamma = -0.5\sqrt{I} \tag{35}$$

Basing on Eq.(35) and Eq.(30), we can obtain such relationship:

$$\lg \frac{Q_S}{Q} = -\frac{\xi}{2}\sqrt{I} + \lg A \tag{35}$$

To deepen the comprehension on Eq.(35), it would be helpful for us to analyze the concepts of various parameters. An increase in the ξ will result in a low level of adsorption, and therefore ξ shows a character of inhibitor in Eq.(35). However, an increase in ‘A’ would result in a higher level of adsorption (driving force). Clearly, the effect of ionic strength in protein adsorption can be determined by the relative magnitude of inhibitory force and driving force. As in Eq.(35), plotting $\lg\frac{Q_S}{Q}$ versus $\sqrt{I}$ is normally expected to be a straight line (Fig.16). From the slope and intercept, the inhibitory force and driving force can be obtained. As shown in Table 3, a considerable interesting result is that at pH 2.2, the driving force for protein adsorption is larger than the inhibitory force, and while the inhibitory force is beyond the driving force at pH 4.3. Judging from the origin of ξ, we know that ξ can characterize the intermolecular interaction, the larger ξ indicating the weaker interaction. Clearly, the larger driving force at pH 2.2 indicates that the interaction between BSA molecules is relatively weak, or to say, the presence of NaCl at pH 2.2 can decrease the interaction between BSA molecules. As afore-elucidated, this may be due to the suppression of NaCl on the electric double-layer of BSA [38], which can decrease the bilateral electrostatic repulsion between protein molecules. However, the lower driving force at pH 4.3 indicates that the actual interaction between BSA molecules is stronger than that at pH 2.2. The can be attributed to, as already explained [39,40], the shift of the isoelectric point of protein when increasing the ionic strength. As a result, the increase in NaCl concentration would increase the electrostatic repulsion between BSA molecules.

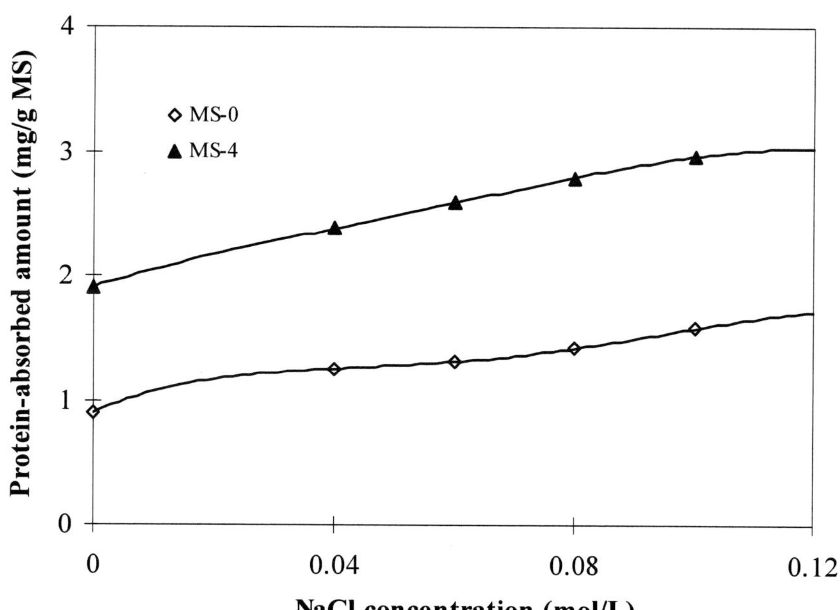

Fig.15a Effect of NaCl (pH 2.2)

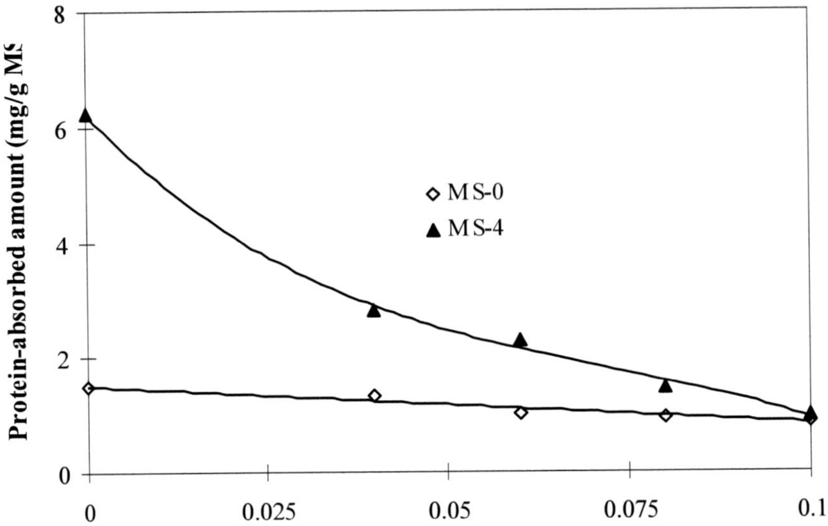

Fig.15b Effect of NaCl (pH 4.3)

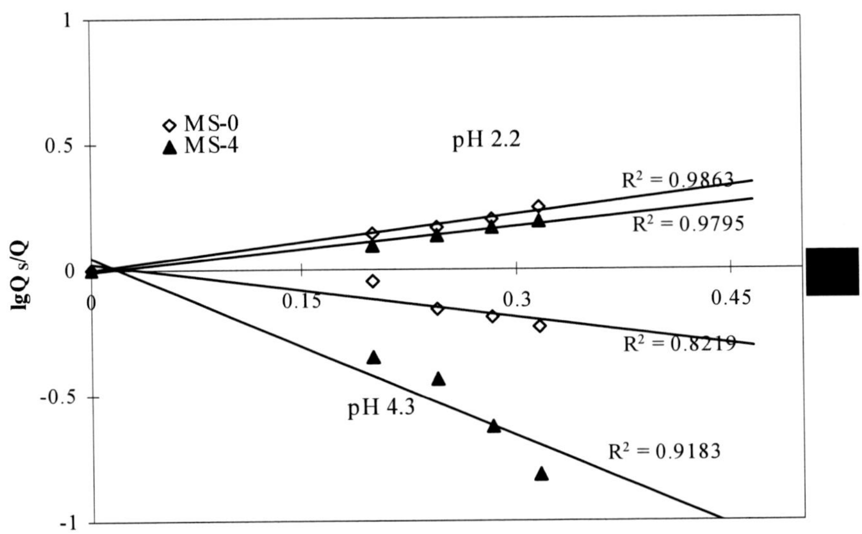

Fig.16 Salt effect curve

Table 3 Driving force and inhibitory force of BSA adsorption

	pH 2.2		pH 4.3	
	MS-0	MS-2	MS-0	MS-4
$-\xi/2$	0.7363	0.5923	-0.7201	-2.3664
$\lg A$	-0.0036	-0.0066	0.0236	0.049

5. Final Remarks

Adsorption protein (especially blood proteins) onto functional microspheres can play important roles in biochemical research. Basing on such process, it would be helpful to understand some bio-phenomena involved the analogous molecular interactions. Although there have much description available in literature, the mechanism, especially intrinsical mechanism is not yet clear to us. As a supplementation, the present research provided some important information involved by protein adsorption process, and specially, the intrinsic information was described emphatically. To help learning the probable process, the origin of classical theories and their derivatives were also shown in present study.

Like many studies, the adsorption process was conducted with BSA as model and sulfonated-microspheres as the matrix. Langmuir, Freundlich and derivative models have been used to fit the experimental date. The result shows that the fitting of derivative model is best and while the worst is obtained from Langmuir fitting. This indicates that the practical adsorption is much complex than the idea situation; the interaction, more or less, can be involved in this process. As a fact, with the adsorption goes on, the more BSA molecules are adsorbed, and the earlier adsorbed macromolecules will inevitably hinder the latter adsorption. Also it is virtually difficult for BSA to exist completely as monomolecular form in common condition. Indeed, even if the commercially available crystal BSA contains still about 10% dimmer and a small amount of oligomer. Judging from pH effect, we know pH can play important roles in protein adsorption process. This may be due to the dependencies of electrostatic interaction and protein conformation on pH conditions. Near the pI, the maximal adsorption-amount can be related to protein conformation that is more folded than at other pH regions. The quick decrease in adsorption amount of protein under a higher or lower pH condition can be related to the bilateral electrostatic repulsion between protein molecules, since they are in same electric charge. Further investigation shows that the adsorption amounts of different pH conditions are actually correlative; also the role of electrostatic interaction can be characterized by kinetic method. As in many literatures, salt can also affect the adsorption. At pH 2.2, the increase in NaCl concentration results in a slight increase of protein-absorbed amount, and while the less adsorption-amount is obtained at pH 4.3. The increase in adsorption amount of protein at pH 2.2 may be due to the suppression of NaCl on the electric double-layer of BSA, as already elucidated, which can decrease the bilateral electrostatic repulsion between protein molecules. The decrease near pI can be attributed to the shift of the isoelectric point of protein when increasing the ionic strength. As a result, the increase in NaCl concentration would increase the electrostatic repulsion between BSA molecules, thereby decreasing the adsorption amount. It is noted that many theories concerning salt-effect have been introduced in literature. However, so far, few workers in biochemical field have employed it to investigate a specific adsorption process. In present study, the theory shows that the effect of ionic strength in protein adsorption can be determined by the relative magnitude of inhibitory force and driving force. At pH 2.2, the driving force for protein adsorption is actually larger than the inhibitory force, and while the inhibitory force is beyond the driving force at pH 4.3. Clearly, the change in the driving force may be responsible for the change of protein-absorbed amount.

It is also necessary to point out that some theories are somewhat tentative; further work is necessary and is currently underway. Other aspects that, have not, as yet, been addressed in the article, are also expected to be significant.

REFERENCES

[1] Norman E, Williams P, Illum L. Influence of block copolymers on the adsorption of plasma proteins to microspheres. *Biomaterials*, 1993, 14(3), 193-202.

[2] Glen SK, You HB, Harry C, Jan FJ, Sung WK. Release of proteins via ion exchange from albumin-heparin microspheres. *J. Contr. Release*, 1992, 22(2), 83-93.

[3] Lu HB, Homola JI, Campbell CT, Nenninger GG, Yee SS, Ratner BD. Protein contact printing for a surface plasmon resonance biosensor with on-chip referencing. *Sensors and Actuator B*, 2001, 74(3), 91-99.

[4] Butler JE. Solid supports in enzyme-linked immunosorbent assay and other solid-phase immunoassays. *Methods*, 2000, 22(1), 4-23.

[5] Panphilov AS, Bekish OJL. Updating of magnetosensitive microspheres from styrene technology for immunomagnetic cell separation. *Immunol. Lett.*, 1997, 56(2), pp462.

[6] Brahim S, Narinesingh D, Guiseppi-Elie A. Kinetics of glucose oxidase immobilized in p(HEMA)-hydrogel microspheres in a packed-bed bioreactor. *J. Mol. Catal. B*, 2002, 18(3), 69-80.

[7] Kellar KL, Iannone MA. Multiplexed microsphere-based flow cytometric assays. *Experiment. Hematol.*, 2002, 30(11), 1227-1237.

[8] Andrade JD. *Surface and interfacial aspects of biomedical polymers, vol.2- Protein adsorption*, Plenum, New York, 1985, 1-80.

[9] kawaguchi H, Amagasa H, Hagiya T, Kimura N, Ohtsuka Y. Interaction between proteins and latex particles having different surface structures. *Colloids Surf.*, 1985, 13(2), 295-311.

[10] Tuncel A, Purvis AD, Piskin CRL. Cibacron Blue F3G-A-attached monosize poly(vinyl alcohol)-coated polystyrene microspheres for specific albumin adsorption. *J. Chromatogr. A,* 1993, 634(2), 161-168.

[11] Peula JM, de las Nieves FJ. Adsorption of monomeric bovine serum albumin on sulfonated polystyrene model colloids 1. Adsorption isotherms and effect of the surface charge density,. *Colloid surf. A,* 1993, 77(3), 199-208.

[12] Haynes CA. Norde W. Globular proteins at solid/liquid interfaces. *Colloid surf. B*, 1994, 2(6), 517-603.

[13] Haynes CA, Sliwinsky E, Norde W. Structural and electrostatic properties of globular proteins at a polystyrene-water interface. *J. Colloid interf. Sci.*, 1994, 164(2), 263-514.

[14] Brazin KN, Fulton DB, Andreotti AH. A specific intermolecular association between the regulatory domains of a tec family kinase. *J. Mol. Biol.*, 2000, 302(3), 607-623.

[15] Ortega LG, Mccotter MD, Henry GL, Mccormack SJ, Thomis DC, Samuel CE. Mechanism of interferon action: biochemical and genetic evidence for the intermolecular association of the RNA-dependent protein kinase PKR from human cells. *Virology*, 1996, 215(1), 31-39.

[16] Welzel PB. Investigation of adsorption-induced structural changes of proteins at solid/liquid interfaces by differential scanning calorimetry. *Thermochim. Act.*, 2002, 382(2), 175-188.

[17] Slomkowski S, Basinska T, Miksa B. New types of microspheres and microsphere-related materials for medical diagnostics. *Polym. Adv. Tech.*, 2002, 13(11), 905-918.

[18] Ortega-Vinuesa L, Hidalgo-Álvarez R. Sequential adsorption of F(ab')2 and BSA on negatively and positively charged polystyrene latexes. *Biotech. Bioeng.*, 1995, 47(6), 633-639.

[19] Lee JH, Yoon JY, Kim WS. Continuous separation of serum proteins using a stirred cell charged with carboxylated and sulfonated microspheres. *Biomed. Chromatgr.*, 1998, 12(6), 330-334.

[20] Yoon JY, Kim JH, Kim WS. Interpretation of protein adsorption phenomena onto functional microspheres. *Colloids Surf. B*, 1998, 12(1), 15-22.

[21] Yoon JY, Kim JH, Kim WS. The relationship of interaction forces in the protein adsorption onto polymeric microspheres, *Colloids Surf. A*, 1999, 153(3), 413-419.

[22] Gu R, Li CS, Liu CX. Preparing mono-sized PMMA microspheres by dispersion polymerization. *Chin. J. Shandong Normal Univ.* (Nat. Sci.), 1999, 14(4), 470-473.

[23] Yoon JY, Park HY, Kim JH, Kin WS. Adsorption of BSA on highly carboxylated microspheres—quantitative effects of surface functional groups and interaction forces. *J. Colloid Interf. Sci.*, 1996, 177(2), 613-620.

[24] Li ZJ. *Kinetic bases of heterogeneous catalytic reactions*. Beijing: Beijing University Press, 1990, p39-42.

[25] Yang CH. Statistical mechanical study on the Freundlich Isotherm Equation. *J. Colloid Interf. Sci*, 1998, 208(2), 379-387.

[26] Johnson RD, Arnold FH. The temkin isotherm describes heterogeneous protein adsorption. *Biochimica et Biophys. Acta. (BBA) - Protein Structure and Molecular Enzymology,* 1995, 1247(2), 293-297.

[27] Song HC, Strubinger JR, Parcher JF. Tracer pulse chromatographic method for the determination of nitrogen BET isotherms and surface areas. *J. Chromatogr. A*, 1990, 518(1), 319-328.

[28] Wassell DTH, Hall RC, Embery G. Adsorption of bovine serum albumin onto hydroxyapatite. *Biomaterials*, 1995, 16(9), 697-702.

[29] Jiling University (Ed.). *Basis of catalysis*. Beijing: Chinese Scientific Press, 1980, p1-30.

[30] Hu H, Trejo M, Nicho ME, Saniger JM García-Valenzuela A. Adsorption kinetics of optochemical NH_3 gas sensing with semiconductor polyaniline films. *Sensors and Actuators B*, 2002, 82(1), 14-23.

[31] Cortés J. Gas-solid adsorption isotherms: the art and science of determining monolayer capacity. *Adv. Colloid Interf. Sci.*, 1985, 22(2), 151-176.

[32] Fu XC, Shen WX, Yao TY. *Physicochemistry*. Beijing: Chinese High Education. Press, 1998, p937-942.

[33] Yin G, Liu Z, Zhan J, Ding FX, Yuan NJ. Impacts of the surface charge property on protein adsorption on hydroxyapatite. *Chem. Eng. J.*, 2002, 87(2), 181-186.

[34] Suzawa T, Shirahama H. Adsorption of plasma proteins onto polymer lattices. *Adv. Colloid Interf. Sci.*, 1991, 35(1), 139-172.

[35] Kondo A, Oku S, Murakami F, Higashitani K. Conformational changes in protein molecules upon adsorption on ultrafine particles. *Colloids Surf. B*, 1993, 1(3), 197-201.

[36] Matsumoto H, Koyama Y, Tanioka A. Interaction of proteins with weak amphoteric charged membrane surfaces: effect of pH. *J. Colloid Interf. Sci.,* 2003, 264(1), 82-88.

[37] Lapizco-Encinas BH, Pinto NG. Effectiveness of the H-root method for determining adsorption isotherms of protein–salt systems in open micro-channels. *J. Chromatogr. A,* 2004, 1036(1), 61-72.

[38] Sharma S, Agarwal GP. Interactions of proteins with immobilized metal ions: role of ionic strength and pH. *J. Colloid Interf. Sci.*, 2001, 243(1), 61-72.

[39] Revilla J, Elaïssari A, Carriere P, Pichot C. Adsorption of bovine serum albumin onto polystyrene latex particles bearing saccharidic moieties. *J. Colloid Interf. Sci.*, 1996, 180(2), 405-412.

[40] Zhang Z, Dalgleish DG, Goff HD. Effect of pH and ionic strength on competitive protein adsorption to air/water interfaces in aqueous foams made with mixed milk proteins. *Colloids Surf. B*, 2004, 34(2), 113-121.

[41] Fu XC, Shen WX, Yao TY. *Physicochemistry*. Beijing: Chinese High Education. Press, 1998, p541-572.

[42] Wuhan University (Ed.). *Analytic chemistry*. Beijing: Chinese High Education. Press, 1995, p39-44.

In: New Frontiers in Polymer Research
Editor: Robert K. Bregg, pp. 31-53
ISBN: 1-59454-956-7

Chapter 2

METAL-CONTAINING POLYURETHANES, POLY(URETHANE-UREA)S AND POLY(URETHANE-ETHER)S

R. Jayakumar[1*], S. Nanjundan[2] and M. Prabaharan[1]
[1]Department of Polymer Engineering, University of Minho, Campus of Azruem, Guimaraes-4800 058, Portugal
[2]Department of Chemistry, Anna University, Chennai-600 025, India

ABSTRACT

Polyurethanes are becoming increasingly important as engineering materials because they have excellent abrasion resistance and the properties of both elastomers and plastics. The growth of science and technology of polyurethanes leads to the development of new materials with more desirable properties. Such kinds of materials include metal-containing polyurethanes, poly(urethane-urea)s and poly(urethane-ether)s with isocyanate structural units combining the properties of enhanced thermal stability, fire retardancy, flexibility and solubility. Ionic diols containing metal salts are used as important starting materials for the synthesis of metal-containing polyurethanes in which the metal is firmly incorporated in the backbone of the polymer chain. Incorporation of metal into polyurethanes has led to wide application as aqueous thickeners, impregnates, textile sizers, adhesives, additives, resins and catalysts. The aim of this work is to present an overview of the various methods of preparation and properties of metal-containing polyurethanes and their copolymers.

Keywords: Metal-containing polyurethanes: Poly(urethane-urea)s: Poly(urethane-ether)s

---

[*] jayakumar77@yahoo.com

1. INTRODUCTION

Polyurethanes (PUs) represent an important class of thermoplastic and thermoset polymers as their mechanical, thermal, and chemical properties can be tailored by the reaction of various polyols and polyisocyanates. The PUs includes those polymers which contain a significant number of urethane groups (-HN-COO-), regardless of what the rest of the molecule may be. Usually the PUs are obtained by the combination of polyisocyanate with reactants which have at least two hydroxyl groups, eg., polyethers, castor oil and simple glycols. Other reactive groups such as amino and carboxyl groups may also be present. Thus a typical `polyurethane' may contain ester, ether, amide, and urea groups in addition to urethane groups. The general structure of a linear polyurethane derived from a diol (HO-R-OH) and a diisocyanate (OCN-R'-NCO) can be represented as follows;

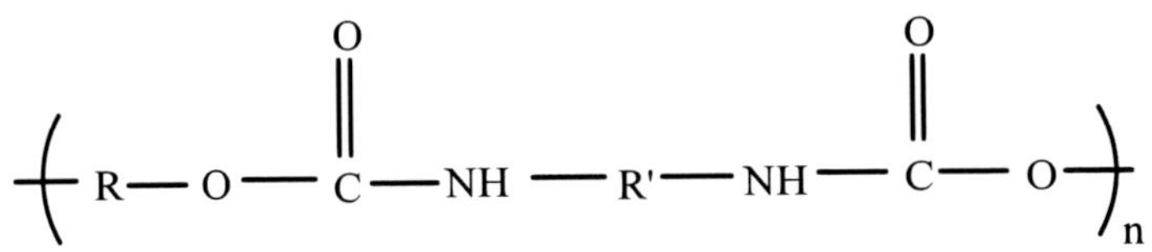

The properties of PUs can be varied in numerous ways according to their demand. The functionality of the hydroxyl compound as well as the isocyanate can be increased to three or more in order to form branched or crosslinked polymers. Other structural changes can also be made at which, the nature of R (polyether, polyester, and glycol) and R' (diisocyanates) might be altered by changing its molecular weight and type. For these reasons, the crosslinking and chain flexibility properties of PUs and its intermolecular forces can be varied widely and independently. PUs, in general, present good adhesive characters due to their elastomer properties, enhanced by soft segments of the polyol and by the polar character of the urethane groups [1, 2]. These materials are widely used as coatings, foams, different kinds of plastics and elastomers, and rocket motor liners for the securing of composite propellant gain within the rocket motor [2-6].

Metal-containing polymers represent a broad classification of polymers with inorganic salt groups attached to the polymer chain. They are generally referred as ′polyelectrolytes' when they have sufficient ionic charge to be soluble in water and as ′ionomers' when the concentration of ion-containing groups is too low for water solubility. Unlike homogeneous polymer systems, polymers containing ionic group interact to form ion rich aggregates contained in the non-polar polymer matrix. The resulting interactions strongly influence polymer properties and applications, which have made this as a fertile area of research and development. Metal-containing polymers are the most versatile and useful of the new polymeric materials currently being developed. These polymers have been investigated for several decades, attracting scientific and commercial attention. These polymers have a wide variety of properties that lead to different applications as aqueous thickeners, impregnants, textile sizers, adhesives, additives [7, 8], resins [9-11], catalysts [12] and in the biomedical field [13, 14]. The aim of this review is to focus on the various methods of preparation and properties of metal-containing polyurethanes (PUs), poly(urethane-urea)s (PUUs), and poly(urethane-ether)s (PUEs).

2. METAL-CONTAINING PUS

PUs possess excellent mechanical and physical properties, high combustion resistance, and wear resistance with broad temperature range good blood and tissue compatibility [2]. Modification of PUs by incorporating metal and functional groups are used extensively to improve various typically desired properties of materials, such as enhanced thermal stability, fire retardancy, flexibility and solubility. These properties lead to variety of applications. Ionic diols containing ionic linkages between carboxylate anion (COO^-) and metal cation (M^{2+}) are of interest and are very important starting materials for the synthesis of metal-containing PUs in which the metal is firmly incorporated into the backbone of the polymer chain [15-27]. Various kinds of ionic diols such as divalent metal salts of mono (hydroxyethyl) phthalate, mono (hydroxybutyl) phthalate, mono (hydroxyethoxyethyl) phthalate, mono (hydroxypentyl) phthalate, mono (hydroxybutyl) hexolate are reported as starting materials. In addition, diols containing phosphorous or sulphur group are also used as important monomers for the synthesis of PUs having fire resistance and blood contacting properties. The DMF and DMSO are found to be good solvents for the synthesis of metal-containing PUs. However, the reaction rate of urethane formation in DMF is found to be very high [28]. Di-*n*-butyltin dilaurate (DBTDL) is reported as an effective catalyst for the synthesis of PUs. The reaction of diisocyanates with diols catalyzed by DBTDL takes place via the formation of a ternary complex between the catalyst and the reagents [29].

Matsuda [9] was the first to report the synthesis of PUs using Mg and Ca salts of mono(hydroxyethyl)phthalate with hexamethylene diisocyanate (HMDI) or tolylene 2, 4-diisocyanate (TDI). These polymers were found to be glassy materials or white powders, based on the species of diisocyanates and glycols. It was found that the viscosities of the prepared PUs in DMF were decreased markedly with increasing of metal content. The decomposition temperatures were lowered by introduction of metals into the PUs. However, up to 50% weight loss, the decomposition rate of the Mg and Ca-containing PUs was lower than those of the corresponding PUs not containing metal (Table 1). Moreover, the thermal stability of these polymers was little influenced by the species of metals. From these facts, it was observed that the existence of metals promotes the thermal decomposition of urethane linkage.

Table 1. Thermal stability of Mg and Ca-containing PUs and blank PUs

Diisocyanate	Diol	Temperature of 10% weight loss, °C	Temperature of 50% weight loss, °C
TDI	$(HEP)_2Mg$	192	332
TDI	$(HEP)_2Ca$	192	338
TDI	EG	248	314
HMDI	$(HEP)_2Mg$	200	303
HMDI	$(HEP)_2Ca$	200	310
HMDI	EG	295	340

Durairaj and Rao [15] have synthesized Mn and Pb-containing PUs from Mn and Pb salts of mono(hydroxyethyl)phthalate, HMDI and TDI. The general reaction scheme for the synthesis of metal-containing PUs from the various monomers is shown in Figure 1. Here also, the decomposition temperatures of metal-containing PUs were found to be significantly lower than those of metal-free polymers of similar structure. The rates of decomposition of Mn and Pb-containing PUs were lower than those of metal free PUs. It was reported that the inherent viscosities of the PUs in DMSO at 30 °C were low, ranging from 0.043 to 0.067. The viscosities of the PUs were found to be low as compared to the analogues without metal due to the fact that in more polar solvents like DMSO, the ionic links present in the polymer chain dissociate resulting in reduction of the molecular weight of the polymer when the concentration is low. Similar kind of properties were reported for Co and Cu-containing PUs, which were prepared from Co and Cu salts of mono(hydroxyethyl)phthalate with TDI [16].

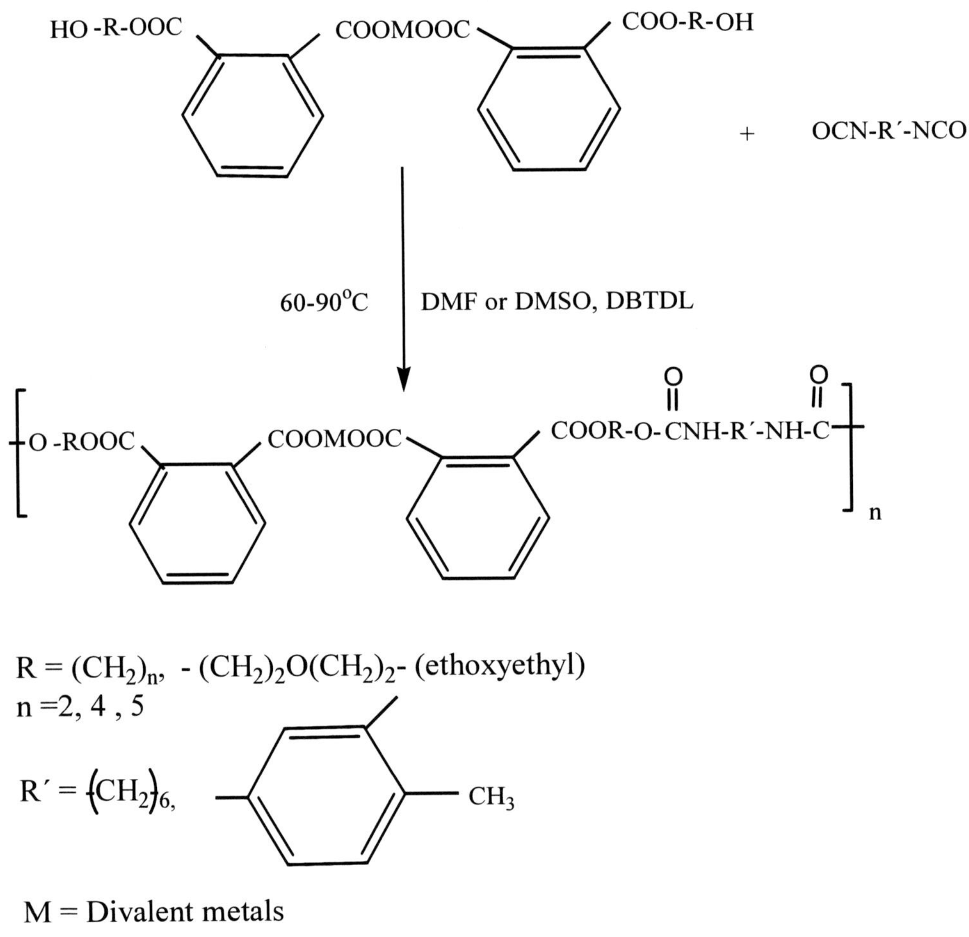

Figure 1. Synthesis of metal-containing PUs based on divalent metal salts of mono(hydroxyl-ethyl/hydoxybutyl/hydroxypentyl/hydroxyethoxyethyl)phthalate.

PUs prepared from Zn, Cd and Hg salts of mono(hydroxyethyl)phthalate, HMDI, TDI and 4, 4'-diphenylmethane diisocyanate (MDI) were found to be yellow to brown in color powder [17, 18]. It was reported that the yields of these metal-containing PUs were consistently good in all cases. The PUs based on MDI gave slightly higher yields than those based on TDI due to the higher reactivity of MDI compared to TDI. The X-ray diffraction studies of HMDI based Zn, Cd and Hg-containing PUs was also discussed by the authors in the aspect of orientation. The relative crystallinity values given in Table 2 revealed that the HMDI-based PUs had a better crystalline fit, resulting from close packing in the hard segments and their orientation. The order of crystallinity observed (based on relative crystallinity) for these metal-containing PUs was as follows: HMDI-based > MDI-based > TDI-based. Similar studies were also carried out by Arun Prasath et al. for Ca, Mn, Pb and Zn-containing PUs which were prepared from HMDI and TDI with Ca, Mn, Pb and Zn salts of mono(hydroxybutyl)phthalate [19, 20].

Table 2. Relative crystallinity of Zn, Cd and Hg-containing PUs

Polymers	Relative crystallinity (%)	Nature of phase
Zn $(HEP)_2$-MDI-PU	55	Semicrystalline
Zn $(HEP)_2$-TDI-PU	35	Amorphous
Zn $(HEP)_2$-HMDI-PU	69	Crystalline
Cd $(HEP)_2$-MDI-PU	44	Amorphous
Cd $(HEP)_2$-TDI-PU	32	Amorphous
Cd $(HEP)_2$-HMDI-PU	73	Crystalline
Hg $(HEP)_2$-HMDI-PU	70	Crystalline

A series of metal-containing PUs were synthesized by the reaction of HMDI or TDI with Ca, Cd, Pb, Cu, Mn and Zn salts of mono(hydroxyethoxyethyl)phthalate and mono(hydroxypentyl)phthalate and characterized [21-23]. The viscosities of this metal-containing PUs were found to be low due to the chain dissociation of polymers in the solvent. The polymers were found to be insoluble in most of the organic solvents and soluble in highly polar solvents like DMF, DMSO and DMAc. The authors observed that the TDI-based PUs showed a higher thermal stability than HMDI-based PUs due to the presence of stiff phenylene rings in the polymer chain. The XRD patterns (Figure 2) showed that the HMDI-based PUs were partially crystalline, whereas TDI-based PUs were amorphous in nature due to the folding of hexamethylene group present in the backbone of the PUs. The elemental analysis showed that the experimentally determined percentage values of carbon, hydrogen and metal content were within the calculated values. It was reported that all these metal-containing PUs were exhibit antibacterial activities and can be used as antibacterial coatings.

Metal-containing copolyurethanes were synthesized by the polyaddition reaction of HMDI or TDI with 1:1 mixtures of Ca, Cd, Pb and Zn salts of mono(hydroxypentyl)phthalate or mono(hydroxyethoxyethyl)phthalate and diethylene glycol (DG) and with 1:1 mixtures of Ca and Zn salts of mono(hydroxybutyl)phthalate and DG [24-26] (Figure 3 & 4). The antibacterial activity of these polymers was investigated by agar diffusion method and found that the Ca-containing copolyurethanes exhibited some antibacterial activity. This work also reported that the antibacterial activity of these metal-containing copolyurethanes was not only due to the metal ions but also the isocyanate group of metal-containing copolyurethanes. The Cd-containing copolyurethanes exhibited strong inhibitory effects on all the bacteria, but the copolyurethanes containing other metals exhibited different antibacterial activities depending on the bacteria. The elemental analysis showed that the experimentally determined percentage values of carbon and hydrogen were slightly higher and those of metals were lower than the calculated values. From this result, it was inferred that the reactivity of ionic diol was lesser than that of DG

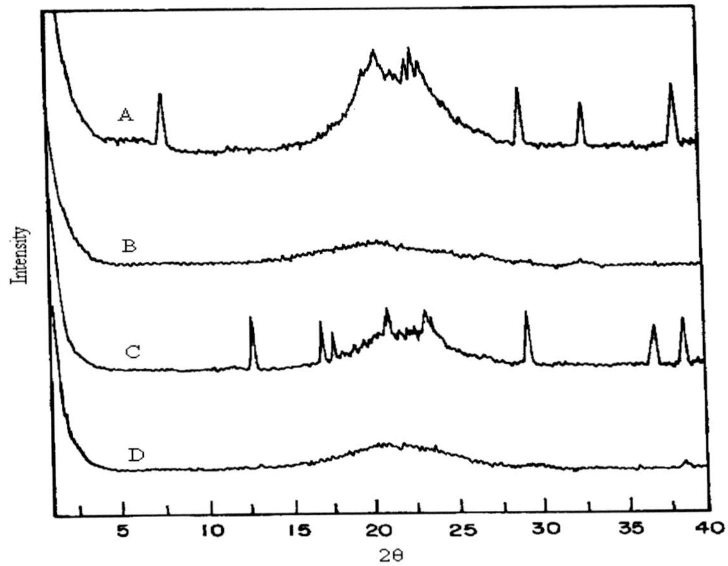

Figure 2. XRD patterns of (A) HMDI-based PU, (B) TDI- based PU, (C) HMDI- based PU, and (D) TDI-based PU.

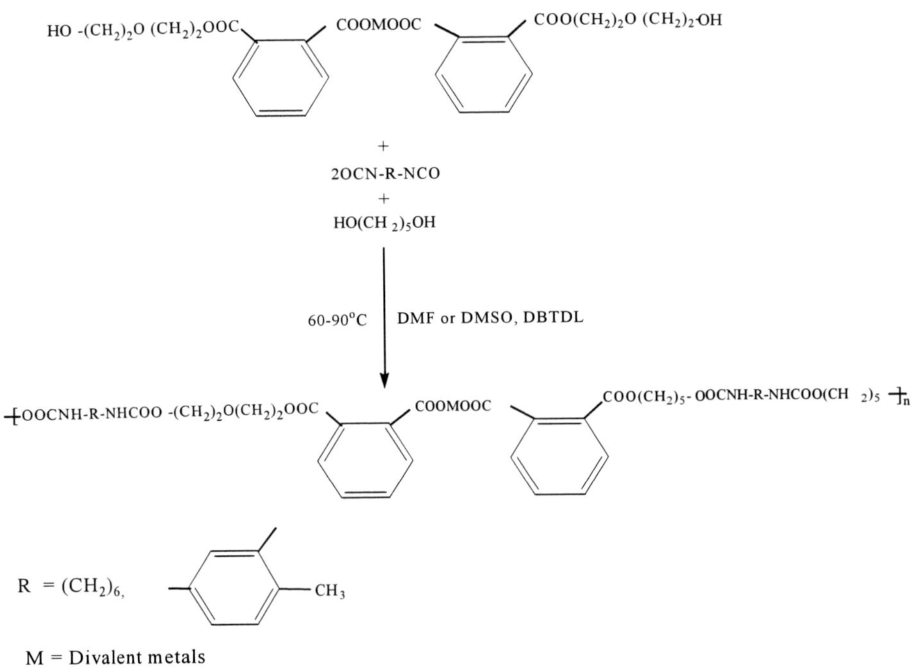

Figure 3. Synthesis of metal-containing copolyurethanes based on mono(hydroxyethoxyethyl)phthalate.

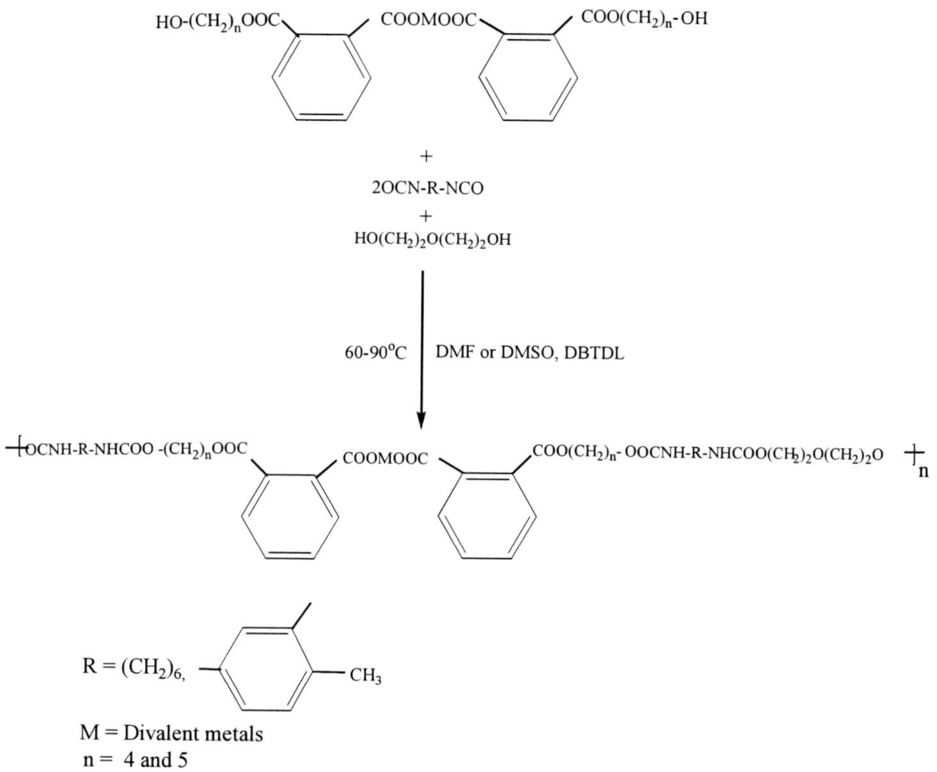

Figure 4. Synthesis of metal-containing copolyurethanes based on mono(hydoxybutyl/hydroxylpentyl)phthalate

Lipatov and Kosyanchuk [30] prepared linear and crosslinked Cu-containing PUs based on oligo(tetramethylene glycol), HMDI and dioxybenzoic acid. The results obtained were compared with data on the structure of PUs based on aromatic diisocyanates. In polymers with weakly pronounced microphase separation (PUs based on TDI), the metal ion crosslinks served as the centers of structure formation. In a PU based on 4, 4′-diphenyl-methane diisocyanate, the crosslinking inhibited microphase separation to yield a polymer with loose structure. In a polymer based on HMDI, the microphase separation was stronger and resulted in a large degree of ordering of the structural elements

Flame-retardant Ca, Mn and Pb-containing PUs were synthesized by the solution polymerization of HMDI with Ca, Mn and Pb salts of mono(hydroxybutyl)hexolate (Figure 5) [27]. The flame-retardant property of these polymers was investigated by measuring limiting oxygen index (LOI) and compared with metal-containing PUUs. Due to the higher chlorine content, the metal-containing PUs showed higher flame retardancy than the corresponding metal-containing PUUs. Though the Pb-containing PUs have less chlorine than the Ca and Mg-containing PUs, their flame-retardant properties were found to be similar to those of the other metal-containing polymers due to the lower percentage of combustible materials available in Pb-containing polymers than in Ca and Mg-containing polymer.

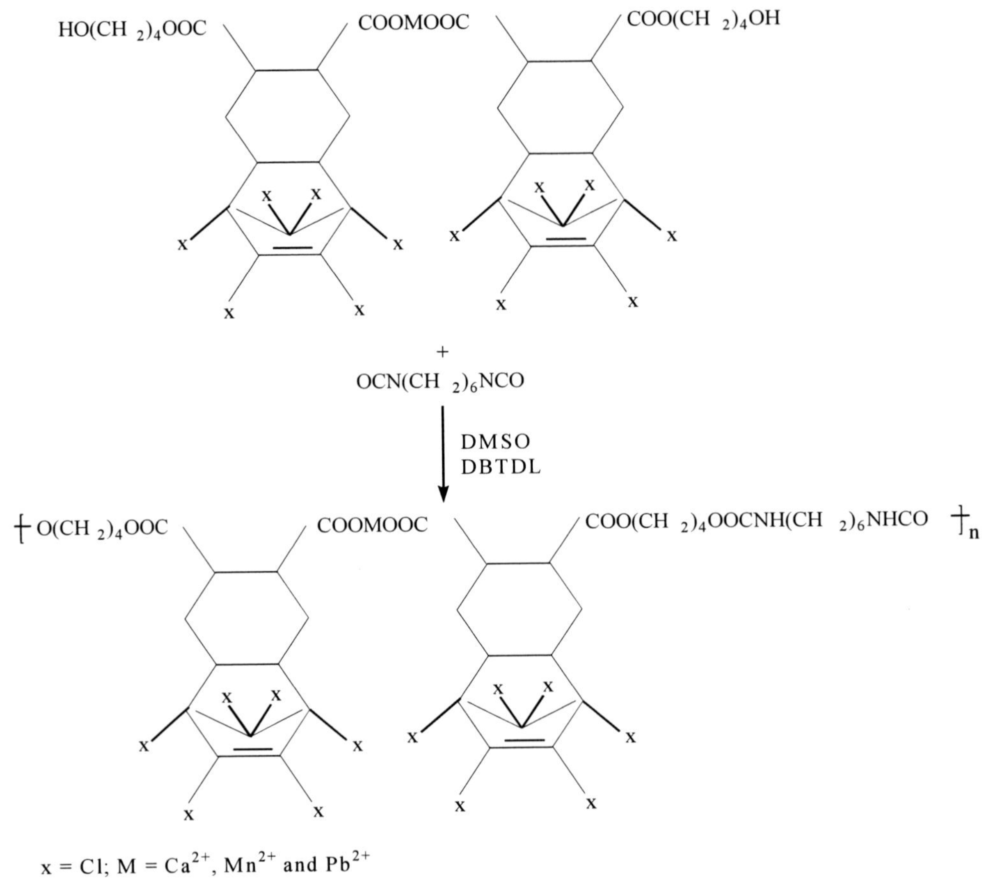

Figure 5. Synthesis of metal-containing PUs based on mono(hydoxybutyl)hexolate.

The inclusion of phosphorus group in a polymer chain can impart interesting properties of the polymer. This polymer is widely used in fire resistance [31-33], to enhance process ability [34, 35], and provide plasticizing properties [36]. Rendering polymers as a flame retardant by using of phosphorus has received much attention. Liu et al. [37] synthesized phosphorous-containing PUs by reacting phosphorous-containing diisocyanates, bis (4-isocyanatophenoxy) phenyl phosphine oxide and bis (3-isocyanatophenyl) phenyl phosphine oxide with various diols. The thermal analysis of the phosphorous-containing PUs showed that incorporating phosphorous into the polymers increased exothermicity during their decomposition. According to these results, the phosphorous-containing PUs were less thermally stable than conventional PUs. Char yield and LOI measurements demonstrated that the flame retardancy of PUs markedly improved by the incorporation of phosphorous group. Similarly, flame retardant properties of PUs resulting from the interaction of an organic polyisocyanate and polyol material comprising a polymeric halogenated phosphorous diol were reported by Wadim et al [38]. PUs containing phosphonate group were also prepared by the reaction of 1, 2-dihydroxpropyl phosphonic acid and a prepolymer based on 4, 4'-diphenylmethane diisocyanate and poly(propylene glycol) with 1, 4-butane diol as the chain extender [39]. DSC studies showed that the T_g values of the polymer were increased with increasing the ionic content which indicated a decrease in segmental motion of the soft segments due to the hydrogen bonding of ether oxygen's of the soft segment with the phosphonic acid groups in ionic diols (Figure 6). This result showed that the introduction of phosphonic acid group influenced the thermal properties of the parent PUs.

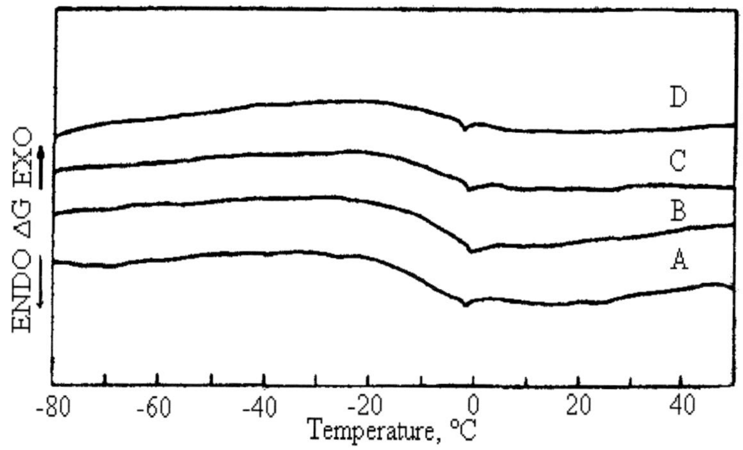

Figure 6. DSC traces of phosphorous-containing PUs (A) 0% phosphorous, (B) 20% phosphorous, (C) 60% phosphorous, and (D) 100% phosphorous.

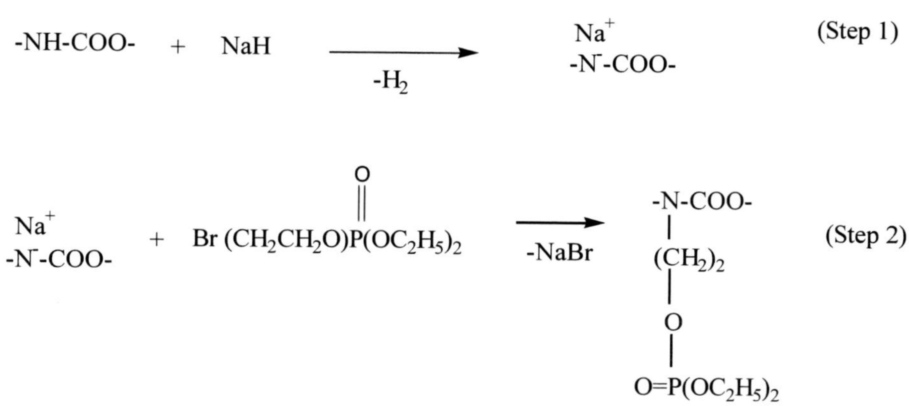

Figure 7. Synthesis of phosphorous-containing PUs based on bisphenol-AF.

Troev et al [40] synthesized phosphorous-containing PUs by reacting phosphoric acid diesters (RO_2) P (O) X (X=H, CH_3, Ph) with hydroxy carbamates. These polymers were found to be readily soluble in highly polar solvents like DMF, DMSO and water. The results showed that these polymers have low temperature degradation due to the cleavage of P-O-C bonds. These polymers were reported as additives and carriers of biologically active compounds. The phosphorous-containing PUs based on bisphenol-AF were synthesized by N-alkylation in a two step process (Figure 7) [41]. First, the polymer was metalated with sodium hydride; then the prepared urethane polyanion was treated with diethyl 2-bromoethyl phosphate. The results of TGA revealed that the thermal stability of phosphorous-containing PUs was decreased due to the thermal dissociation of the phosphorous moiety in the PUs. When the degree of N-substitution increased, the T_g and polarity of the PUs decreased due to weakened intermolecular interactions between the polymer chains caused by the decrease in the number of hydrogen bonds between them. The phosphorous-containing PUs exhibited enhanced fire resistance because phosphorous simultaneously promoted carbonization of the polymer and inhibited combustion. The XRD patterns (Figure 8) of these polymers showed that with increasing phosphorous content, the degree of crystallinity of the PUs decreased. The authors noted that this observation was due to decreasing the number of hydrogen bonds and increasing the content of phosphorous groups that act as plasticizers.

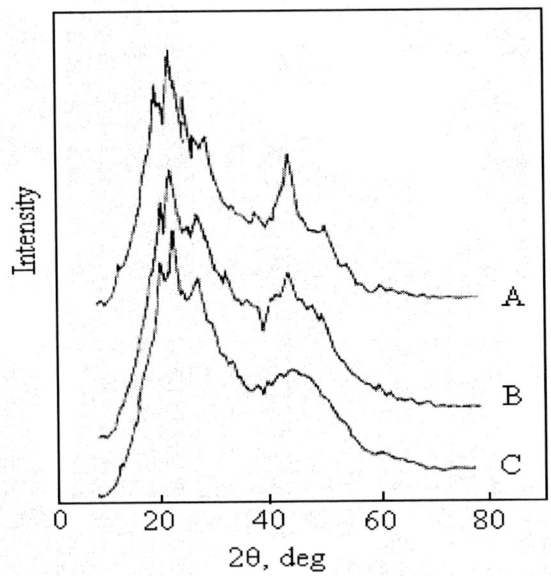

Figure 8. Effect of phosphorous content on XRD patterns of polyurethanes (A) 0%, (B) 1% and (C) 2%.

The incorporation of ion-containing groups on to a polymer back bone has been one method used to improve the blood contacting properties of a variety of polymers. The blood-contacting properties of various ionomers have indicated that the presence of sulphonate ionic groups can produce favorable blood-contacting responses including antithrombotic activity, reduced complement activation, and low platelet deposition and activation [13]. Silver et al. [13] studied the anti-coagulant effects of sulphonated PUs. For this work, the base PU was synthesized by reaction of methylene bis(p-phenyl isocyanate), 1, 4-butane diol, and polytetramethylene oxide in a 3:2:1 mole ratio. After that, the derivatives of this material were made by replacement of 30, 40 and 80 mol % of the urethane hydrogens with propyl sulphonate groups. The results showed that the PUs had excellent blood contacting properties. These polymers were investigated to determine their influence on thrombus formation. These polymers were shown to delay clotting times in the following ways: by direct complex formation between the polymer and thrombin; by interference with fibrin polymerization; and by complex interactions between polymer, thrombin, plasma antiproteases and fibrinogen in plasma. A series of elastomeric sulphonated PUs based on TDI and hydroxyl terminated polybutadiene, poly(propyleneoxide), and polytetramethylene oxide were prepared by Ding et al [42, 43]. The choice of cation was found to have little effect on the T_g of the materials, but had a significant effect on their modulus-temperature properties and a profound effect on their stress-strain behavior at room temperature. The rubbery plateau modulus was found to vary with cation, presumably due to differing levels of isolated ionic groups in the matrix. The addition of water to the casting solvent caused more ionic groups to be dispersed in the matrix, lowering the rubbery plateau modulus. Small-angle X-ray scattering studies indicated that as the polyol molecular weight was increased, the aggregate number density decreased and the aggregate size increased. It was found that the poly(propyleneoxide)-based ionomers formed smaller and more regularly spaced ionic aggregates than TDI- polytetramethylene oxide or hydroxyl terminated polybutadiene-based ionomers. Ding et al. [44] also reported the sulphonated PUs based on poly(tetramethylene oxide) of molecular weight 1000, neutralized with Ca^{2+}, Ni^{2+}, Zn^{2+}, Cd^{2+}, Cs^{2+} and Ea^{3+}. The extended X-ray absorption fine structure (EXAFS) spectroscopy showed that the degree of local order correlates inversely with the strain at which stress crystallization begins, indicating that the cohesiveness of the ionic aggregates controls the large-strain behavior of the materials by suppressing ion-hopping. Upon hydration, the materials lose their mechanical strength. In this work, the local structure in the Ca^{2+} neutralized material was insensitive to hydration, while that in the Ni^{2+} neutralized material changed dramatically.

Novel soft-segmented ionic PUs (linear and crosslinked) were synthesized by grafting the sodium sulfonate onto the chain of poly(ethylene oxide) with molecular weights of 400, 600, 800, and 1000 and characterized by Wang et al [45]. The effect of plasticizer (poly(ethylene oxide) monomethyl ether and poly(ethylene oxide) dimethyl ether) on the ionic conductivity of the polyurethane was investigated. The ionic conductivity of both linear and crosslinked soft segmented PUs was found to be enhanced by the addition of the plasticizer. The ionic conductivity of a polymer electrolyte was substantially reduced with an increase in the terminal hydroxyl group content in the plasticizer. It was found that crosslinking of sodium sulfonate-side chain poly(ethylene oxide) enables improvement of the film stability and slightly enhances the ionic conductivity. Wei and Yu [46] prepared a series of polyether polyurethane ionomers having different contents of sodium sulfonate groups in the soft segments. These ionomers exhibited good film-forming ability, typical elastomeric behaviors, and high interdomain connectivity. The PUs with polyethylene oxide soft segments had better ionic conductivity at room temperature than formerly synthesized PUs with anions in the hard segment domains. It was reported that increase in the ionization level increased the compatibility of the hard and soft segments and the glass transition region of the soft segment became broader.

Three series of new thermoplastic, high molecular weight, segmented thio PUs were synthesized by a one-step melt polymerization from thiodiols (bis[4-(2-hydroxyethyl)thiomethylphenyl]methane, bis[4-(3-hydroxypropyl)thiomethylphenyl] methane, and bis[4-(6-hydroxyethyl)thiomethylphenyl]methane (BHHM)), HMDI and 20-80 mol % of poly(oxytetramethylene)glycol by Kultys and Pikus [47]. All these obtained polymers were colorless solids that possessed partially crystalline structures; those obtained with 40-80 mol % poly(oxytetramethylene)glycol were elastomers. An increase in the poly(oxytetramethylene)glycol content decreased hardness, modulus of elasticity, and tensile strength, whereas elongation at break increased. BHHM-based PUs obtained in the melt polymerization showed the best tensile properties due to the incorporation of diphenyl methane units with active methylene groups in their structure. The PUs exhibited definite glass transitions (-70 to -59°C) that were nearly independent of the hard-segment content up to about 50 wt% (40-80 mole % poly(oxytetramethylene)glycol), indicated the existence of mainly microphase-separated soft and hard segments.

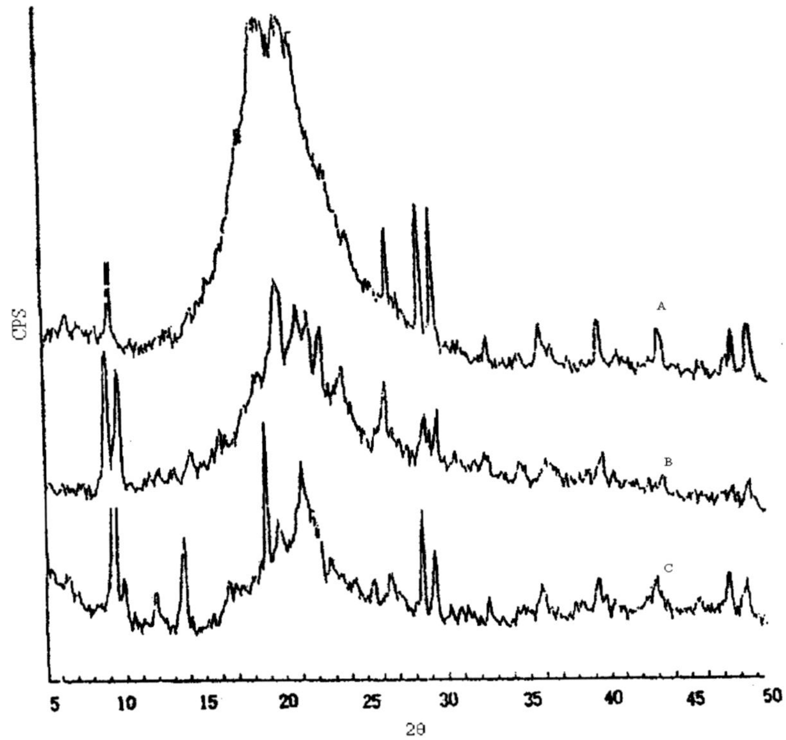

Figure 9. WAXD patterns for the samples (A) Ni-PU, (B) Cu-PU, and (C) Co-PU.

Mechanical properties of PUs crosslinked with Ni, Cu, and Zn ions and their complexes with crown ethers were studied by Kosyanchuk et al [48]. It was shown that the mechanical properties of the PUs depend on the nature of the metal in the network junction. The behavior of the PUs containing crown ethers in the network junctions was primarily determined by the ability of macrocyles to form complexes with metal ions, which results in weakened interaction between the metal ion and the polymer chain. PUs with ligating sites for metal complexation was synthesized by Acharya et al [49]. In this work, the –COOH group was incorporated into the polyurethane chain by the reaction of a tri-functional monomer, bis(hydroxymethyl) propionic acid with HMDI and TDI, respectively. The pristine tri-functional monomers themselves formed stable metal complexes with the metals and so were chosen to be incorporated into the polyurethane chain. The free –COOH group was used for metal complexation using non-toxic metals like Zn and Ag. These PUs were found to be useful as antibacterial agents and for potential bio-medical applications. The relationship between structure and morphology on one hand and molecular mobility on the other hand in segmented polyurethane based on oligo-oxytetramethylene glycol with metal ions Cu and Ni incorporated in the chains was studied by Georgoussis et al [50]. These systems were considered as potential ionic conductors and models for the investigation of single anionic conductivity. The various techniques employed in this study provided several indications that the degree of microphase separation into hard-segment microdomains and soft segment microphase was higher in the system with Ni than with Cu ions. Chen et al. [51] reported the coordination PUs based on novel diols bis[(2-hydroxyphenyl)methylene]hydroxylethylene-amino]-copper(II), cobalt(II), and nickel (II). The geometric structures of the metal-Schiff base, which plays an important role on the properties of these materials, were studied by EXAFS spectroscopy and wide angle X-ray diffraction techniques (WAXD). EXAFS study indicated that the coordination structures of the samples were determined by the nature of the central metal ions and were not affected by the composition and morphology of the polymer backbone. Among the samples, the geometric structure of the Ni^{2+} Schiff base coordination polymer showed less regularity than that of Cu^{2+} and Co^{2+} analogues. The metal-Schiff base coordinated polyurethanes with regular microstructure exhibited interesting liquid crystalline character. The WAXD profiles (Figure 9) showed that the metal-Schiff base with different metal ions resulted in different coordination structures. Cu and Co-containing polymer trended to form a more regular microstructure than Ni-containing polymer.

3. Metal-Containing PUUs

PUUs are a kind of important copolymers. They are composed of a class of elastomers exhibiting superior extensibility, toughness and extensively used ranging from textile fibers to medical prosthesis [52]. Introduction of urea group into the polymer backbone is expected to improve the solubility of the polymer without decreasing the thermal stability significantly. Incorporation of metal into PUUs has emerged and finds wide applications in various fields. Keeping this fact in mind, many researchers have synthesized and characterized various metal-containing PUUs using the different types of monomers.

Matsuda [53] synthesized Ca and Mg-containing PUUs by the polyaddition reaction of HMDI with mixtures of Ca and Mg salts of mono(hydroxyethyl)phthalate and bisureas containing two hydroxyl groups which were obtained from monoethanolamine and diisocyanates. The general reaction scheme for the synthesis of metal-containing PUUs from the various moneomers is shown in Figure 10. The obtained PUUs were found to be yellowish powder. Elemental analysis data showed a close agreement between observed and calculated values of metal content of the PUUs. The viscosities of these metal-containing PUUs were found to be decreased markedly with increase in metal content. From thermal studies, it was found that the decomposition temperature of these metal-containing PUUs was decreased with increase in metal-content.

The PUUs obtained by the reaction of TDI with Mn, Cu, Co and Pb salts of mono(hydroxyethyl)phthalate and bisurea were found to be brown to green in colour [54]. TGA studies indicated that the decomposition temperature of PUUs was decreased by introducing metal into the polymers. These polymers exhibited different behavior in thermal analysis due to the presence of different metals in the PUUs. The order of the degradation reactions, the activation energies of degradation were calculated for all these metal-containing PUUs by the Murray and White method. The activation energies given in Table 3 showed that the temperature change was less in the first stage than in the second stage, whereas the weight loss and activation energy were higher in the first stage than in the second stage for all the

PUU systems. The authors reported that the first stage may be considered to be urethane-urea chain scission followed by the formation of isocyanate and hydroxyl components and the second stage may be involved cyclization and secondary degradation. Based on the activation energy, the thermal stability of these metal-containing PUUs followed in the order, Mn > Pb > Cu > Co. As metals were present in the main chain of PUU by ionic linkage, the thermal stability of these transition metals was explained by the presence of odd electrons, in the outermost orbital, which probably enhanced the ionic hold of the metal in the polymer chain. Hence, Mn exhibited a thermally more stable ionic link in the back bone chain of PUU by the support of the d^5 orbital.

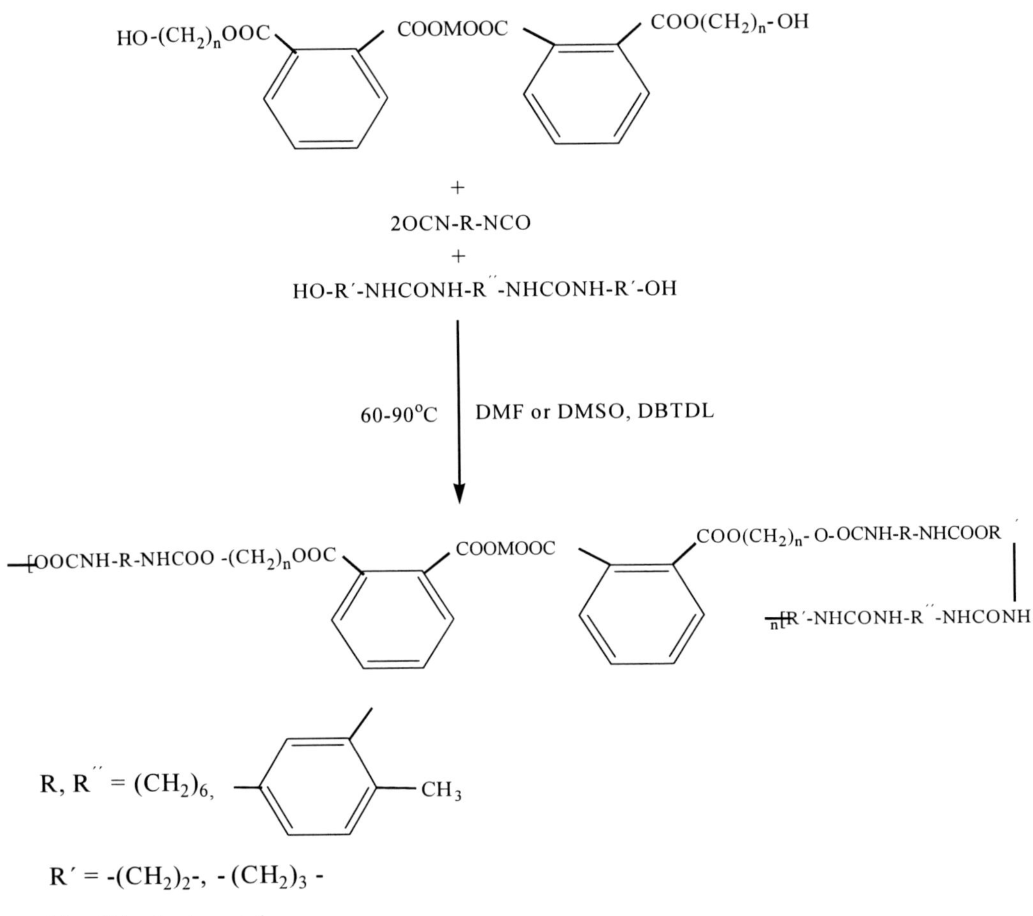

R′ = $-(CH_2)_2-$, $-(CH_2)_3-$

M = Divalent metals

n = 2, 4 and 5

Figure 10. Metal-containing PUUs based on divalent metal salts of mono(hydroxyethyl/hydoxybutyl-/hydroxylpentyl)phthalate.

Ca, Mn, Pb and Zn-containing PUUs were synthesized by the reaction of Ca, Mn, Pb and Zn salts of mono(hydroxybutyl)phthalate and 1:1 mixtures of toluylene 2,4-bis(ω, N-hydroxyethyl)urea and mono(hydroxybutyl)phthalate, hexamethylene bis(ω,N-hydroxyethyl)urea and mono(hydroxybutyl)phthalate with HMDI or TDI [19, 20]. Thermal studies showed that these metal-containing PUs had higher initial decomposition temperature than that of PUUs. The PUUs were found to be more stable than PUs as the former was stabilized by more hydrogen bonding. It was observed that the existence of metal promotes the thermal decomposition of urethane and urea and with increase in metal content of the polymers the stability decreases. From analytical methods, the metal content in PUs or PUUs were found to be less than the value calculated based on equal reactivity of co-monomers. This showed that the reactivity of diol and hexamethylene bis(ω,N-hydroxyethyl)urea are higher than that of mono(hydroxybutyl)phthalate towards diisocyanates.

Table 3. Degradation and activation energies of PUUs

Polymers	Stage 1			Stage 2		
	Temperature (K)	Weight loss (%)	E_a (KJ mol^{-1})	Temperatur e (K)	Weight loss (%)	E_a (KJ mol^{-1})
Cu-PUU	468-598	60.0	42.83	598-798	19.8	25.37
Mn-PUU	468-548	48.2	71.18	548-713	31.7	39.68
Pb-PUU	448-573	46.0	66.99	573-823	33.2	32.23
Co-PUU	448-573	63.9	34.75	573-783	16.0	25.37

Zn-containing PUUs were synthesized by the reaction of HMDI or TDI with 1: 1 mixture of Zn salt of mono(hydroxyethoxyethyl)phthalate or mono(hydroxpentyl)phthalate or mono(hydroxybutyl)phthalate and bisureas such as hexamethylene-bis(ω, N-hydroxyethyl) urea, tolyelene-2, 4-bis(ω, N-hydroxyethyl) urea, hexamethylene-bis-(ω, N-hydroxypropyl) urea and tolyelene-2, 4-bis(ω, N-hydroxypropyl) urea (Figure 11) [26, 55, 56]. The TDI-based PUUs showed higher initial decomposition temperature than the HMDI-based PUUs due to the presence of stiff phenylene rings in the main chain of the TDI-based PUUs. From the analytical methods it was found that the metal content in PUUs were lower than the calculated values, which indicated that the reactivity of bisureas was higher than that of metal containing ionic diols relative to the diisocyanates. Similarly, Ca-containing PUUs were synthesized from HMDI or TDI and 1: 1 mixtures of calcium salt of mono(hydroxyethoxyethyl)phthalate and bisureas by Jayakumar et al [57, 58]. DSC studies showed that these PUUs prepared from HMDI-based bisures have lower T_g values than those derived from TDI-based bisureas due to flexible hexamethylene group. The viscosities of these polymers were found to be very low due chain dissociation in DMSO solvent. The X-ray diffraction studies showed that the HMDI-based polymers were partially crystalline and TDI-based polymers were amorphous in nature.

Flame-retardant metal-containing PUUs were prepared by reacting HMDI with 1:1 mixtures of Ca, Mn and Pb salts of mono(hydroxybutyl)hexolate and bisureas such as hexamethylene bis [N′-(1-hydroxy-2-methyl-prop-2-yl)urea] and tolylene 2, 4-bis [N′-(1-hydroxy-2-methyl-prop-2-yl)urea] [27]. These metal-containing PUUs exhibited somewhat lower LOI values than other flame-retardant polymers such as PVC and polyphosphoramide esters. The flame retardancy of these polymers was due to the flame being cooled by the hexachloropenta diene which was formed by a Retro-Diels-Alder reaction and by various other rearranged products formed during decomposition. The flame retardancy was also due to the modification of the combustion mechanisms by the chlorine radicals formed. The PUUs based on tolylene 2, 4-bis [N′-(1-hydroxy-2-methyl-prop-2-yl)urea] showed slightly higher flame retardancy than those derived from heamethylene bis [N′-(1-hydroxy-2-methyl-prop-2-

yl)urea] because of the presence of aromatic rings in the main chain of the former. In general, the thermal stability of these metal-containing PUUs followed in the order Ca > Mn > Pb.

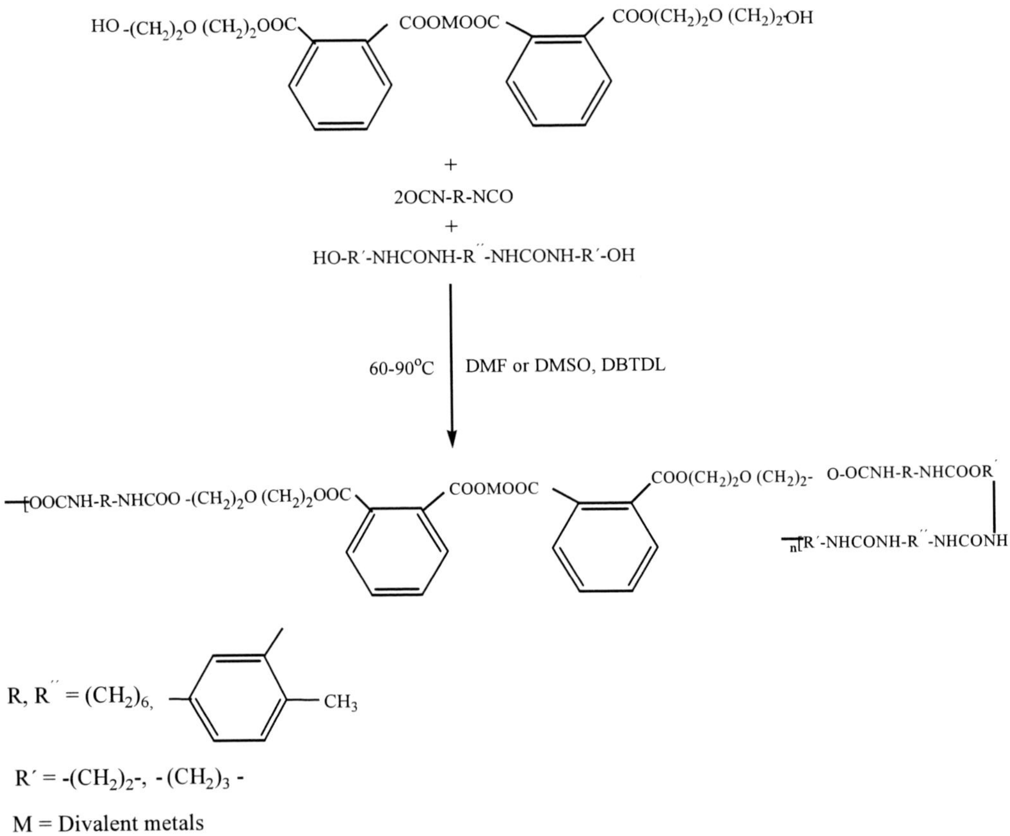

Figure 11. Synthesis of PUUs based on divalent metal salts of mono(hydroxyethoxyethyl)phthalate.

Halato-telechelic PUUs were prepared by the polyaddition reaction of isocyanate-terminated adducts of diethylene glycol with TDI [59]. The reactions were carried out by using Mg or Ca salts of p-aminobenzoic acid and 4, 4'-diaminodiphenylmethane dissolved in DMF. The obtained PUUs were white solids when the metal content was low, and those with high metal content were slightly brown solids, or slightly brown or brown glassy materials. Further, those with a low metal content and the parent polymers were soluble in DMF and DMSO. Generally, they became insoluble in these solvents as the metal content increases. The reduced viscosity of these PUUs in DMF or DMSO increased as the concentration increased. In addition, it was reported that the increase in viscosity was higher in DMF than in DMSO due to the lower depression of ionic association in DMF because of its lower dielectric constant (Є=36.7) than in DMSO (Є= 46.6). The viscosity behavior of the halato-telechelic poly-ureas and PUUs at low concentrations was discussed theoretically by using the equation (1).

$$[\eta_n] - = [\eta]\, x/100\text{-}x \qquad (1)$$

Where $[\eta]$ is the intrinsic viscosity, $[\eta_n]$ is the intrinsic viscosity when viscosity-average molecular weight increased by a factor n, and x is the degree of association of ionic groups.

Equation (1) showed that the difference between $[\eta_n]$ of partially associated polymers and $[\eta]$ of completely dissociated ones is proportional to $[\eta]$. Hence, the difference between $[\eta_n]$ and $[\eta]$, became larger as the molecular weight between ionic groups, i.e. $[\eta]$ increases. Equation (1) also indicated that even if the molecular weight between ionic groups is constant, an increase in degree (x) of ionic association, caused by increasing the added amount of $MgCl_2$, leads to an increase in the difference between $[\eta_n]$ and $[\eta]$. Thus, equation (1) clearly suggested that the viscosity behaviors of the halato-telechelic poly-ureas and PUUs at low concentrations are attributable mainly to partial reassociation of dissociated ions.

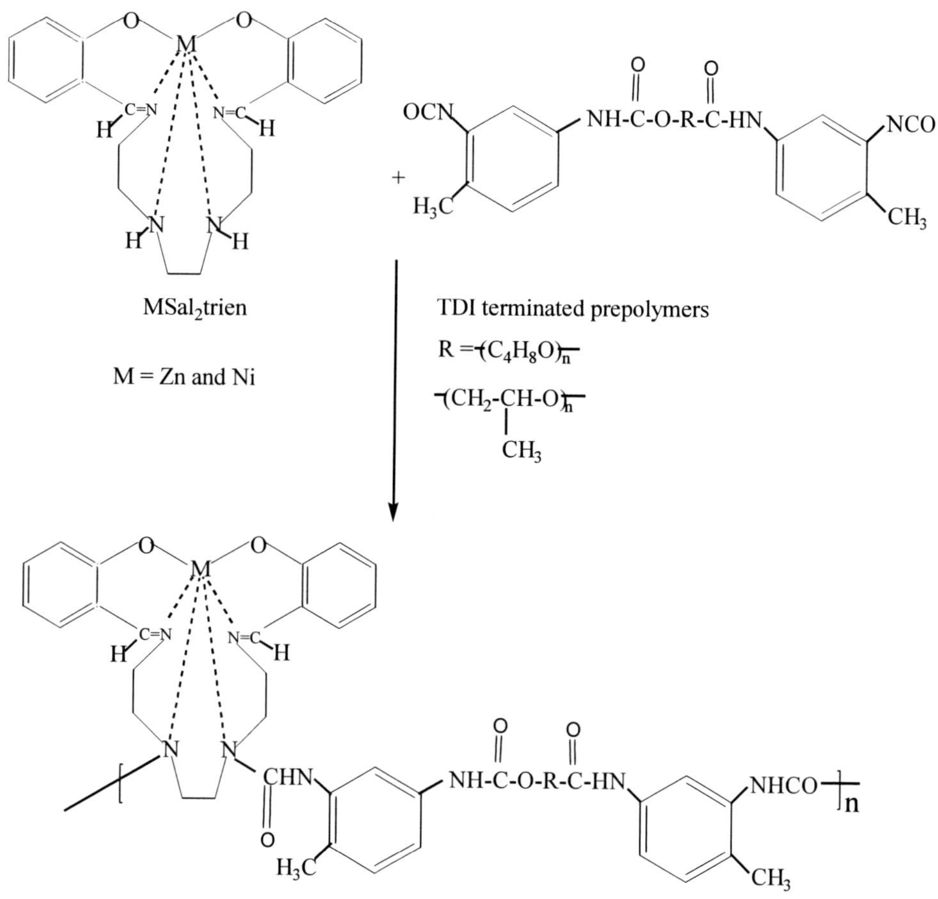

Figure 12. Synthesis of metal-containing PUUs from $MSal_2$trien and poly(1,4-butanediol) or poly(propyleneglycol).

Ni and Zn-containing PUUs (Figure 12) and copolyurethane-ureas (Figure 13) were synthesized by the reaction between hexadentate Schiff base metal complexes ($MSal_2$trien, where M = Ni and Zn, Sal=salicyladehyde and trine= triethylenetetramine) and isocyanate-terminated prepolymers. The NCO-terminated prepolymers used were TDI-terminated poly(1,4-butanediol), TDI-terminated poly(propylene glycol) and the prepolymers synthesized from MDI and diols. The diols employed were polycaprolactone diol and poly(tetramethylneoxide). Copolyurethane-ureas were obtained by the reaction between MSal2trien, poly(1,4-butanediol) or poly(propylene glycol) and MDI [60]. The results showed that the inherent viscosity of polymers increased with increasing wt% of $MSal_2trien$ in the polymer. Ni and Zn-containing polymers showed good thermal stability compared to the polymers without $MSal_2trien$ in the backbone and found that thermal stability increased with increasing wt % of $MSal_2trien$ in the polymer. The presence of metal in these polymers showed influence on LOI value. In comparison between the polymers obtained from the same metal complex, it was found that the LOI values in most polymers increased with increasing wt % of $MSal_2trien$ in the polymers. Ni and Zn-containing polymers obtained at the same amount of MSal2trien showed the same LOI value.

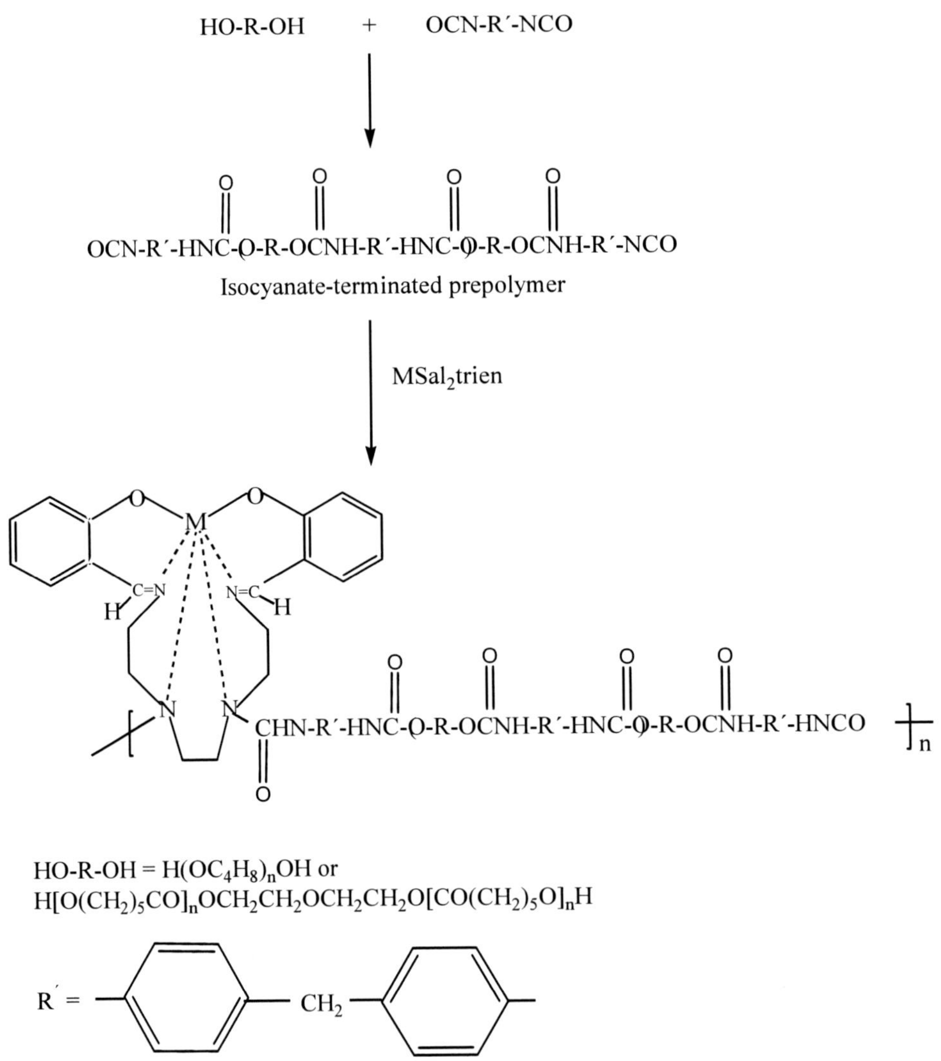

Figure 13. Synthesis of metal-containing PUUs from $MSal_2trien$ and polycaprolactone diol or poly(tetramethylneoxide) and MDI.

4. Metal-Containing PUEs

PUs with flexible polyether backbone have generally good retention properties at low temperatures which make them suitable candidates for use in adhesives, coatings etc., under cryogenic conditions [6]. PUEs have been proven to have sufficient biocompatibility and biodegradability for short-term use in a variety of blood contact applications, such as leaf-ventricular-assist devices for experimental artificial hearts and blood pumbs [61]. Due to its potential important, a great attention has been paid to synthesis various metal-containing PUEs and to study its properties and applications.

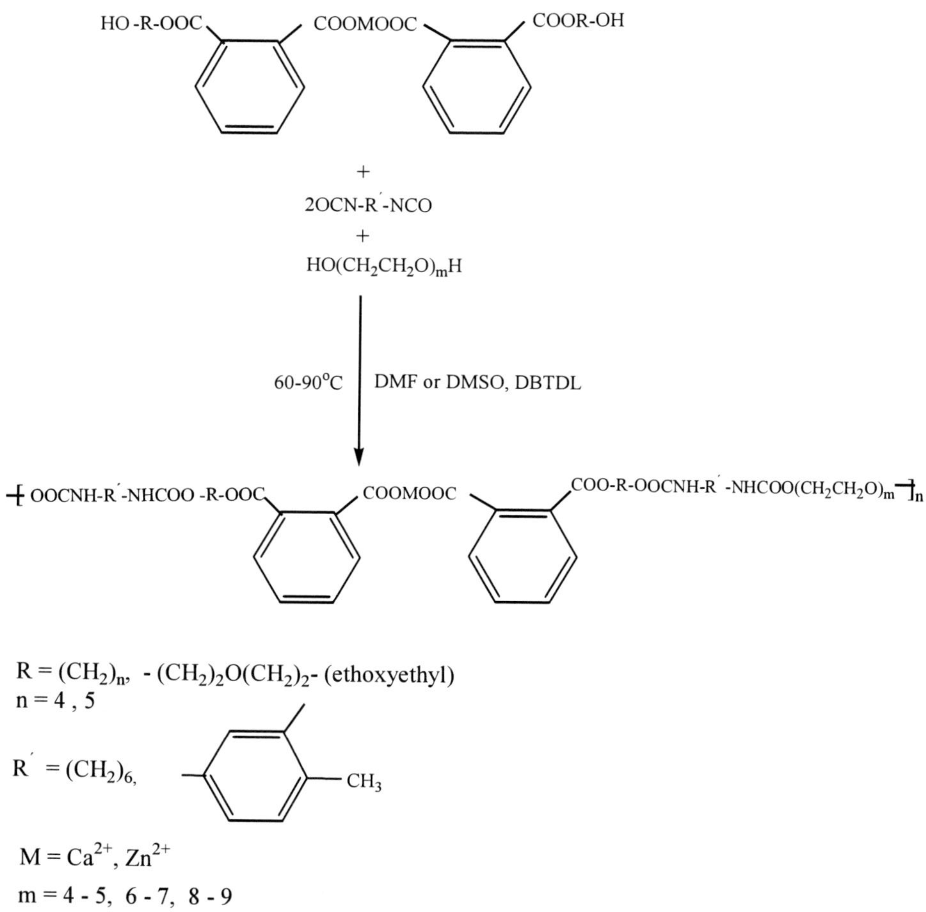

Figure 14. Synthesis of metal-containing PUEs based on Ca and Zn salts of mono(hydoxybutyl/-hydroxylpentyl/hydroxyethoxyethyl)phthalate.

Ca and Zn-containing PUEs of different compositions were synthesized from Ca and Zn salt of mono(hydroxyethoxyethyl)phthalate or mono(hydroxypentyl)phthalate with poly-ethylene glycol (PEG_{300} or PEG_{400}) and HMDI or TDI using the molar ratios of 3:1:4, 2:2:4, and 1:3:4 by Jayakumar et al [62-64]. The general reaction scheme for the synthesis of metal-containing PUEs from the various monomers is shown in Figure 14. The viscosity of PUEs increased with the soft segment content and the unit length and decreased with the Ca content due to dissociation into low molecular weight fragments in solution. The initial decomposition temperatures of TDI-based PUEs were higher than that of HMDI-based PUEs due to the presence of rigid aromatic rings. The T_g values of PUEs increased with the Ca content and decreased with the soft segment content and length. Two T_g values were observed for PUEs for certain compositions due to the presence of heterogeneity, suggesting that phases with different mixing ratios of hard segments and soft segments were present. It was generally observed that TDI-based PUEs had higher T_g values than HMDI-based PUEs. It was observed that the coating properties of PUEs such as, tensile strength, elongation at break, tear strength, water vapour permeability, flexing endurance, cold crack resistance, abrasion resistance, color fastness, and adhesive strength were comparable to the standard values. Therefore, it was concluded that PUEs are suitable for coating applications.

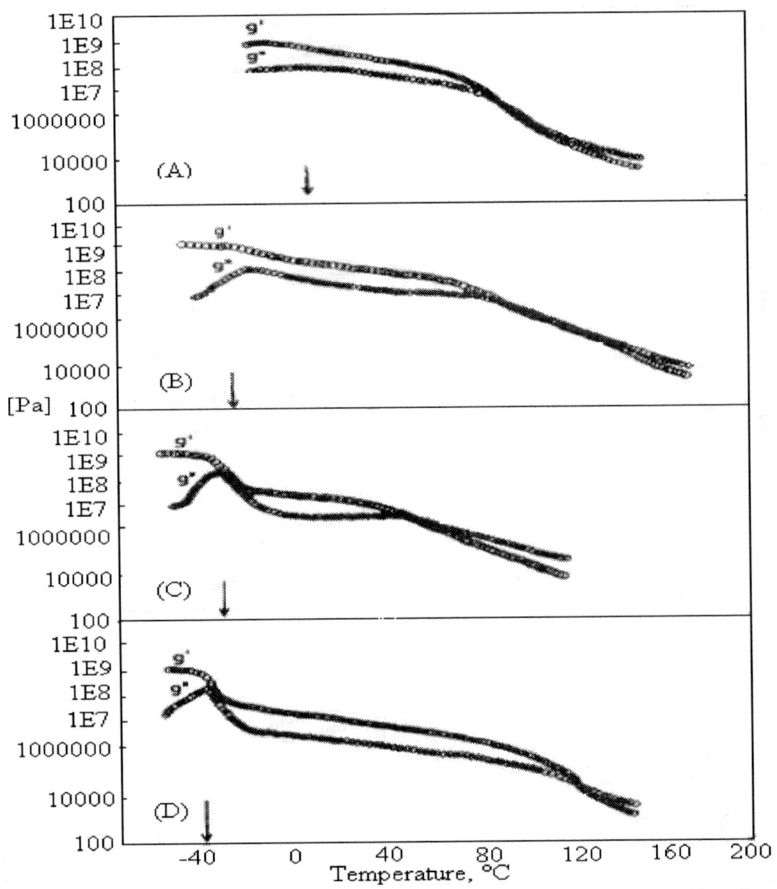

Figure 15. Plot of storage modulus and loss modulus vs. temperature of Ca-(hydoxybutyl)phthalate/ PEG_{400}/HMDI (A) 3:1:4 ratio, (B) 2:2:4, (C) 1:3:4, and (D) Blank PUEs.

The effect of Ca content on the properties of PUEs prepared from the mixture of Ca salt of mono(hydroxybutyl)phthalate, PEG_{200} or PEG_{400} and HMDI or TDI with different molar ratios was studied by Arun Prasath et al [65] (Figure 15). Figure 17 shows the dynamic mechanical analyses of PUEs. This figure showed that at any given temperature, modulus (g' and g'') increased with increase in the ionic content in the polymers. The magnitude of decrease in the modulus (g' and g'') with increase in temperature was found to be less for the polymers with higher ionic content. The polymers without Ca showed lower modulus at any given temperature. The higher value of the storage modulus (g') observed for polymers having higher amount of ionic content, which were found to be related to the formation of physical crosslinks in the copolymer system. As the ionic concentration in the polymer increased the ionic cluster forming capacity also increased which in turn increased the modulus. A rubbery plateau was noted for all the PUEs above the first T_g. The composition of ionic and soft segment content in the polymer system had influenced on the plateau in the rubbery state. A soft rubbery state with lower plateau value of g' was observed for the system with higher amount of soft segment or without ionic species. A hard rubbery state with higher plateau value of g' was observed for the system with higher ionic content. Similar type of thermal and dynamic mechanical properties were reported for Ca-containing poly(urethane-ester)s and poly(urethane-ether-ester)s [66].

5. Conclusions

Metal-containing PUs are the most versatile and useful polymeric materials. In this review, the various methods of preparation and properties of metal-containing PUs and their copolymers, in which the metal ions were located on the main chain, were discussed. The results showed that the incorporation of metal ions into the polymer chain greatly influenced the mechanical, thermal and fire retardancy properties. PUUs and PUEs had sufficient biocompatibility and biodegradability properties, which led to a variety of blood contact applications. In conclusion, metal-containing PUs and their copolymers are very promising materials for the different applications such as, antibacterial agents, aqueous thickeners, impregnants, textile sizers, adhesives, additives, resins and catalysts. Results of these studies will likely determine the future investigation of new metal-containing polymers for the various other applications.

References

[1] Keiji, I. *Handbook of Polyurethane Resins*; Nikkan Kosyo Shimmbun, Tokyo, Japan, 1987; Chapter 1.

[2] Hepburn, C. *Polyurethane Elastomers*; 2nd Ed., Elsevier, Amsterdam, 1991.

[3] Woods, G. *The ICI polyurethane book*, John Wiley & Sons Ltd., Chichester, 1990; Chapter 1.

[4] Comstock, M. J. in *Urethane Chemistry and Applications; ACS Symposium Series*, No. 172, American Chemical Society, Washington, D. C., 1981; Chapter 1.

[5] Meckel, W.; Goyert, W.; Wieder, W., in *Thermoplastic elastomers*; Legge, N. R.; Holden G.; Schroeder H. E. Hanser Gardener Publications, New York, N. Y., 1987, Chapter 2.

[6] Nair, P. R.; Nair, C. P. R; Francis, D. J. *Eur. Polym. J.*1997, 13(1), 89-95.

[7] Matsuda, H. *J. Appl. Polym. Sci.* 1976, 14 (7), 1783-1795.

[8] Matsuda, H. *J. Macromol. Sci., Pure Appl Chem.* 1883, A19 (1), 1-6.

[9] Matsuda, H. *J. Polym. Sci., Part A. Polym. Chem. Ed.* 1974, 12 (2), 455-468.

[10] Matsuda, H. *J. Macromol. Sci., Pure Appl Chem.* 1977, A11 (2), 431-477.
[11] Matsuda, H. *J. Appl. Polym. Sci.* 1978, 22 (12), 3371-3386.
[12] Matsuda, H. *J. Appl. Polym. Sci.* 1978, 22 (9), 2603-2610.
[13] Okkema, A. Z.; Cooper, S. L. *Bio. Mat.* 1991, 12 (8), 668-676.
[14] Silver, J. H.; Hart, A. P.; Williams, E. C.; Cooper, S. L.; Charef, S.; Labarre, D.; Jozefowicz, M. *Bio. Mat.* 1992, 13 (6), 339-344.
[15] Durairaj, B.; Rao, K. V. *Eur. Polym. J.* 1980, 16 (4), 941-944.
[16] Kothandaraman, H.; Rao, K. V.; Ragavan, A.; Chandrasekaran, V. *Polym. Bull.* 1983, 13 (2), 353-356.
[17] Rajalingam, P.; Radhakrishnan, G.; Devanathan, S. B.; Selvi, K. T.; Rao, K. V. *Acta Polym.* 1989, 41 (11), 560-564.
[18] Rajalingam, P.; Radhakrishnan, G.; Vasudevan, C.; Selvi, K. T.; Rao, K. V. *Polym. Commun.* 1990, 31(1), 243-246.
[19] Arun Prasath, R.; Nanjundan, S. J. Macromol. Sci., *Pure Appl. Chem.*1998, A35 (5), 821-842.
[20] Arun Prasath, R.; Jayakumar, R.; Nanjundan, S. *J. Macromol. Sci., Pure Appl. Chem.* (2000), A37 (5), 469-488.
[21] Jayakumar, R.; Lee Y. S.; Rajkumar, M.; Nanjundan, S. *J. Appl. Polym. Sci.* 2004, 91 (1), 288-295.
[22] Jayakumar, R.; Rajkumar, M.; Nagendran R.; Nanjundan, S. *J. Macromol. Sci., Pure Appl. Chem.* 2001, A38 (9), 869-888.
[23] Jayakumar, R.; Rajkumar, M.; Nagendran, R.; Nanjundan, S. *J. Appl. Polym. Sci.* 2002, 85 (6), 1194-1206.
[24] Jayakumar, R.; Lee, Y. S.; Nanjundan, S. *Reac. Func. Polym.* 2003, 55 (3), 267-276.
[25] Jayakumar, R.; Lee, Y. S.; Nanjundan, S. *Int. J. Polym. Ana. Ch.* 2004 (In press).
[26] Arun Prasath, R.; Nanjundan, S. *Eur. Polym. J.* 1999, 35 (11), 1939-1948.
[27] Arun Prasath, R.; Vijayanand, P. S.; Nanjundan, *S. Polym. Int.* 2000, 49 (11), 1464-1472.
[28] Nishijima, Y.; Fukushina, M. *Kobunshi Kagaku*, 1973, 30 (1), 13-20.
[29] Lipatova, T. E.; Bakalo, L. A.; Sivotinskaya, A. L. *Technomic publication*, Section II, Chapter 17, (1975) 99-102.
[30] Lipatov, Y. S.; Kosyanchuck, L. F. *Vyso.Soedin. Seriya A & Seriya B.* 1998, 40 (12), 2022-2028.
[31] Banks, M.; Ebdon, J. R.; Johnson, M. *Polymer*, 1993, 34 (21), 4547-4556.
[32] Nair, C. P. R.; Clouet, G.; Guilbert, Y. *Polym. Deg. Stab.* 1989, 26 (4), 305-331.
[33] Sato, M.; Yokoyama, M. *Eur. Polym. J.* 1979, 15 (1), 75-79.
[34] Ignatious, F.; Sein, A.; Cabasso, I. Smid, J. *J. Polym. Sci. Part A: Polym. Chem.* 1993, 31 (1), 239-247.
[35] Banerjee, S.; Palit, K.; Maiti, S. *J. Polym. Sci. Part A: Polym. Chem.* 1994, 32 (2), 219-227.
[36] Maiti, S.; Banerjee, S.; Palit, K. *Prog. Polym. Sci.* 1993, 18 (2), 227-261.
[37] Liu, Y. L.; Hsiue, G. H.; Lan, C. W.; Chiu, Y. S. *J. Polym. Sci. Part A: Polym. Chem.*1997, 35 (9), 1769-1780.
[38] Wadim, B. Uniroyal Inc., US 3945954, 19760323, US 519441, 19741031, (1976).
[39] Kakati, D. K.; Gosain, R.; George, M. H. *Polymer*, 1994, 35 (2), 398-402.

[40] Troev, K.; Tsevi, R.; Bourova, T.; Kobayashi, S.; Uayama, H.; Roundhill, D. M. *J. Polym. Sci. Part A: Polym. Chem.* 1996, 34 (4), 621-631.
[41] Liaw, D. J.; Lin, S. P. *J. Appl. Polym. Sci.* 1996, 61 (2), 315-319.
[42] Ding, Y.; Register, R. A.; Yang, C. Z.; Cooper, S. L. *Polymer.* 1989, 30 (7), 1204-1212.
[43] Ding, Y.; Register, R. A.; Yang, C. Z.; Cooper, S. L. *Polymer.* 1989, 30 (7), 1213-1220.
[44] Ding, Y.; Register, R. A.; Yang, C. Z.; Cooper, S. L. *Polymer.* 1989, 30 (7), 1221-1226.
[45] Wang, X.; Li, H.; Tang, X.; Chang, F. C. *J. Polym. Sci. Part B: Polym. Phys.* 1999, 37 (8), 837-845.
[46] Wei, X.; Yu, X. *J. Polym. Sci. Part B: Polym. Phys.* 1997, 35 (2), 225-232.
[47] Kultys, A.; Pikus, S. *J. Polym. Sci. Part A: Polym. Chem.* 2001, 39 (10), 1733-1742.
[48] Kosyanchuk, L. F.; Lipatov, Y. S.; Babich, V. F.; Perepelitsyna, L. N. *Polym. Sci. Series.* 2003, A 45 (7), 644-649.
[49] Acharya, V.; Prabha, C. R.; Murthy, C. N. *J. Polym. Mat.* 2003, 20 (1), 83-89.
[50] Georgoussis, G. Kanapitsas, A.; Pissis, P.; Savelyev, Y. V.; Veselov, V. Y.; Privalko, E. G. *Eur. Polym. J.* 2000, 36 (6), 1113-1126.
[51] Chen, L.; Yu, H.; Yang, C. Z. *Polym. Adv. Tech.* 1997, 8 (6), 335-338.
[52] Yamazaki, E.; Hanahata, H.; Hiwatari, J.; Kitahama, Y. *Polym. J.* 1997, 29 (10), 811-820.
[53] Matsuda, H. *J. Polym. Sci. Part A: Polym. Chem.* 1974, 12 (2), 469-481.
[54] Rajalingam, P.; Radhakrishnan, G. *Polymer.* 1992, 33 (10), 2214-2216.
[55] Jayakumar, R.; Arun Prasath, R.; Nanjundan, S. *J. Macromol. Sci., Pure Appl. Chem.* 2002, A39 (8), 853-877.
[56] Jayakumar, R.; Radhakrishnan, S.; Nanjundan, S. *Reac. Func. Polym.* 2003, 57 (1), 23-31.
[57] Jayakumar, R.; Lee, Y. S.; Nanjundan, S. *J. Appl. Polym. Sci.* 2003, 90 (13), 3488-3496.
[58] Jayakumar, R.; Nanjundan, S. *J. Polym. Sci. Part A: Polym. Chem.* 2004, 42 (8), 1809-1819.
[59] Matsuda, H.; Takechi, S. *J. Polym. Sci. Part A: Polym. Chem.* 1991, 29 (1), 83-91.
[60] Chantarasiri, N.; Damrongkosit, T.; Jangwong, W.; Sridaeng, D.; Suebphan, S. *Eur. Polym. J.* 2004, 40 (8), 1867-1874.
[61] Hayashi, K.; Takano, H.; Matsuda, T.; Umezu, M. *J. Biomed. Mater. Res.* 1985, 19 (2), 179-193.
[62] Jayakumar, R.; Lee, Y. S.; Nanjundan, S. *J. Polym. Sci. Part A: Polym. Chem.* 2003, 41 (18), 2865-2878.
[63] Jayakumar, R.; Lee, Y. S.; Nanjundan, S. *J. Appl. Polym. Sci.* 2004, 92 (2), 710-721.
[64] Jayakumar, R.; Nanjundan, S. *Polym. J.* 2003, 35 (9), 734-739.
[65] Arun Prasath, R.; Nanjundan, S.; Pakula, T.; Klapper, M. *Eur. Polym. J.* 2004, 40 (8), 1767-1778.
[66] Arun Prasath, R.; Nanjundan, S.; Pakula, T.; Klapper, M. *Polym. Deg. Sta.* 2004, 85 (2), 911-923.

In: New Frontiers in Polymer Research
Editor: Robert K. Bregg, pp. 55-74
ISBN: 1-59454-956-7

Chapter 3

SOY PROTEIN ADHESIVES AS FIBERBOARD PACKAGING SEALANTS

Zhikai Zhong[1], X. Susan Sun[1], Donghai Wang[2] and Jo Ann Ratto[3]
[1]Department of Grain Science and Industry;
[2]Dept. of Biological & Agricultural Engineering; Kansas State University, Manhattan, KS 66506
[3]U.S. Army Natick Soldier Center; Kansas Street, Natick, MA 01760

ABSTRACT

Fiberboard has been used widely in the fields of construction, insulation, soundproofing, decoration, and packaging. The most popular adhesives for the fiberboard industry are petroleum-based and starch-based adhesives. The petroleum-based adhesives impact the environment with toxicity issues and problems during manufacturing, transportation, and utilization. On the other hand, the starch-based adhesives are environmentally friendly but usually have low adhesion strength and water resistance. Therefore, the development of an environmentally friendly adhesive with high performance is greatly needed. Modified soy protein has great potential as an adhesive because of its natural structure.

This chapter summarizes the research results on soy protein adhesives for fiberboard packaging applications. Soy protein adhesives can be obtained by modifying soy material with at least 50% protein. Variables affecting soy protein adhesion performance include modifier and its concentration, soy protein concentration, pressing conditions, and assembly time. Under optimum processing conditions, soy protein-based adhesives gave appropriate working viscosity, adhesion strength, and water resistance as a fiberboard packaging sealant. After modification, soy protein-based adhesives can provide dry shear strength of about 1.9 MPa and wet shear strength of about 0.4 MPa for solid fiberboard. For corrugated fiberboard, the dry shear strength was about 1.0 MPa and wet shear strength was about 0.2 MPa. Both the dry and wet solid and/or corrugated fiberboard exhibited 100% cohesive failure within fiberboard. The percentage of water-soluble-mass of the modified soy protein adhesive had a value as low as 1.7%, which is low compared to the requirement. Both the solid and corrugated fiberboard packaging boxes sealed with the soy protein adhesives could be easily transported with no damage or alteration in the box after subjecting the boxes to the standard tests of exposure to rain and impact tests. The adhesives developed are geared toward packaging applications, however, they have the potential to be used for many other applications.

INTRODUCTION

Fiberboard is a category of sheet products derived from wood fibers and used for a variety of applications depending on its density. Fiberboards are classified according to the density into insulation board (density = 0.16 ~ 0.50 g/cm^3), medium-density fiberboard (MDF) (density = 0.64 ~ 0.80 g/cm^3), and hardboard (density = 0.88 ~ 1.12 g/cm^3). Insulation board products are mainly used for sheathing, roof decking, roof insulation, ceiling tile, and sound deadening board. It is also used for automotive and furniture industries. Hardboard is mainly used in construction, and MDF is mainly used for industrial core stock (furniture panels). For container and packaging materials, one or more pieces of corrugated or solid fiberboard are used. The pieces are creased, slotted, joined, and folded to a container. Compared to other packaging materials, fiberboard boxes offer many advantages such as lightweight, high strength, and ease of assembly and transportation.

Currently, there are two major types of adhesives used in the fiberboard packaging industry [1]. One is petroleum-based synthetic adhesive, such as urea-formaldehyde resin, phenol-formaldehyde resin, polyolefin-based or polyamide-based hot-melt adhesives, and ethylene vinyl acetate based adhesives. The other one is a renewable material-based adhesive, such as a starch-based adhesive. The synthetic adhesives usually provide high adhesion strength and high water resistance, but they are not biodegradable, making them difficult to dispose of. They can cause unpleasant odors and even produce toxic materials during manufacturing, transportation, and utilization. Migration of some adhesive components from the final products glued with the synthetic adhesives is observed [2-4], and this is always alerted seriously, especially for food packaging. Also, petroleum resources are naturally limited. The starch- and dextrin-based adhesives are biodegradable and environmentally friendly, but they have low adhesion strength and water resistance [5]. Therefore, there is a need for the development of environmentally friendly adhesives with comparable properties to existing petroleum adhesives.

Water resistance is the main challenge for biopolymer-based adhesives because of hydrophilic nature of most biopolymers. To be considered as a water-resistant adhesive, it should have a water-soluble-mass of 2% or less tested according to ASTM standards D 5570 [6]. From a practical standpoint, the packaging box sealed with the adhesive should be able to be transported without the flaps opening after subjecting to the rain cycle at medium intensity (50 mm/h) and the drop tests according to ASTM standards D 1974 and D 4169 [7, 8]. The water resistance of biopolymer-based adhesives can be improved by modification. Researches on soy protein-based adhesives are therefore also focused on modification of soy protein. The history of the soy protein-based adhesive originates in the 1920s, and it was formulated mainly for plywood applications. By the late 1920s, plywood adhesives formulated with soy flour had become a viable product because it provided many advantages: 1) readily available from inexpensive and abundant renewable resources and agricultural processing by-products; 2) easy to handle because of their relatively low viscosity; 3) able to be processed by both hot and cold presses; and 4) veneer with moisture levels as high as 20-35% could be glued without splitting [9]. By the 1960s, soy protein-based adhesives were substantially replaced by the rapidly developed synthetic adhesives, which had better glue strength and water resistance. Recently, however, increasing concerns about environmental problems and supply of the petroleum have promoted the development of environmentally friendly adhesives from renewable, inexpensive resources. Therefore, as a biodegradable adhesive, soy protein gained attention again, and some efforts have been made to improve their adhesive strength and water resistance [10-15]. Some researchers have explored the feasibility of using soy flour for adhesives because the ingredient costs for adhesives based on soy protein isolate are higher than commercially available phenol-formaldehyde and urea-formaldehyde adhesives. However, soy protein isolate-based adhesive has high adhesion strength that cannot be obtained with soy flour based adhesive [15].

Soybean consists of about 20% oil, 40% protein, and 35% carbohydrate. Soy protein has two major components: glycinin (11S) and conglycinin (7S). They account for two-thirds of the soy protein. Soy protein is made of 20 amino acids with side groups from hydrophobic to hydrophilic, and has a typical globular structure (Fig. 1).

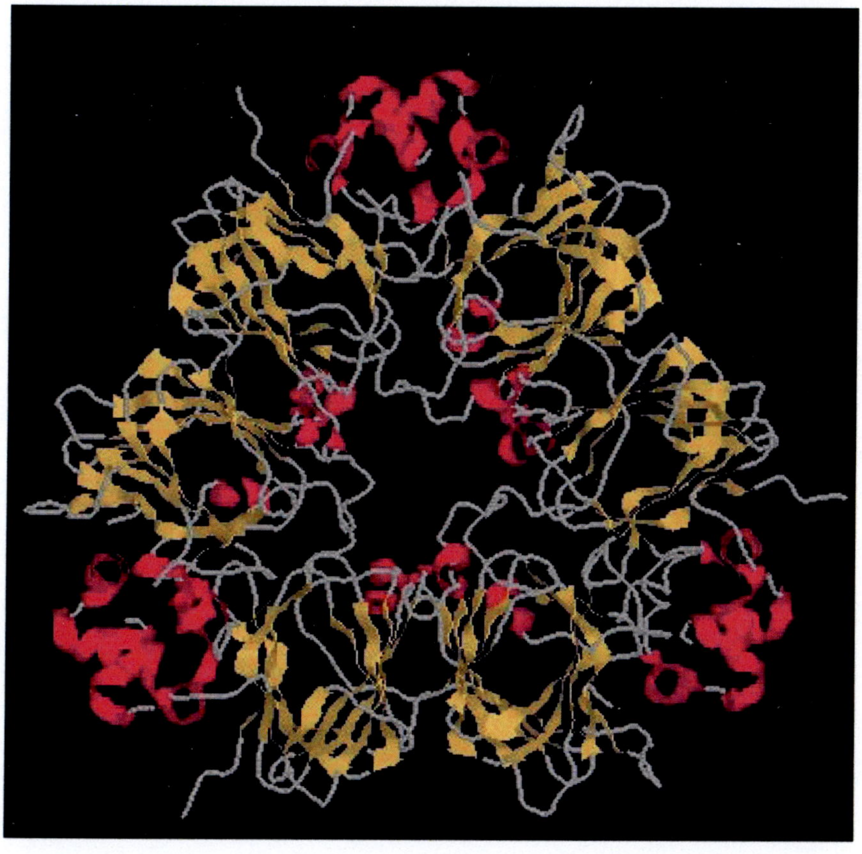

Figure 1. The β-conglycinin domains show typical globular structure of soy protein. The diagram also shows the α–helix and β–sheet secondary structures with different shape/color. (Courtesy from April Zheng, Kathy Mitchell, and X. Susan Sun)

Most hydrophilic residues of the soy protein are exposed outside and most hydrophobic residues were embedded. This kind of configuration and the globular structure have made native soy protein does not having acceptable adhesion and/or water resistance.

Fiberboard is a porous substrate, and made of cellulosic fibers. To achieve good adhesion performance with fiberboard, the soy protein-based adhesive should wet and spread to the fiberboard substrate, and penetrate into the fiberboard porous structure. The exposed hydrophilic residues of the soy protein could interact with hydrophilic group of cellulose, such as the hydroxyl group, but the interaction might be limited by the globular structure. Also, the exposed hydrophilic residues easily interact with water, resulting in lower water resistance. Therefore, modification of the soy protein is an important approach to improve its adhesion performance.

A common method used by researchers is to modify soy protein with heat or denaturation agents such as alkali, enzyme, or detergent. During modification, soy protein is expected to change its conformation, and hence, interactions between soy protein and the substrate are improved while interactions between soy protein and water are diminished. Hettiarachchy et al. [11] use an alkali-modification method and obtain adhesives with improvement in both adhesion strength and water resistance. Alkali treatment breaks strong internal hydrogen bonds of the coiled protein molecules and cause soy protein molecules to unfold so that all their polar groups are available for adhesion to wood [16]. In order to reduce the viscosity of soy protein adhesives, salts and disulfide bond-cleaving agents are used [12]. Sun and Bian [13] found that urea-modified soy protein adhesives are more water-resistant than those modified by alkali. Soy protein also was blended with other proteins, such as animal blood and casein, to achieve useful adhesive characteristics, including ease of process, high adhesion strength, or high water resistance [16].

Sodium dodecyl sulfate (SDS) and guanidine hydrochloride (GuHCl) are well known protein denaturation agents. They denature protein through a totally different mechanism: SDS unfolds protein through hydrophobic interaction while GuHCl unfolds protein through hydrogen bonding interactions. Therefore, the resulted adhesives are expected to have different thermal, rheological, and adhesion properties [14,15].

The objective of this chapter is to review what has been explored in our laboratory on the research and development of soy protein-based adhesives for fiberboard packaging. SDS and GuHCl were the chemicals used to modify soy protein for improved adhesion and water resistance. Variables (modification parameters and processing parameters) affecting soy protein adhesion performance are examined. Water resistance of the soy protein adhesives is investigated in details using a variety of methods.

Characterization of Soy Protein Adhesives

1. Thermal Properties

Soy protein has a typical globular structure, which is based on peptide chains by forming secondary structure (α-helix and β-sheet), tertiary structure, and quaternary structure (Fig. 1). These structures are destroyed completely or partially upon heating or chemical treatment. Since the structure of the protein relates to the final adhesion performance, it is essential to measure thermal properties such as denaturation temperature (T_d) and denaturation enthalpy (ΔH_d) to understand and characterize the soy protein isolate (SPI) adhesives. SPI usually has two denaturation transitions located at approximately 74°C and 90°C (11% soy protein content), corresponding to denaturation of 7S and 11S components of SPI, respectively [17]. Modified SPI with SDS and GuHCl exhibited different thermal responding compared to the unmodified SPI as showed in Table 1: Both the T_d and ΔH_d decreased as SDS concentration increased; Whereas, the T_d increased, and the ΔH_d increased first and then decreased as GuHCl concentration increased.

Table 1. Denaturation temperature (T_d) and enthalpy of denaturation (ΔH_d) of unmodified, and SDS-, GuHCl-modified SPI adhesives. Data presented in this table are from references [17,20]

	11S component			7S component	
	T_d (°C)	ΔH_d (J/g SPI)		T_d (°C)	ΔH_d (J/g SPI)
Unmodified					
	88.8	8.4		74.0	1.6
SDS conc. (wt%)					
0.5	90.1	6.3		74.1	1.1
1.0	90.5	4.7		73.3	1.1
3.0	87.9	3.0		72.6	0.1
5.0	84.9	1.2		ND	ND
GuHCl Conc. (M)					
0.5	97.8	9.0		77.5	2.3
1.0	95.8	5.7		77.6	3.1
2.0	91.8	1.1		75.6	1.6
4.0	ND	ND		ND	ND

ND: not detected.

It is well know that SDS is an anionic detergent, consisting of a SO_3^- anionic head and a -$(CH_2)_{11}CH_3$ long hydrophobic tail. During modification, the hydrophobic tail of SDS penetrated into and interacted with the apolar core of soy protein to open its globular structure, leading to unfolding and partial denaturation of SPI. As a result, SDS continually decreased the T_d and ΔH_d of SPI [17]. However, 5.0% SDS-modified SPI stilled showed a thermal denaturation transition for 11S component (Table 1), meaning that SDS could only partially denature SPI [18,19].

On the other hand, GuHCl denatures protein through hydrogen bonding mechanism as mentioned previously. At first, guanidine molecules hydrogen-bonded to the hydrophilic outlier of soy protein globule. As this was proceeding, numerous guanidine molecules bonded to the soy protein and this may have disturbed interaction forces that stabilize the protein globular structure. As a result, the globular structure opened slightly. Then, more guanidine molecules would hydrogen-bonded to the exposed hydrophilic groups of SPI. Therefore, during the first stage of GuHCl modification, SPI-GuHCl might have structure intermediate between native and highly unordered states as suggested by Tanford [19]. These intermediate states may be more thermally stable and give higher T_d and ΔH_d [20]. At higher GuHCl concentration (4 M), the SPI was unfolded and denatured completely (Table 1).

The above hypothesis was also supported by circular dichroism (CD) studies. Figure 2 showed CD spectra reading at 222 nm as affected by SDS (Fig. 2a) and GuHCl (Fig. 2b) modification. Protein α-helix secondary structure dominates an intense negative band at about

222 nm in the CD spectra [21]. Increasing in the CD spectra intensity at 222 nm (Fig. 2a) suggested that more α-helix structure was formed as SDS content increased. Tanford [19] has indicated that the protein was only partially unfolded after denaturation by detergents even at high concentrations and that existed ordered regions or newly formed ones or possibly even isolated hydrophobic side chains could interact with the hydrophobic tail of detergent molecules to form micellelike regions. These micelle-like regions should have similar structure as α-helix. The intensity for the GuHCl-modified SPI did not change significantly as GuHCl concentration increased up to 1 M, but the increase was significant at 2 ~ 6 M (Fig. 2b). Therefore, the GuHCl only destroyed the protein α-helix structure significantly at a concentration of 2 M or higher.

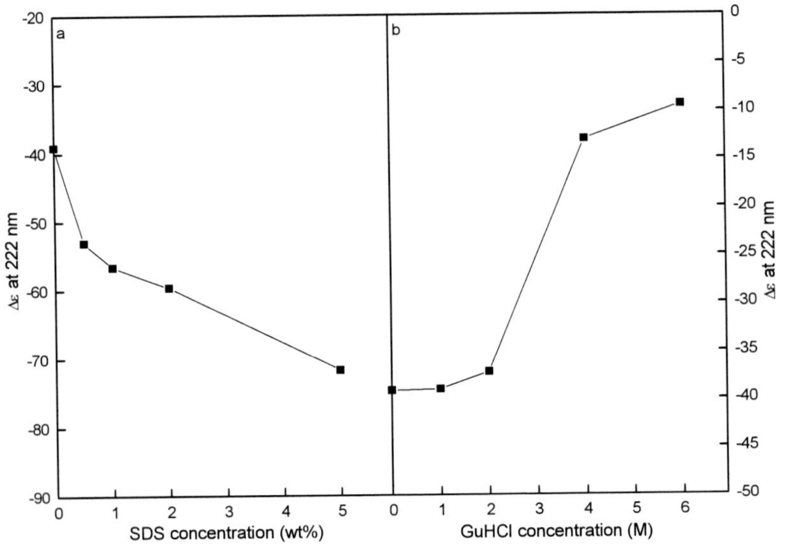

Figure 2. Effects of SDS and GuHCl modification on CD reading at 222 nm. α-Helix content increases as the reading decreases.

2. Rheological Properties

Rheological properties affect the application of adhesives significantly, especially the spreading and penetrating of the adhesives into substrate. Appropriate viscosity range of adhesives is required. Very low viscosity would result in an over-penetration of adhesive, and very high viscosity would lead to inability to spread the adhesive and minimal penetration.

Table 2 summarizes viscosities of the unmodified, and SDS- or GuHCl-modified SPI adhesives. The unmodified SPI adhesive had a viscosity of 12 mPa.s. SDS-modification increased the viscosity greatly, especially at SDS content of 3 wt% or higher [17]. As discussed above, SDS continued to denature SPI, resulting in an unfolded and loose protein structure, which could increase the viscosity. GuHCl, however, did not unfold and destroy soy protein globular structure at low concentration. In fact, the bonding of GuHCl molecules onto SPI at low concentration might interfere with interactions of the SPI molecules with water, which would decreased the viscosity. When GuHCl content was 2 M or higher, SPI was unfolded significantly, and hence the viscosity was increased. However, because of a large amount of GuHCl molecules, the viscosity was not increased as sharply as SDS-modified SPI. The GuHCl molecules may act as a dilution agent [20].

Table 2. Viscosity of unmodified, and SDS or GuHCl-modified SPI adhesives at 26°C and 100.44 sec^{-1} of shear rate (SPI conc. = 10.7 wt%). Data presented in this table are from references [17,20]

SDS conc. (wt%)	0	0.5	1.0	3.0	5.0
Viscosity (mPa·s)	12.0	36.1	69.2	358	>600
GuHCl conc. (M)	0	0.5	1.0	2.0	4.0
Viscosity (mPa·s)	12.0	7.4	10.7	13.4	41.7

Adhesion Performance of Soy Protein

1. Variables Affecting Shear Strength

The most common method to apply soy protein adhesive is to brush and to spread it onto fiberboard substrate [17, 20, 22]. Before assembly, the brushed fiberboard is at room atmosphere. Soy protein molecules penetrate into the fiberboard's porous structure, and water evaporated during this time. After that, the fiberboard assembly was pressed in a temperature-regulated press (Hot Press). Molecule penetration and water evaporation continue, and protein structure may also change because of the high temperature. This process continues until the soy protein adhesive is set. Therefore, any factor that might affect any of these steps would play a role in the final properties of the bonded fiberboard. These parameters, such as modifier concentration, protein concentration, assembly time, and hot-press conditions (temperature, time, and pressure), must be controlled and optimized.

Modifier concentration: Concentration of modifier has two roles on the final adhesion properties. First, modification of the protein depends on modifier concentration. Secondly, modifier, as an additive, may interfere with the adhesion performance of the soy protein adhesive directly through its interactions with soy protein molecules and fiberboard substrate.

As shown in Table 3, both SDS- and GuHCl-modification improved the dry shear strengths; however, the dry shear strength decreased at higher modifier concentration. The soaked shear strength had the same trend against modifier concentration. With modification, soy protein molecules unfolded from their compact and ordered globular structure to a loose and random structure, which would increase contact area between the protein molecules and the fiberboard substrate, enhance penetration of the protein molecules, and increase possibility of protein molecule entanglement. Those reasons contributed to the increased shear strength.

Table 3. Effects of modifier concentration on shear strength of solid fiberboard bonded with soy protein-based adhesives. Data presented in this table are from references [17,20]

	Unmodified	3% SDS modified	5% SDS modified	1M GuHCl modified	5M GuHCl modified
Dry shear strength (MPa)	1.71 ± 0.05	1.97 ± 0.07	1.39 ± 0.05	1.93 ± 0.08	1.82 ± 0.04
Soaked shear strength (MPa)	1.46 ± 0.03	1.68 ± 0.07	1.07 ± 0.08	1.80 ± 0.09	1.37 ± 0.06

The mechanism for the decreasing of shear strength at higher modifier concentration was different for SDS and GuHCl. For SDS, the reason was because of the very high viscosity.The 5% SDS-modified SPI had a viscosity bigger than 600 mPa.s (Table 2). The high viscosity caused difficulty for the adhesive in wetting and penetrating the fiberboard substrate [17]. However, it was not likely due to the same reason for GuHCl because the viscosity was low (Table 2). The major reason should be related to the amount (weight basis) of GuHCl in the 4.0 M GuHCl-modified SPI adhesives. It is much more than the SPI content, and pure GuHCl solution does not have any detectable bonding ability to fiberboard. The larger amount of GuHCl molecules might interfere with the bonding interactions of soy protein macromolecules with fiberboard substrate [20].

SPI concentration: It is understandable that concentration of soy protein should affect adhesion properties because the concentration directly controls the bonding interaction strength. At lower SPI concentrations, the interaction between the SPI adhesive and fiberboard substrate increased greatly as SPI content increased. Therefore, both the dry and soaked shear strengths increased as SPI concentration increased (Table 4). However, high SPI concentration might have two reverse effects: the relative amount of modifier to SPI decreased, which limited the modification (unfolding and denaturation) of SPI; the viscosity may increase, especially for the SDS-modification. Hence, the shear strength decreased as the SPI concentration further increased (Table 4). For GuHCl modification, the first reverse effect is explained by the concentration of 0.16 (ratio of SPI/GuHCl solution) giving a viscosity of only 44.4 mPa.s (Table 4), which was not high enough [20].

Press temperature: Processing temperature is another important role affecting properties of the soy protein-based adhesives for fiberboard. As showed in Table 5, high processing temperature greatly improved the adhesion strength for both unsoaked and soaked specimens. The improvement was more significant in the temperature range of 50 ~ 100°C. Because SPI had denaturation temperatures of about 75°C and 90°C for components of 7S and 11S (Table 1), respectively, the high temperature would denature SPI significantly, resulting in an unfolded and loose structure. This effect was similar to that of the modification. High press temperature may also (1) increase molecular mobility and enhance penetration of SPI molecules; (2) improve water evaporation and increase the rate of curing for SPI molecules; and (3) increase the chemical reactions between SPI and fiberboard [17, 20, 22].

Table 4. Effect of SPI concentration on viscosity and shear strength of 1M GuHCl-modified soy protein adhesives. Data presented in this table are from references [20]

SPI/1M GuHCl solution (wt/wt)	6/100	8/100	10/100	12/100	14/100	16/100
Viscosity (mPa.s)	3.7	6.0	10.6	16.7	29.2	44.4
Dry shear strength (MPa)	1.57 ± 0.08	1.74 ± 0.07	1.93 ± 0.08	2.00 ± 0.05	1.94 ± 0.05	1.93 ± 0.05
Soaked shear strength (MPa)	1.14 ± 0.10	1.62 ± 0.10	1.80 ± 0.09	1.87 ± 0.08	1.93 ± 0.06	1.88 ± 0.08

Table 5. Effect of press temperature on shear strength of solid fiberboard bonded with soy protein-based adhesives. Data presented in this table are from references [17,20]

Press temperature (°C)	25	50	75	100	120
3% SDS modified					
Dry shear strength (MPa)	1.06 ± 0.07	1.24 ± 0.06	1.66 ± 0.09	1.71 ± 0.05	1.64 ± 0.09
Soaked shear strength (MPa)	0.01	0.29 ± 0.08	0.69 ± 0.08	1.35 ± 0.07	1.31 ± 0.06
1M GuHCl modified					
Dry shear strength (MPa)	1.48 ± 0.07	1.68 ± 0.05	1.85 ± 0.07	1.93 ± 0.08	1.94 ± 0.03
Soaked shear strength (MPa)	0.13 ± 0.05	0.28 ± 0.08	1.52 ± 0.09	1.80 ± 0.09	1.85 ± 0.08

Effects of time: There were two times that need to be mentioned: assembly time and press time. The assembly time is the time that the adhesive-coated substrate rested at room temperature prior to assembling. The press time is the time that the assembled substrate pressed in the hot-press. During assembling, the SPI molecular chains penetrated into the fiberboard through the porous structure, and some water evaporated. During pressing, the SPI penetration into the fiberboard and the chemical/physical interactions were enhanced because of the physical contact. Water continued to evaporate, and SPI became more entangled upon curing. Therefore, enough time was required. However, very long assembly and press times could decrease the shear strength because of too much water evaporation and starving glue line formation resulted from over-penetration of SPI molecules.

Table 6 summarizes the effects of the assembly time and press time on the shear strength of soy protein-based adhesives. For the SDS-modified SPI adhesives, the shear strength increased first and then leveled off as assembly time and press time increased. No decrease in the shear strength was observed and this might be due to the short experimental times [17]. For the GuHCl-modified SPI adhesives, optimum values were observed for both assembly time (10 min) and press time (5 min) [20].

Press pressure: Processing pressure could also affect the final adhesion performance. When the soy protein-based adhesive was applied to the fiberboard substrate, it was spread and it wetted the surface. The protein molecules interact with the substrate by physical and/or chemical forces. The molecular chain could also penetrate into the fiberboard porous structure, enhancing interactions between the soy protein and the substrate. Increased press pressure improved contact at the interface between the adhesive layer and the fiberboard, helped the wetting and penetrating of the adhesives to substrate, and hence might increase the shear strength [22].

Table 6. Effects of time on dry shear strength of solid fiberboard bonded with various SPI adhesives. Data presented in this table are from references [17,20]

Press time (min) (1)	1	2	5	10	15
3% SDS modified SPI (MPa)	1.12 ± 0.07	1.49 ± 0.09	1.71 ± 0.05	1.97 ± 0.07	-
1M GuHCl modified SPI (MPa)	1.89 ± 0.07	1.99 ± 0.08	2.00 ± 0.05	1.90 ± 0.05	1.84 ± 0.07
Assembly time (min) (2)	2	5	10	15	20
3% SDS modified SPI (MPa)	1.63 ± 0.05	1.71 ± 0.05	1.81 ± 0.07	1.83 ± 0.06	-
1M GuHCl modified SPI (MPa)	1.81 ± 0.06	1.92 ± 0.05	2.00 ± 0.05	1.99 ± 0.07	1.90 ± 0.05

(1) The samples are prepared at 100°C and 2 MPa with an assembly time of 5 min for 3% SDS modified SPI and an assembly time of 10 min for 1M GuHCl modified SPI.

(2) The samples are prepared at 100°C and 2 MPa for 5 min.

Drying-treatment: Drying is one of the processing parameters that needs to be controlled in preparing the soy protein and the adhesive. Drying is expected to affect the physical and/or chemical status of the protein and, hence, affects final performance [23]. Dried adhesive powder is also easy for handle, transportation, and storage.

The 3% SDS-modification denatured most of the soy protein, leading to only a small peak during differential scanning calorimetry heating (Table 7). Drying treatment did not further denature the SPI, indicated by the lack of significant differences between the drying-treated and control samples for both the denaturation temperature and denaturation enthalpy (Table 7). Also, it seemed that drying treatment temperature did not affect the denaturation behavior of SPI; although the drying treatment temperature of 90°C was higher than the denaturation temperature of the control SPI adhesive. This was probably because water evaporation was rapid during drying treatment at 90°C, resulting in low moisture content of the sample and great increase in denaturation temperature [24]. Therefore, the remaining ordered conformation of SPI did not have enough time to unfold.

However, drying-treatment did affect the rheological property of the SDS-modified SPI adhesives. Viscosity increased greatly as drying treatment temperature increased (Table 7). During adhesive preparation, it was also observed that the drying-treated SDS-modified SPI powder was easier to disperse in water than the unmodified one did. Although drying treatment did not further denature the remaining ordered conformation of SPI molecules, it may change interactions among SPI molecules and physical states of denatured SPI molecules. This increase in molecule entanglement results in higher viscosity and easier dispersion, which is confirmed by the shear strength studies.

Table 7. Effects of drying-treatment on thermal and rheological properties of the SDS-modified SPI adhesives. Data presented in this table are from references [23]

Drying-treatment temperature (°C)	Denaturation temperature (°C)	Denaturation enthalpy (J/g SPI)	Viscosity (mPa·s)
Control	87.0	4.2	470
30	87.3	4.7	770
50	86.9	4.8	1945
70	87.0	4.5	>5500
90	87.0	4.8	4090

Table 8 gives representative results of the effects on drying-treatment on adhesion properties for the SDS-modified SPI. In general, the soaked sample had similar shear strength as the unsoaked sample, indicating most bonding strength was recovered after the wet specimen was dried. The wet sample had much lower shear strength since the wet fiberboard itself had low strength. Fiberboard bonded with drying-treated adhesives has higher shear strength than fiberboard bonded with the untreated adhesives. Drying-treatment temperature has only minor effects on shear strength (Table 8). Comparing to the control, drying treatment increases the percentage of cohesive failure within fiberboard significantly; 100% cohesive failure within fiberboard is found for all samples bonded with the adhesives drying-treated at 70 and 90°C (Table 8). This meant that the modified SPI adhesives have superior water resistance than the adhesive used to manufacture the fiberboard.

Table 8. Effects of drying-treatment on corrugated fiberboard bonded with the SDS-modified SPI adhesives (1). Data in parenthesis indicated the average cohesive failure within fiberboard. Data presented in this table are from references [23]

Drying-treatment temperature (°C)	Unsoaked shear strength (MPa)	Soaked shear strength (MPa)	Wet shear strength (MPa)
Control	0.65 ± 0.09 (22)	0.65 ± 0.11 (43)	0.17 ± 0.06 (40)
30	0.94 ± 0.13 (81)	0.96 ± 0.09 (24)	0.18 ± 0.04 (0)
50	1.03 ± 0.07 (100)	1.05 ± 0.05 (100)	0.25 ± 0.04 (84)
70	0.91 ± 0.14 (100)	1.01 ± 0.13 (100)	0.25 ± 0.03 (100)
90	1.01 ± 0.07 (100)	0.99 ± 0.04 (100)	0.22 ± 0.01 (100)

(1) The unsoaked shear strength is obtained from control, unsoaked sample; the soaked shear strength is obtained from the sample that is soaked in tape water for 24 h and then dried for 5 days; the wet shear strength is obtained from the sample that is soaked in tape water for 24 h and then tested immediately.

Therefore, drying treatment did improve water resistance of the modified SPI adhesives. Drying treatment may enhance hydrophobic interactions among SPI molecules and increase SPI molecule entanglements. The treated SPI is easier to disperse into water and has higher viscosity. The increased molecular entanglement also enhanced the interactions between SPI and fiberboard substrate, as well as the interactions among SPI molecules, which would improve water resistance [23].

2. Percentage of Water-Soluble-Mass

Water resistance is an essential parameter for an adhesive being used for a package closing or sealing, which may be exposed to a variety of humid environments. The water resistance of adhesive can be characterized by percentage of water-soluble-mass, which is measured according to the ASTM Test Method D5570 [6] (Scheme 1):

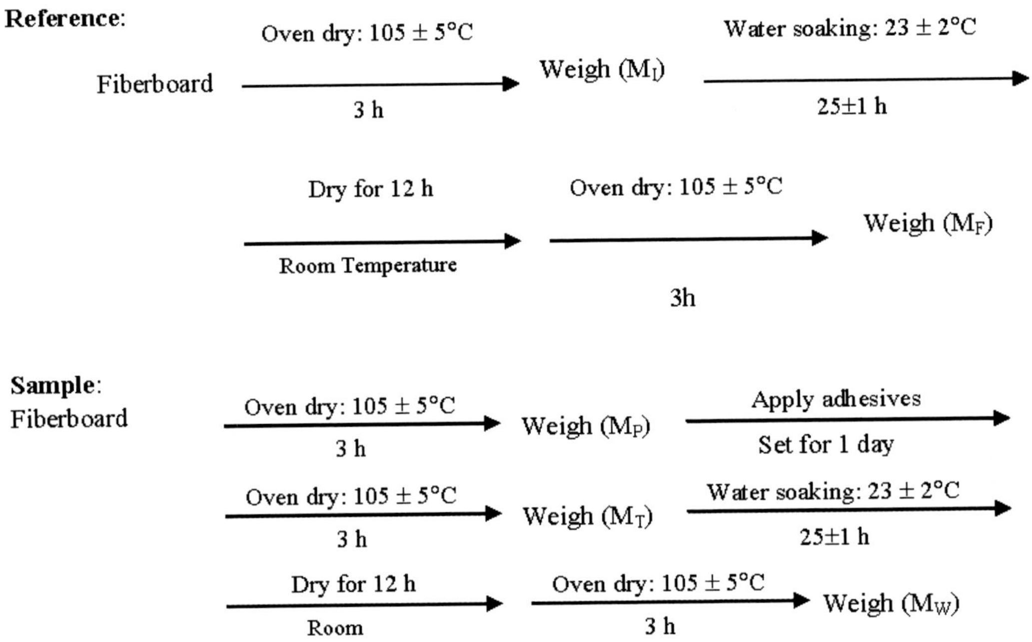

Scheme 1. Measurement of percentage of water-soluble-mass.

Water-soluble-mass (M_R) in **Reference** was: $M_R = M_I - M_F$

The mass of adhesive (M_A), which was applied onto **Sample**, was calculated by:

$$M_A = M_T - M_P$$

The percentage of water-soluble-mass of adhesive in **Sample** was:

$$S_A\% = \frac{M_T - M_W - M_R}{M_A} \times 100$$

The unmodified SPI had the water-soluble-mass of 5.9% (Table 9). According to ASTM D5570 [6], a water-soluble-mass of up to 2% could be attributed to incidental materials in the adhesive and experimental variation. Therefore, data in Table 9 suggested that some components of the unmodified SPI adhesive were removed by water during the soaking. However, the 1% SDS-modified SPI adhesive had an average water-soluble-mass of 1.7% (Table 9), indicating that the adhesive had adequate water resistance. The 1M GuHCl-modified SPI adhesive had a much higher water-soluble-mass (39.3%), which could be attributed to the loss of GuHCl molecules because the 1M GuHCl-modified SPI adhesive had about 42 wt% GuHCl (dry-base) [23].

Table 9. Percentage of water-soluble-mass of various SPI-based adhesives. Data presented in this table are from references [23]

	Percentage of water-soluble-mass, S_A (%)
Unmodified SPI	5.9 ± 5.0
1% SDS-modified SPI	1.7 ± 2.4
1M GuHCl-modified SPI	39.3 ± 0.4

3. Evaluation of Water Resistance for Fiberboard Packaging Box

The water resistance of the soy protein-based adhesives can be further evaluated on packaging boxes. As stated in ASTM D 1974 [7], successful completion of the following performance tests provide a demonstration of water resistance: (a) Subject three samples of box closure to the rain cycle of Test Method D 951 [25] at medium intensity (50 mm/h) for 6 h; (b) The boxes shall immediately be subjected to the drop tests of Practice D 4169 [8]; (c) The sample boxes need to be picked up and moved without the flaps opening.

For this purpose, a rain simulation system (Fig. 3) was designed and constructed according to ASTM Method D 951 [25]. The system consisted of six parts: Test Room, Sprays, Flow Control Valves, Circulating System, and Water Temperature Control. One part of the cold water was heated through an Electric Water Heater, and then was mixed with the other part of cold water in the Thermostatic Mixing Valve. This valve together with temperature measure system controlled water temperature. The water, with desired temperature, then flowed through pipes and other control valves, and finally sprayed in the Test Room. The Spray Nozzles were designed and located so that the droplets are falling from gravitational force only when they strike the specimens. The Circulating System together with Flow Control Valves is designed so that the system can provide continuous water spraying at proper intensity. Overall, the system provided a test room area of 150 × 120 cm^2, which was suitable for six 30 cm × 30 cm boxes. The system could simulate raining from medium intensity (50 ± 13 mm/h) to high intensity (100 ± 25 mm/h) and with temperature from 23°C to 50°C.

The water resistance of various soy protein-based adhesives is tested by spray method on both solid and corrugated fiberboard-packaging boxes. The boxes are filled with wood veneers to certain weight, and then sealed with the adhesives by securely bonding at least 50% of the contact area of the flaps according to ASTM D1974, Closure Method 2A3 (Fig. 4). After subjecting the samples to rain at medium intensity (50 mm/h or 2 in./h) for 6 h, the boxes are immediately subjected to drop tests according to ASTM D 4169, Element A at assurance Level II [8]. The weight of the box, which is measured before subjecting it to raining, is used to determine the drop height (381 mm or 15 in.). The final (sixth) drop is conducted from twice the specified drop height. Table 10 listed the specifications of the drop test. After the test, the box is immediately examined for flaps opening and handling.

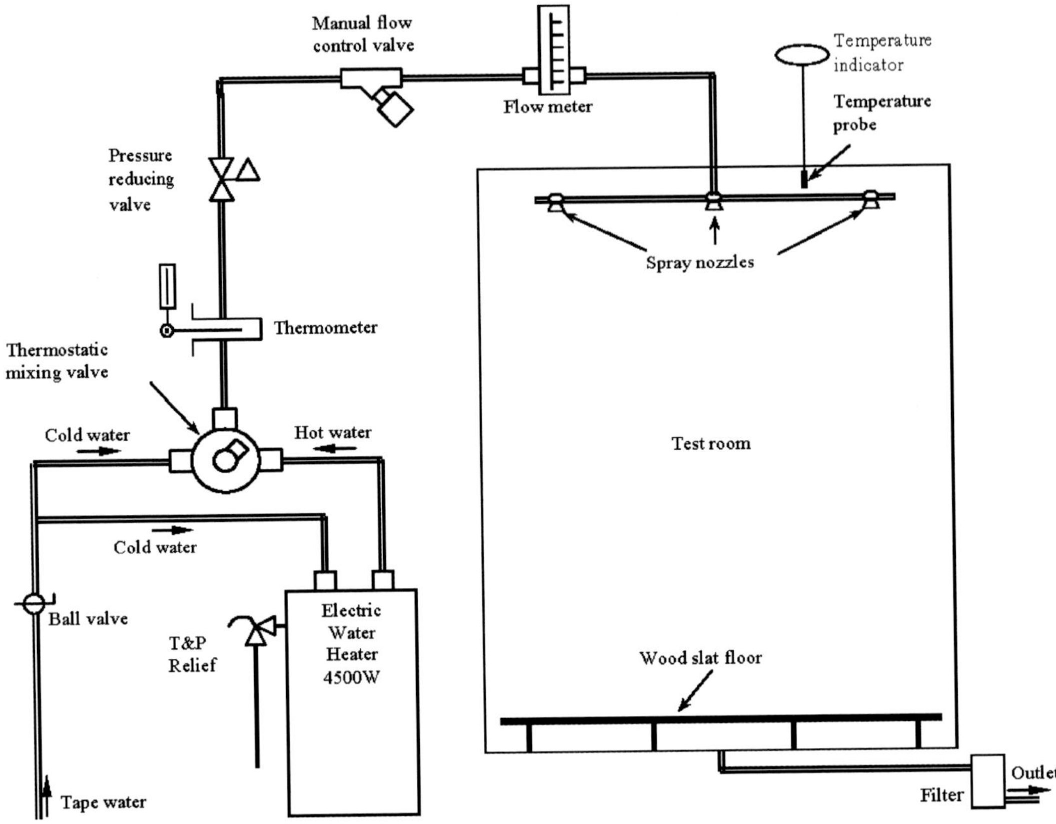

Figure 3. Schematic diagram of rain simulation system.

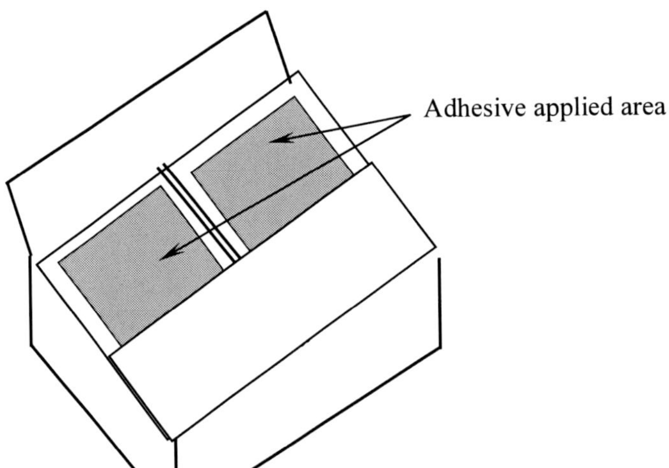

Figure 4. ASTM D1974 Box Closure Method 2A3.

Table 10. Specifications of the drop test

Drop height (mm)	Number of drops at specified height	Impact orientation
381	One	Top
381	Two	Adjacent bottom edges
381	Two	Diagonally opposite bottom corners
762	One	Bottom

After the dropping tests, most solid fiberboard boxes are still sealed as shown in Figure 5. No flaps opened and the boxes are conveyable, indicating that the adhesives are water resistant. To open the boxes, one had to use both hands with force. The opened boxes clearly showed cohesive failure within fiberboard (Fig. 6). However, the corrugated fiberboard boxes showed linerboard opening after the dropping tests (Fig. 7). This is because of low water resistance of the adhesive used to make the corrugated fiberboard. In our opinion, it was not the flap opening because the other linerboard of the corrugated fiberboard was still bonded well to the remaining box (Fig. 7). The corrugated fiberboard packaging boxes were still conveyable after the tests. This also indicated that the soy protein-based adhesives had superior water resistance than the adhesive used to make the fiberboard.

Figure 5. Solid fiberboard packaging box after the dropping tests, sealed by soy protein-based adhesives.

Figure 6. Solid fiberboard packaging box (after the dropping tests), sealed by soy protein-based adhesives, and opened by hand with force.

Figure 7. Corrugated fiberboard packaging box after the dropping tests, sealed by soy protein-based adhesives.

Tables 11-13 summarize the data from the tests. It can be seen that the water resistance results are also affected by the processing parameters, such as pressure and temperature. For the solid fiberboard boxes, most of the soy protein-based adhesives gave satisfactory results (no flaps opening and conveyable) except the 1% SDS-modified adhesive with low press pressure. But the 1% SDS-modified adhesive worked well at higher press pressure (Table 11). For the corrugated fiberboard boxes, all the soy protein-based adhesives passed the specifications even at low press pressure (Table 12). Although the control, unmodified SPI adhesive gave satisfactory results if it was processed at 100°C, it lead to non-conveyable box after tests if processed at room temperature (Table 13). The unmodified SPI adhesive also had high percentage of water-soluble-mass (Table 9) as discussed before. As discussed previously, high processing temperatures could also be treated as a kind of modification because it could denature potentially the soy protein, enhance molecule entanglement, and increase the possibility of chemical reactions. Therefore, temperature is critical for most soy protein-based adhesives. To use at room temperature, latex-like soy adhesive has been developed by our group [26]. The latex-like soy adhesive is extracted from soy flour using a multi-step precipitation method, and has excellent adhesion performance. The latex-like soy adhesives could work well at room temperature and low press pressure (Table 13).

Table 11. Results of water resistance of the soy protein-based adhesives on packaging box made from solid fiberboard (ASTM D1974 and D4169). Assembly time = 10 min

Sample	Press Conditions	#	Weight (kg)	Results
Control	100°C/0.5MPa/5min	1	7.97	No flap opening/conveyable
		2	7.98	No flap opening/conveyable
		3	7.90	No flap opening/conveyable
1M GuHCl Modified	100°C/0.5MPa/5min	1	7.98	No flap opening/conveyable
		2	7.98	No flap opening/conveyable
		3	7.90	No flap opening/conveyable
1% SDS Modified	100°C/0.5MPa/5min	1	7.96	Partial flap opening/conveyable
		2	7.99	Flap opening/non-conveyable
		3	7.91	No flap opening/conveyable
1% SDS Modified	100°C/2MPa/5min	1	7.96	No flap opening/conveyable
		2	7.99	No flap opening/conveyable
		3	7.91	No flap opening/conveyable

Table 12. Results of water resistance of the soy protein-based adhesives on packaging box made from corrugated fiberboard (ASTM D1974 and D4169). Assembly time = 10 min

Sample	Press Conditions	#	Weight (kg)	Results
Control	100°C/0.5MPa/5min	1	7.73	No flap opening/conveyable
		2	7.74	Partial flap opening/conveyable
		3	7.66	No flap opening/conveyable
1M GuHCl Modified	100°C/0.5MPa/5min	1	7.72	No flap opening/conveyable
		2	7.74	No flap opening/conveyable
		3	7.66	No flap opening/conveyable
1% SDS Modified	100°C/0.5MPa/5min	1	7.72	No flap opening/conveyable
		2	7.74	No flap opening/conveyable
		3	7.65	No flap opening/conveyable
1% SDS Modified	100°C/2MPa/5min	1	7.63	No flap opening/conveyable
		2	7.71	No flap opening/conveyable
		3	7.73	No flap opening/conveyable

Table 13. Results of water resistance of the soy protein-based adhesives on packaging box made from corrugated fiberboard and pressed at different temperature (ASTM D1974 and D4169). Assembly time = 10 min

Sample	Press Conditions	#	Weight (kg)	Results
Control	100°C/0.5MPa/5min	1	7.73	No flap opening/conveyable
		2	7.74	Partial flap opening/conveyable
		3	7.66	No flap opening/conveyable
Control	23°C/0.5MPa/5min	1	7.70	Partial flap opening/non-conveyable
		2	7.72	No flap opening/conveyable
		3	7.66	Flap opening/non-conveyable
Latex-like soy adhesive	100°C/0.5MPa/5min	1	7.73	No flap opening/conveyable
		2	7.75	No flap opening/conveyable
		3	7.64	No flap opening/conveyable
Latex-like soy adhesive	23°C/0.5MPa/5min	1	7.70	No flap opening/conveyable
		2	7.72	No flap opening/conveyable
		3	7.67	No flap opening/conveyable

CONCLUSION

Soy protein is an abundant protein that can be successfully used as an adhesive. Modification is required to achieve satisfactory adhesion performance. This study revealed that modifications may follow different mechanisms and result in various structures and properties. For example, the SDS-modification following a hydrophobic interaction mechanism does not completely destroy soy protein secondary structure. The GuHCl-modification undergoes a hydrogen-bonding interaction mechanism and denatures the soy protein completely at high concentration.

Adhesion performance of the soy protein-based adhesives on fiberboard are affected by many parameters, such as modifier, concentration, temperature, pressure, time, and pre-treatment. The 1% SDS or 1M GuHC modifications and concentrations proved to be the most promising for the adhesive performance. The hot-press temperature of 100°C, pressure of 2MPa and press time of 5 minutes give optimal results for the adhesive performance. Temperature appears to be the most critical parameter for most soy protein-based adhesives. High processing temperatures could also be treated as a kind of modification because it could denature potentially the soy protein, enhance molecule entanglement, and increase the possibility of chemical reactions. However, for the latex-like soy adhesive there is better processing ability when it is applied at room temperature and low pressure.

In this study, the soy protein-based adhesives have higher water resistance than some commercially available adhesives used to manufacture the fiberboard. The water resistance results are also affected by the processing parameters, especially press pressure. For the solid fiberboard boxes, most of the soy protein-based adhesives gave satisfactory results except the 1% SDS-modified adhesive with low press pressure of .5 MPa. For the corrugated fiberboard boxes, all the soy protein-based adhesives passed the specifications even at low press pressure. Overall, this research has produced adhesives derived from modified soy proteins that compete with existing adhesives, but have improved water resistantance and are environmentally friendly.

ACKNOWLEDGEMENT

The project was mainly supported by the Navy's Waste Reduction Afloat Protects the Sea (WRAPS) Program. Part of the work was also supported by the U.S. Army Research Office (#DAAD19-01-1-0570)

REFERENCES

[1] Estes, G. M. & Kuespert, D. (1977). The Future of Adhesives Use in Packaging and Packaging Materials. *Adhesives Age*, 20, 35-38.

[2] Hse, C-Y. (1995). Environmental Concerns Related to Wood Adhesives: Developments and Trends. *IUFRO XX World Congress*, August 6-12, Tampere, Finland.

[3] Sellers, T. & Haupt, R. A. (1995). New Developments in Adhesives and Gluing in North America. *IUFRO XX World Congress*, August 6-12, Tampere, Finland.

[4] Gruner, A. & Piringer, O. (1999). Component Migration from Adhesives Used in Paper and Paperboard Packaging for Foodstuffs. *Packaging Tech. Sci.* 12, 19-28.

[5] Kennedy, H. M. (1998). Starch- and Dextrin-Based Adhesives. In R. W. Hemingway, A. H. Conner, & S. J. Branham (Eds.), *Adhesives from Renewable Resources (ACS Symposium Series, No.385*, pp. 326-336). Washington, D.C.: American Chemistry Society.

[6] ASTM D 5570 (1994). Standard Test Method for Water Resistance of Tape and Adhesives Used as a Box Closure. In *Annual Book of ASTM Standards* (Vol. 15.09, pp. 799-800). Philadelphia, PA: ASTM.

[7] ASTM D 1974 (1994). Standard Practice for Methods of Closing, Sealing, and Reinforcing Fiberboard Boxes. In *Annual Book of ASTM Standards* (Vol. 15.09, pp. 233-248). Philadelphia, PA: ASTM.

[8] ASTM D 4169 (1994). Standard Practice for Performance Testing of Shipping Containers and Systems. In *Annual Book of ASTM Standards* (Vol. 15.09, pp. 551-561). Philadelphia, PA: ASTM.

[9] Burnett, R. S. (1951). Soybean Protein Industrial Products. In K.S. Markley (Ed.), *Soybeans and Soybean Products* (Vol. 2, pp. 1003-1053). New York: Interscience Publishers.

[10] Kalapathy, U., Hettiarachchy, N. S., Myers, D., & Hanna, M. A. (1995). Modification of Soy Proteins and Their Adhesive Properties on Woods. *J. Am. Oil Chem. Soc.*, 72, 507-510.

[11] Hettiarachchy, N. S., Kalapathy, U., & Myers, D. (1995). Alkali-Modified Soy Protein with Improved Adhesive and Hydrophobic Properties. *J. Am. Oil Chem. Soc.*, 72, 1461-1467.

[12] Kalapathy, U., Hettiarachchy, N. S., Myers, D., & Rhee, K. C. (1996). Alkali-Modified Soy Proteins: Effect of Salts and Disulfide Bond Cleavage on Adhesion and Viscosity. *J. Am. Oil Chem. Soc.*, 73, 1063-1066.

[13] Sun, X. S. & Bian, K. (1999). Shear Strength and Water Resistance of Modified Soy Protein Adhesives. *J. Am. Oil Chem. Soc.*, 76, 977-980.

[14] Huang, W. & Sun, X. S. (2000). Adhesive Properties of Soy Proteins Modified by Urea and Guanidine Hydrochloride. *J. Am. Oil Chem. Soc.*, 77, 101-104.

[15] Huang, W. & Sun, X. S. (2000). Adhesive Properties of Soy Proteins Modified by Sodium Dodecyl Sulfate and Sodium Dodecylbenzene Sulfonate, *J. Am. Oil Chem. Soc.*, 77, 705-708.

[16] Lambuth, A. L. (1994). Protein Adhesives for Wood. In A. Pizzi, & K.L. Mittal (Eds.), *Handbook of Adhesive Technology* (Chap. 13, pp. 259-281). New York: Marcel Dekker, Inc.

[17] Zhong, Z., Sun, X. S., Fang, X., & Ratto, J. A. (2001). Adhesion Strength of Sodium Dodecyl Sulfate-Modified Soy Protein to Fiberboard. *J. Adhesion Sci. Technol.*, 15, 1417-1427.

[18] Zhong, Z. & Sun, X. S. (2001). Thermal and Mechanical Properties and Water Absorption of Sodium Dodecyl Sulfate-Modified Soy Protein (11S). *J. Appl. Polym. Sci.*, 81, 166-175.

[19] Tanford, C. (1968). Protein Denaturation. In C. B. Anfinsen, M. L. Anson, J. T. Edsall, & F. M. Richards (Eds.), *Advanced in Protein Chemistry* (pp. 121-283). New York: Academic Press, Inc.

[20] Zhong, Z., Sun, X. S., Fang, X., & Ratto, J. A. (2002). Adhesion Strength of Guanidine Hydrochloride-Modified Soy Protein for Fiberboard Application. *Intl. J. Adhesion & Adhesives*, 22, 267-272.

[21] Johnson, W. C. (1990). Protein Secondary Structure and Circular Dichroism: A Practical Guide. *Proteins: Structure, Fuction, and Genetics*, 7, 205-214.

[22] Zhong, Z., Sun, X. S., Fang, X. & Ratto, J. A. (2001). Adhesion Properties of Soy Protein with Fiber Cardboard. *J. Am. Oil Chem. Soc.*, 78, 37-41.

[23] Zhong, Z., Sun, X. S., Wang, D., & Ratto, J. A. (2003). Wet Strength and Water Resistance of Modified Soy Protein Adhesives and Effects of Drying Treatment. *J. Polym. Environ.*, 11, 137-144.

[24] Zhong, Z. & Sun, X. S. (2000). Thermal Behavior and Nonfreezing Water of Soybean Protein Components. *Cereal Chem.*, 77, 495-500.

[25] ASTM D 951 (2000). Standard Test Method for Water Resistance of Shipping Containers by Spray Method. In *Annual Book of ASTM Standards* (Vol. 15.09, pp. 123-125). West Conshohocken, PA: ASTM.

[26] Sun, X. S. & Zhu, L. (2004). Latex Based Adhesives Derived from Soybeans. *Patent pending.*

In: New Frontiers in Polymer Research
Editor: Robert K. Bregg, pp. 75-97

ISBN: 1-59454-956-7

Chapter 4

ADHESION PHENOMENA OF POLYMERS AT A LOCAL SCALE

Olivier Noel[*], Maurice Brogly and Jacques Schultz

Université de Haute Alsace (UHA)-Institut de Chimie des Surfaces et Interfaces (ICSI)-CNRS UPR 9069-BP 2488-MULHOUSE CEDEX-France

ABSTRACT

Study of the adhesion phenomena at a molecular scale is fundamental to obtain a better control of the properties and the performances of the materials for numerous modern applications in various domains (as biotechnology, microelectronics…). In this paper, the Atomic Force Microscope (AFM) was used to perform force measurements in quasi-static mode to investigate surface properties of model systems at the nanoscale. In particular, in the experimental part of the paper, we describe the systematic calibration procedure of AFM measurements that was performed to produce quantitative data. The influence of capillary condensation of atmospheric water vapor at the tip-sample interface on the measured forces is also discussed. In the main part of the paper, we focused on two types of model systems. The first one was composed of a rigid substrate (silicon plates), which was chemically modified by molecular self-assembling (SAMs) to display different surface properties (methyl and amine functional groups). The second system consists of model polymer networks (cross-linked polydimethylsiloxane or PDMS) of variable mechanical properties, whose surfaces were chemically modified with the same groups as before with silicon substrates (methyl and amine functional groups). Quantitative assessment values were obtained on the model polymer networks from the AFM force curve measurements in considering that viscoelasticity and adhesion can influence the experimental data. The comparison of the force curves obtained from the two model systems shows that the viscoelastic or mechanical contribution dominates in the adhesion on polymer substrates. Finally, a relationship, which expresses the separation energy at a local scale as a function of the energy dissipated within the contact zone, on one hand and the surface properties of the polymer on the other hand, is proposed.

---

[*] Université du Maine- Laboratoire de Physique de l'Etat Condensée –CNRS-UMR 6087; Avenue Olivier Messiaen, 72085 Le Mans CEDEX 9-France

1. Introduction

The comprehension of the adhesion phenomena of materials has always been fundamental for numerous industrial applications in different and various domains as biotechnology, electronics, automotives... However, adhesion has only been fully considered as a science since the fifties[1] considering the diversity of the concepts, of the ideas and, of the pluridisciplinarity of the involved phenomena (molecular interactions, physics and chemistry of materials, rheology…). Adhesion science relates to as well phenomena involved within the contact as the study of the mechanisms of the contact separation.

The first researches reported in the literature have only considered surface properties within the contact. However, the first measurements performed by using macroscopic techniques (peel tests[2], double cantilever beam tests[3], tack or JKR tests[4,5]…) showed that bulk properties cannot be dissociated from surface properties. Thus, we usually measure the adherence (that takes into account both contributions) rather than the adhesion itself (that is related to only the intermolecular forces). Historically, Gent and Schultz[6] are the first researchers who have proposed an empirical law that takes into consideration both mechanical and chemical contributions. This law has been confirmed by Andrews and Kinloch[7] experimental observations and is still developed by many other researchers (Schultz and Carré[8], Barquins and Maugis[9], de Gennes[10]…).

Nowadays, no theory of adhesion can explain all the involved mechanisms. An approach of the phenomena at a molecular scale is necessary to understand the various couplings involved in the adhesion phenomena and to obtain a better control of the properties and of the performances of the materials for the adhesion problems[11]. This is also amplified by the emergence of new nano-materials used in competitive applications as the bio-nanotechnologies or microelectronics.

In this paper, we propose to quantify the different contributions (the adhesive and mechanical ones) included in a surface force measurement, at a local scale (nanometer scale), using the same technique (in-situ measurements). To achieve this objective, we have used the Atomic Force Microscope (AFM)[12]which, is a promising tool for the investigation of material surface properties at the nanoscale. Precise surface force measurements (or adherence measurements) of chemically modified surfaces can be achieved with a nanometer scale probe. In particular, we were interested in the measurements of local attractive or adhesive forces with the AFM, of controlled chemical and mechanical model substrates. As model substrates, we have used grafted silicon wafers (or SAMs) and cross-linked PDMS grafted with the same functionalities as with the silicon wafers. The comparison of the force curves obtained from these two model systems showed that we can established a relationship between viscoelastic properties of the bulk and adhesive properties of the surface on one hand and, the separation energy on the other hand. For the following investigations, all experimental AFM measurements were performed in air, with a commercial apparatus (NanoscopeIIIa D3000, DI) and a V-shaped nitride silicon tip (Si_3N_4).

2. EXPERIMENTAL

A. Materials

As model substrates, we have used silicon wafer and cross-linked PDMS sheets, which are both grafted with identical molecules to display different surface properties.

Silicon wafers (100) (supplied by MAT TECHNOLOGY France) polished on one side were used as substrate for the SAM film grafting. All chemicals used in chemical handling (cleaning, synthesis) were of reagent grade or better (supplied by Aldrich). Bi-distilled and deionized water was used for contact angle measurements.

In this paper, "as received silicon (Si_{wafer})" refers to a silicon wafer previously cleaned in hexane then, in ethanol in an ultrasonic bath. That means a contaminated layer still remains on the surface.

Two organosilane grafts (supplied by ABCR Karlsruhe-Germany) were used. One hydrophobic grafted model surfaces were prepared by using hexadecyltrichlorosilane ($C_{16}H_{42}O_3Si$ purity is 99%) and one hydrophilic grafted model surfaces by using (6-aminohexyl)-aminopropyltrimethoxysilane ($C_{12}H_{30}N_2O_3Si$, purity is 95%).

Other hydrophobic and hydrophilic surfaces on only silicon substrates have been prepared. These are prepared by using 1H,1H,2H,2H-perfluorodecylmethyldichlorosilane ($C_{11}H_7Cl_2F_{17}Si$ or Si_{CF3}) and 2-(carbomethoxy)ethyltrichlorosilane ($C_4H_7Cl_3O_2Si$ or Si_{COOR}).

B. Sample Preparation

B.1. Preparation of Oxidized Silica Surfaces

Before grafting the substrates (silicon wafer) must be chemically modified in order to get a hydrophilic surface (SiO_2). The silicon surface was first cleaned with ethanol and dried with nitrogen before oxidation. Oxidized surfaces were obtained after cleaning by submersing the substrate in a warm piranha (60°C) solution (3:7 v/v 30% H_2O_2 and H_2SO_4 mixture) for about 30 minutes, in order to keep a smooth surface (mean roughness of the substrate after oxidation and determined by AFM is 0.2nm), and then thoroughly rinsed with deionized and twice distilled water. This treatment produces a high hydroxyl group density on the surface (SiOH groups), to which functional silanes will adsorb upon hydrolysis[13]. Just before being grafted with silane the wafers are dried with nitrogen.

Surfaces modified with hydroxyl groups (Si_{OH}) were synthesized with this method and immediately probed in order to avoid contamination of the surface by the environment due to the high reactivity of SiOH groups.

B.2. Functionalized Sams. Experimental Procedure

Vapour–phase molecular self-assembling technique[14] slightly improved in the laboratory [13] was used. One of the advantages of this technique is that the lack of solvent prevents the SAMs from incorporating small solvent molecules contamination and defects. Moreover, a previous study[15] showed that the molecular films prepared with this method, are more homogeneous.

In this method, the oxidized silica surfaces are placed above a previously de-aired solution of 100μL of organosilane - 3mL of paraffin mixture. The vapour-phase deposition of the molecular film on the substrate is performed in a vacuum chamber (50 min at 5×10^{-3} Torr) at room temperature. Si_{CH3}, Si_{NH2} refer to the silicon substrates respectively grafted with methyl, and amine groups.

B.3. Grafted Silicon Wafer Monolayer Characterization

Contact angle measurement is an effective way to characterize surface hydrophilicity and hydrophobicity. Wetting properties with water and glycerol in a thermostated room (20°C) are measured with a Krüss G2 system. The volume of the liquid drop in air lies between 1-2 μL and the measurement chamber was previously saturated with water vapor. Equilibrium static angle, advancing, and receeding angles are reported in Table 1.

Table 1. Wetting properties with water (θe is an equilibrium contact angle, θa is an advancing contact angle, θr is a receeding contact angle) and wetting hysteresis

End groups	θe Water(°) ± 2°	θa Water(°) ± 2°	θr Water(°) ± 2°	Wetting hysteresis (°)
CF_3	106	117	86	31
CH_3	103	111	101	10
Si_{wafer}	78	81	70	11
COOR	71	75	59	16
NH_2	57	59	21	38
OH	6			

It is interesting to note the expected correlation between contact angle measurements and the hydrophobic-hydrophilic character of the end-group of the grafts. Such evolution confirms the formation of SAMs on the silicon wafers. Moreover, the wetting hysteresis is low (except with the Si_{CF3} and Si_{NH2} surfaces). This means that our functionalized surfaces show high compacity and homogeneity at a macroscopic scale.

Surface morphologies of the different types of layers were studied by AFM (Nanoscope III-Digital Instrument) in the TappingMode™. SAMs topographic and phase contrast AFM images (image size: 1 μm × 1 μm) confirmed that no aggregates were formed and showed a complete homogeneous recovery of the grafts. Moreover, the RMS roughness of 0.15 nm deduced from the AFM images, which was the expected value for the substrate roughness, confirmed that our grafts were well ordered and packed at micrometer scales (this is also the case for the Si_{CF3} and Si_{NH2} surfaces).

XPS spectrometry (Leybold LHS 11) with a take-off angle of 15° was also carried out. Comparison of the spectra related to the different types of molecules grafted on the silicon wafer with the spectra obtained on the silicon wafer confirms the effective grafting. Moreover, no peaks related to the chloride element have been detected regarding the spectra related to the chlorinated grafts. This confirms the efficiency of the covalent grafting and of the surface cleaning procedure.

Finally, the thickness of the oxide and organic films were determined by ellipsometry measurements by fitting both the refractive index and the thickness of the film (Table 2).

Table 2. Ellipsometry measurements. Experimental and theoretical[16] thicknesses of the molecular films (for a fully extended graft normal to the surface plane)

End groups	Thicknesses of the grafted organic films (Å)	Theoretical thicknesses (Å)
CH_3	21	22.5
COOR	5	6
NH_2	10	15

These results are in good agreement with the literature for the CH_3 and the NH_2 grafted films[17,18], and confirm that only one monolayer was grafted in these cases. These thicknesses were determined by previously measuring the thickness of the oxide layer formed during the 30 minute piranha treatment. A 3 nm reproducible value was obtained for this oxyde layer thickness.

Even though important effort was made to characterize the grafted surfaces (XPS, ellipsometry, wetting, AFM images), one can suspect that the grafting density and the morphology of the monolayer may affect the comparison of these surfaces in terms of quantitative adhesion force. To prevent such effects, evolution with grafting time of the properties of the monolayers were investigated in previous studies[19]. In the present study, grafting time has been chosen to get stable layers vs. time.

B.4. Preparation and of Cross-Linked Grafted PDMS

PDMS chains were previously cleaned in acetone in order to narrow the polydispersity index. Then, PDMS samples were cross-linked under nitrogen in a glove box using tetrakis(dimethylsiloxy)silane as a cross-linker and a platinum-based catalyst. Cross-linking was achieved by heating the samples at 80°C for 12 hours. Table 3 reports the PDMS chain features before and after cross-linking (the weight molecular mass Mw has been determined by gel permeation chromatography (GPC)). In particular, the classification of PDMS substrates refers to the entanglement mass after crosslinking. This entanglement mass, Mc, has been determined by using the Flory-Erman relationship[20] and experimental swelling measurements in toluene Thus, PDMS 0.8k is the hardest substrate, whereas 34k refers to the softest one.The PDMS sheet thickness is about 2 mm.

Table 3. Weight molecular mass (Mw) and polydispersity index (Ip) before cross-linking and entanglement mass (Mc) after cross-linking

PDMS Classification	Mw before cross-linking ($g.mol^{-1}$)	Ip before cross-linking	Mc after cross-linking($g.mol^{-1}$)
PDMS 0,8k	1500	1.4	800
PDMS 13k	13000	1.5	13000
PDMS 34k	55000	1.7	34000

Before grafting, as for silicon substrates, the PDMS substrates must be chemically modified to get a hydrophilic surface (SiO_2). The PDMS surfaces were first cleaned with toluene for five days and dried under vacuum before oxidation. Oxidized surfaces were obtained by using a water plasma treatment (Plasma power: 80W for 30s). This treatment produces a high hydroxyl group density on the surface (SiOH groups), to which functional silanes will adsorb upon hydrolysis. Grafting procedure is the same as with silicon wafers (vapour–phase molecular self-assembling technique).

B.5. Characterization of the Cross-Linked Pdms (Pdms)

Surface morphologies of the different types of layers were studied by AFM (Nanoscope III-Digital Instrument) in the Contact Mode AFM (for PDMS substrates). SAMs topographic AFM images (image size: 500nm × 500nm) confirmed that no aggregates were formed and showed a complete homogeneous recovery of the grafts (at a local scale). Moreover, the RMS roughness of 0.3nm for PDMS substrates deduced from the AFM images, which were the expected value for the substrate roughness, confirmed that our grafts were well ordered and packed at micrometer scales. It also confirms that the plasma treatment has no effect on the roughness of the substrates.

XPS spectrometry (Leybold LHS 11) with a take-off angle of 15° was also carried out. Comparison of the spectra related to the different types of molecules grafted on the PDMS substrate with the spectra obtained on the raw PDMS sheet confirms the effective grafting. Moreover, as for silicon wafers, no peaks related to the chloride element have been detected regarding the spectra related to the chlorinated grafts. This confirms the efficiency of the covalent grafting and of the surface cleaning procedure.

The thickness of the oxide and organic films were determined by ellipsometry measurements by fitting both the refractive index and the thickness of the film. These results are also coherent with the literature for the CH_3 and the NH_2 grafted films, and confirm that only one monolayer was grafted in these cases.

Finally, macroscopic mechanical tests (uniaxial elongation and dynamical tests) were carried out and have been shown to concord with the Flory theory[21] related to the rubberlike elasticity. This proves that a good control of macroscopic mechanical properties is achieved (this is also confirmed by the reproducibility of the measurements). Young moduli determined by uniaxial elongation were found to lie over one decade (Table 4).

Table 4. Young moduli, E (MPa)

PDMS substrates	E (MPa)
0.8k	2.2 ± 0.4
13k	0.30 ± 0.06
34k	0.13 ± 0.02

Moreover, dynamic mechanical tests have shown that the Young modulus is independent of the strain speed considering the experimental conditions. In term of dissipative energy, DMA also showed that the 0.8k substrate (high modulus substrate) is less dissipative than the 34k substrate (low modulus substrate). We have also checked that the plasma treatment does not affect the mechanical properties of the studied PDMS substrates.

B.6. Surface Energy Determination of Grafted Silicon and Cross-Linked Pdms Sheets

From contact angle measurements with water and glycerol, theoretical surface energy components were calculated from the Good-Girifalco and Fowkes[22] theories. Table 5 shows that the non-dispersive components are negligible for all the surfaces, except for the

Si_{NH2} surface. In this last case, interactions with water through hydrogen bonding can occur, and thus, capillary forces have to be considered. Moreover, the total surface energy increases with the hydrophilicity of the surfaces.

Table 5. Theoretical surface energies: γ_S^d is the dispersive component (due to Van der Waals contributions), γ_S^{Nd} is the non-dispersive (or polar) component of the surface energy, γ_S is the total surface energy

Samples	γ_S^d (mJ.m^{-2})	γ_S^{Nd} (mJ.m^{-2})	γ_S (mJ.m^{-2}) ± 2mJ.m^{-2}
Si_{CF3}	21	0	21
Si_{CH3}	21	1	22
Si_{wafer}	25	7	32
Si_{COOR}	34	9	43
Si_{NH2}	38	15	53
Si_{OH}	39	37	76
PDMSraw	28	0	28
$PDMS_{CH3}$	21	0	22
$PDMS_{NH2}$	38	19	53
0.8k	29	0	29
13k	27	0	27
34k	27	0	27

It is interesting to notice that for a given surface chemistry, the surface energies are independent of the nature of the substrate (silicon wafer or cross-linked PDMS). We can also mention that the surface energies obtained respectively on a silicon wafer and on the raw cross-linked PDMS are very close.

C. Adhesion Force Measurements with AFM

The AFM device can be roughly broken up into three main independent parts (see Figure 1).

The first one is the nanoscale probe made of a cantilever bearing a tip at its end. For convenient reasons, the cantilever is tilted at a 12° angle to the horizontal line. The tip radius, at the apex, is of the order of ten of nanometers. When interacting with the sample, the cantilever bends itself. Its deflection is then detected with an optical device consisting of a laser and a four-quadrant photodiodes detector.

In the case of the DI 3000 microscope, the probe can be vertically moved backward and forwards via a piezoelectric ceramic, while the sample remains motionless.

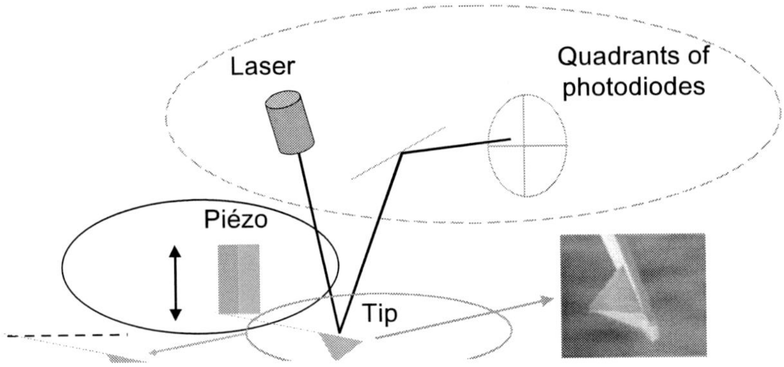

Figure 1. AFM schematic representation (DI 3000)

Force measurements with AFM in the contact mode consist of recording the deflection of the spring (cantilever) while interacting with the sample surface[23]. Provided that the normal spring constant of the cantilever, k, is known, an interacting force (adhesion or adherence force), F_{adh}, can be calculated using Hooke's law:

$$F_{adh} = - k \cdot \Delta z_{pull\text{-}off} \qquad (1)$$

(k is the cantilever stiffness and Δz is the vertical cantilever deflection)

Simultaneously, the tip is approached and retracted from the sample. Thus, it is possible to obtain Deflection-Distance (DD) curves, and Force-Distance (FD) curves.

A schematic representation of a DD curve obtained when probing a hard surface is reported in Figure 2.

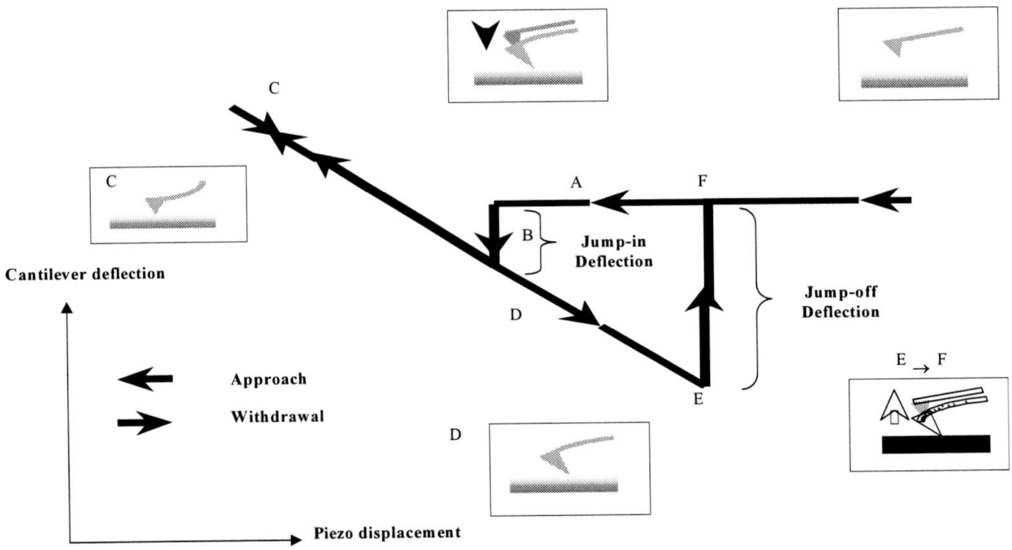

Figure 2. Schematic representation of a Deflection-Distance curve (DD curve).

In zone A, the cantilever is far from the analysed surface and stays at its equilibrium state until it jumps-in the surface (zone B). When in contact (zones C and D), adhesion can be developed. Thus, during the withdrawal cycle, the jump-off (E and F) doesn't occur for a similar piezo displacement as for jump-in. Finally, the cantilever returns to its equilibrium state.

One should notice that jump-in observed during the approach is usually due to mechanical instabilities and is not considered in this paper[23].

A prerequisite is to confirm the thermal stability of both actuator and cantilever. In our laboratory, this is achieved by switching on the AFM for at least 12 hours before starting the experiments. It is also necessary to remove any external vibration and thermal instability. Finally, control of the room's relative humidity is useful[24].

D. AFM Calibration

D.1. Cantilever Calibration

According to equation (1), it is essential to determine the cantilever stiffness, k, to obtain quantitative measurements. To achieve this goal, numerous experimental methods are described in the literature[25-28]. Each of these methods are characterized by their simplicity of implementation and precision[29,30]. We can also distinguish the destructive methods[26,31] from the non-destructive ones[29,32]. In our case, we focused on two non-destructive methods, which have been rigorously studied and compared: the first one is related to the use of rectangular reference cantilevers[33] (supplied by NanoMetrology) and the second is based on the thermal fluctuations measurement of the cantilever[34]. The first method is a very convenient one, as it requires only reference cantilevers, and can be applied to both rectangular and V-shaped cantilevers. The second method requires a spectra analyzer and the opportunity to get a raw signal (before treatment).

In our study, three cantilevers were specially selected to give the same force on a reference silicon wafer. Both calibration methods have then been used and compared, to calibrate the cantilevers. These were triangular shaped cantilevers (supplied by Nanosensor-Germany) and had an effective 0.30 $N.m^{-1}$ spring constant. This value is far from that specified by the supplier (0.58 $N.m^{-1}$).

D.2. Tip Radius Estimation

To deduce thermodynamic parameters from experimental values, it is necessary to have a good estimation of the radius of the tip at the apex[35]. A numerical method[36] is proposed in the literature. However, for convenience reasons, we deduced the radius on the basis of the shape revealed on a scanning electronic microscope (SEM) picture taken at the end of the measurements (Figure 3 (b)).

Both methods have given approximately the same tip radius for all used cantilevers. Our tip radius was estimated to be 50nm ± 5 nm.

We have also tried to image directly the tip in AFM TappingMode™. However, this is time consuming and not precise (Figure 3 (a)).

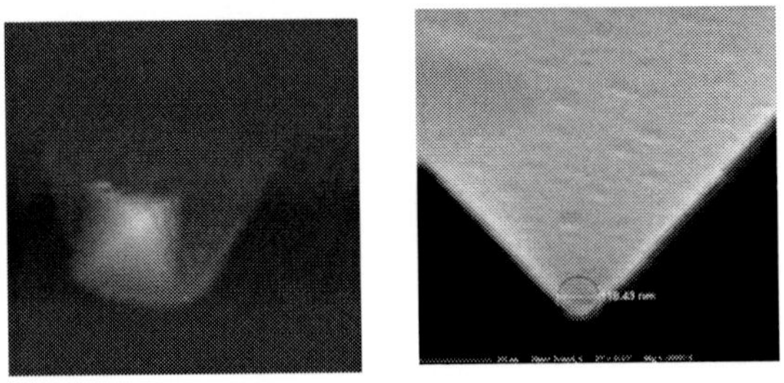

(a) (b)

Figure 3. Tip radius estimation (a) AFM TappingMode™ imaging, (b) SEM picture

Finally, checking regularly and randomly the adhesion force on a reference silicon wafer allows us to verify contamination of the tip, or change of the tip radius during measurements. If organic tip contamination occurs, one can attempt to clean the tip by submersing it in a 2% vol. fluorhydric acid solution[37]. The acid solution reacts with the silicon of the tip and cleans it. After such treatment, one may recalibrate the tip.

D.3. Non-Linearity of the Optical System

First of all, it is necessary to recall that this system measures the angular cantilever deflection rather than the vertical cantilever deflection. The latter is connected to the angular cantilever deflection, Δz, by a linear expression[38]. Thus, to convert the measured angular deflection into the vertical deflection, it is necessary to perform a force curve on a rigid substrate (a silicon wafer for example) and to fix the indentation slope of a force curve (or slope of the force curve in the contact zone (zones C and D in Figure 1)) to unity. However, this conversion requires two conditions to be met. The first one is that the actuator displacement is linear, and the second one is to prevent the tip from sliding or benching on the substrate. These two conditions will be discussed in the following sections.

Secondly, a non-linearity of the optical detector, due to a non-homogeneous spreading of the laser spot on the detector, was first mentioned in the literature by Haugstad et al.[39]. To study this, we have reported the indentation slope (obtained from a rigid substrate) versus the tension recorded by the photo detector (Figure 4). We have also reported the theoretical slope that is expected for a rigid substrate (this one should be equal to 1).

From figure 4, we observe that the domain of linearity of the detector lies between ± 2 V. Finally, if non-linearity is not taken into account, the error on the quantitative results can be significant. In our experiments, all the measurements have been done in the linearity domain.

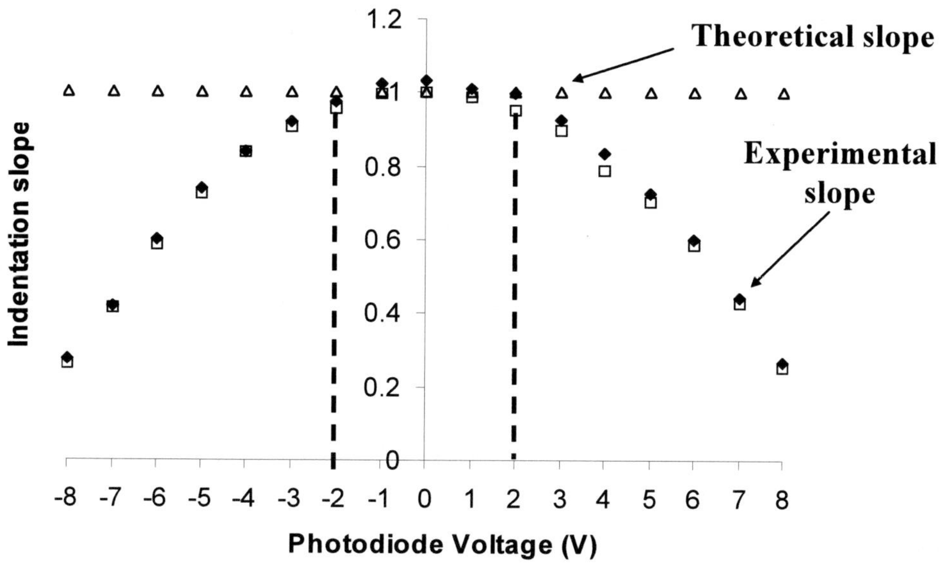

Figure 4 : Non-linearity of the photodetector. The two sets of experimental points (square and rhombus) represent respectively a loading and an unloading experiment.

D.4. Scan Speed of the Cantilever

The piezo actuator displacement is controlled by applying a voltage, which lies between ± 220V. Nevertheless, a linear applied tension will not generate a linear displacement. One might apply a non-linear signal to correct it. However, this effect depends on both actuator scan speed and the total actuator displacement (piezo amplitude), and cannot be completely corrected. In addition, the actuator shows a displacement hysteresis due to the nature of the ceramic that composes the actuator. Michel[40] has shown that the piezo hysteresis can be neglected if the actuator displacement is performed around the 0V applied tension. Finally, to study the non-linearity, we have reported the relative difference of the indentation slope during loading and unloading, versus the scan speed, and for different actuator amplitudes (Figure 5). During the experiment, the actuator is considered thermally stable.

From our results, we observed that a discrepancy appears for very low scan frequencies (below 1 Hz). For higher scan frequencies (more than 20 Hz), the viscosity of the environment could be significant. It is noticeable that the dispersion of the experimental points is minimal for a 10 Hz frequency scan. Thus, this frequency is a good compromise for our actuator.

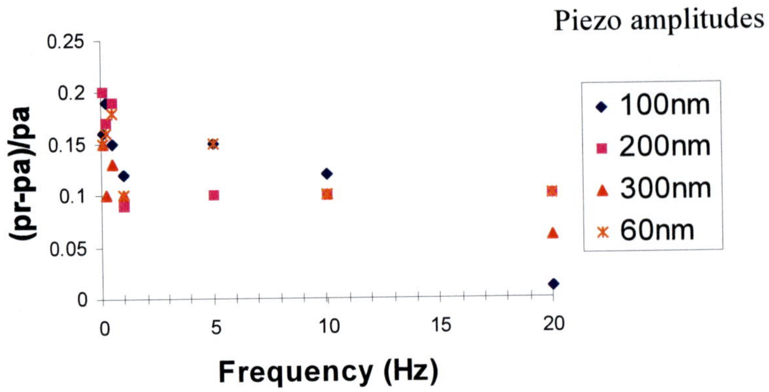

Figure 5. Non-linearity of the actuator.

D.5. Estimation of the Uncertainties of the Physical Values Related to the Experimental Pull-Off Force Measurements

If we consider that the main uncertainties in a pull-off force measurement using AFM are due to the determination of the spring constant and the radius of the tip, we can approximate an error of 10% for the measurement of an adhesion force (due to the uncertainty of the method of calibration of the tip, as specified by ref. [33]), and a relative error of 20% for the determination of the work of adhesion and surface tensions (due to a 10% error on the determination of the tip radius[41]). Nevertheless, even if 20 % of the error is high, one must consider this error as a systematic error. Indeed, all measurements (over 100 for a given surface) were performed with the same tip of the same radius. Experimental results are then significant, and the contribution of the different grafts can be discriminated.

3. Results and Discussion

A. DD Curves From Systems of Controlled Surface Chemistry and Identical Mechanical Properties

A.1.Effect of the Capillary Forces

The tip-sample interaction force was first measured on chemically modified SAMs obtained on rigid substrates (silicon wafers). Figure 6 shows that AFM measurements in our conditions are sensitive and significant to a chemical modification of the layers. When jump-off contact occurs, we measure the corresponding pull-off deflection. Pull-off deflection values ($\Delta z_{pull\text{-}off}$) increase in the following order:

$$\Delta z_{pull\text{-}offSiCF3} < \Delta z_{pull\text{-}offSiCH3} < \Delta_{zpull\text{-}off\text{"}asreceivedSi\text{"}} <$$
$$\Delta z_{pull\text{-}offSiester} < \Delta z_{pull\text{-}offSiNH2} < \Delta z_{pull\text{-}offSiOH}.$$

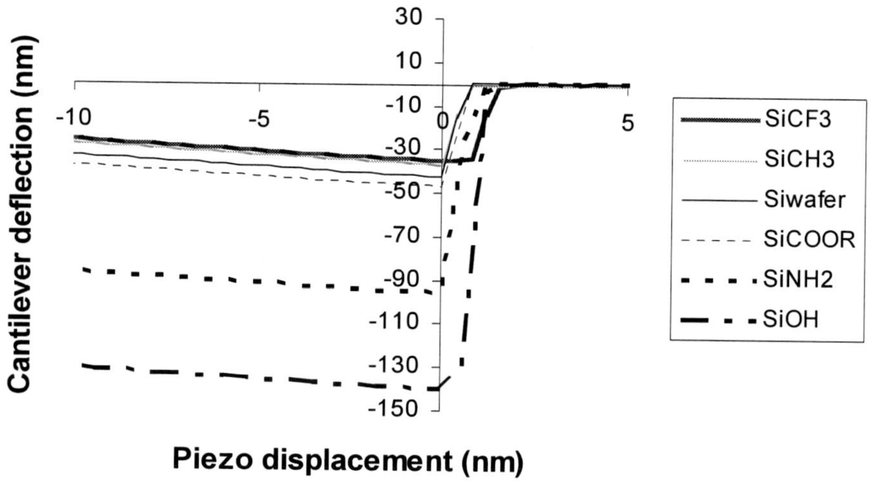

Figure 6. Deflection-Distance force curves on SAMs during the retraction

Knowing the pull-off deflection one can easily deduce the adhesion force, if the cantilever spring constant k is known ($F_{adhesion} = - k . \Delta z_{pull\text{-}off}$). Thus the adhesion force value increases with the hydrophilicity of the surface.

The thermodynamic work of adhesion (W_0) can be calculated from the DMT theory[42] by using the following relationship:

$$W_0 = -\frac{F_{adhesion}}{2\pi R} \text{ (R is the tip radius)} \qquad (2)$$

We can deduce W_0 from the experimental adhesion forces. Figure 7 reports W_0 versus the total surface energy determined as described in section.

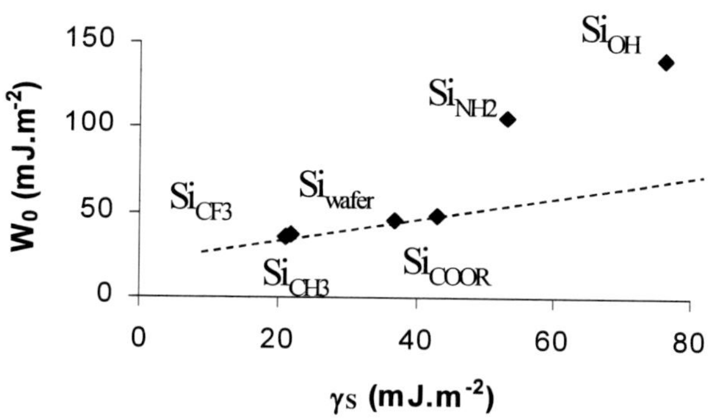

Figure 7. Correlation between thermodynamical work of adhesion and surfaces energy

This shows that W_0 increases linearly (correlation factor is 0.99) with the surface energy of the surfaces. However, singularities are observed for Si_{NH2} and Si_{OH} surfaces (very hydrophilic surfaces). Considering that the measurements are performed in the air, a capillary bridge is formed for these very hydrophilic surfaces and the pull-off mechanism could be different from the other surfaces. This could explain why the corresponding point for Si_{NH2} and Si_{OH} (in figure 7) does not fit the linear extrapolation.

A.2 Discussion on Capillary Force (F_{cap}) Influence in AFM Measurements

A.2.1 General Theory

When performing AFM experiments in air, force measurements include the contribution of Van der Waals (VdW) and capillary forces. In particular, the total adhesion force is given by the following expressions:

$$F_{TheorAdh} = F_{cap} + F_{VdWadh} \tag{3}$$

The theoretical Van der Waals force depends on the medium between both solids in interaction and is given by [43]:

$$F_{VdWadh} = \frac{A \times R}{6 \times d^2} \tag{4}$$

Where, R is the tip radius (equal to 50 nm in our case), d is the separation distance between both surfaces, A is the Hamaker constant, which depends on the medium between the two solids.

The capillary force depends on a meniscus formed between the two surfaces and is given by[44]:

$$F_{cap} = 2 \times \pi \times R \times \gamma \times (\cos(\theta_s) + \cos(\theta_t)) \tag{5}$$

Where γ is the surface free energy of the liquid (0.072 J/m^{-2} for water), θ_s and θ_t are the respective contact angles (with water) of the substrate and the tip.

According to expression (3), capillary forces could be predominant in the adhesion force measurement. Nevertheless, Fisher et al.[45] showed that if ($\gamma_{lv} < \gamma_{sv} \times \cos(\theta)$) the adhesion is dominated by the solid-solid interactions rather than by the capillary forces. For all our systems (except with amine and hydroxyl functional group grafting), the criterion of Fisher et al.[45] is verified.

Thus, we can calculate the "corrected" thermodynamic work of adhesion, $W_{0corrected}$, related to the Van der Waals forces, only:

$$W_0 = \frac{F_{VdW}}{2\pi R} = \frac{F_{adhesion} - F_{cap}}{2\pi R} \quad (F_{cap} \text{ is given by (5)})$$

Figure 8 reports the $W_{0corrected}$ versus the surface energy.

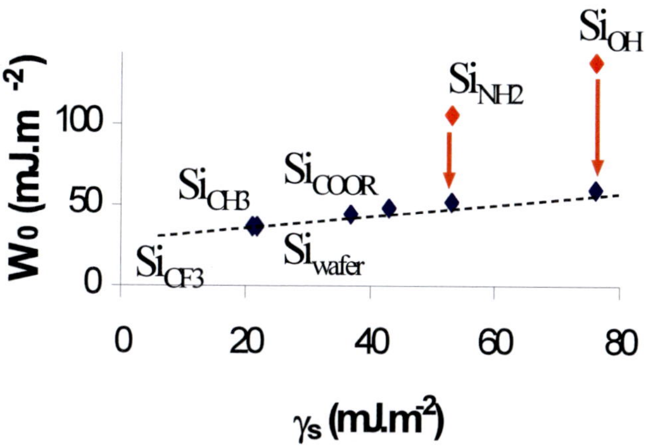

Figure 8: Effect of the capillary forces (red points represents the data obtained in considering the capillary forces (see figure 7))

Figure 8 shows that the singularities observed for the very hydrophilic surfaces are corrected and that the thermodynamic work of adhesion increases linearly with the surface energy when only Van der Waals (and dispersive) forces are considered[46].

B. The Specific Case of Polymers

B.1. DD Curve Measurements

Figure 9 shows the DD curves obtained on a silicon substrate on one hand, and on a PDMS substrate on the other hand. In the following, all the PDMS substrates are at least 1mm thick.

The comparison of the shape of both curves clearly shows the significant contribution of the mechanical properties of the polymer. In particular, compared to silicon wafer, the jump-off contact occurs over a large piezo displacement scale and could correspond to a progressive dewetting of the tip by polymer chains during retracting. The jump-off amplitude is also higher than for silicon wafer. The absolute values of the loading and unloading slopes in the DD representation are much lower than unity in the case of soft polymer systems. That means that the tip indents the polymer.

The jump-in amplitude is also more important for a polymer and cannot be attributed to an instability of the tip only. To explain this feature, we have investigated different assumptions including creep and the formation of a polymer nanoprotuberance during the approach cycle. Influences of creep effects on the AFM measurements have been investigated. In figure 10, we have reported AFM creep experiments[47] on two different PDMS substrates (PDMS 0.8k and PDMS 34k).

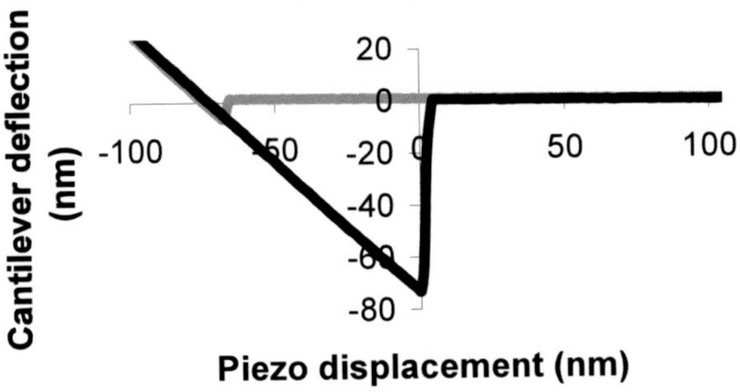

(a)

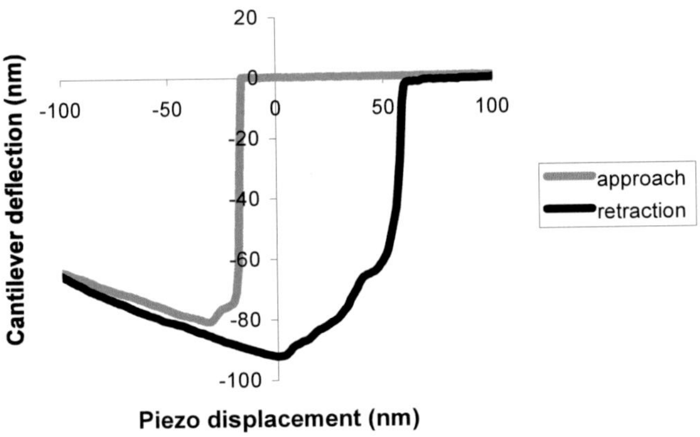

(b)

Figure 9. DD curves obtained on a (a) silicon substrate and on a (b) PDMSsubstrate

Obviously, considering the experimental contact time (t_{exp}<0.1s), the creep effect is not significative in our force curve measurements. However, the comparison between the experimental nanoprotuberance height and the one obtained by a numerical simulation of a nanoprotuberance growth using the model described in reference[48] , showed that the second assumption could be correct. Thus, in a first approximation, the beginning of the indentation is assumed to be at the minimum of the DD curve.

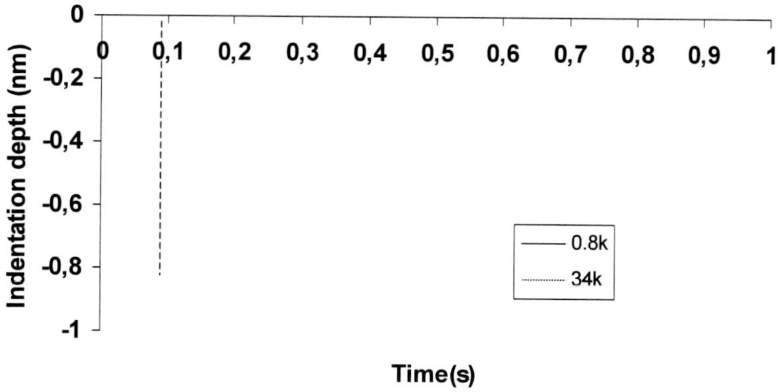

Figure 10: AFM creep experiment.

B.2. Force-Indentation Curves (F-I Curves) Deduced from DD Curves with the AFM

Before monitoring DD curves on PDMS, the actuator and the cantilever were thermally stabilized. The laser spot in contact with the tip was positioned in such a way that tangential forces, due to frictional forces, were minimized. We carried out these measurements on the basis of the above prerequisites so that force measurements could be liable and compared. F-I curves (Figure 11) are deduced from DD curves by assuming that for a given force, the indentation depth is the difference between the experimental deflection value (ds) (blue curve on figure 11) and the one that should be observed if the material was rigid (dr deduced from the slope of 1 for rigid materials-red curve on figure 11).

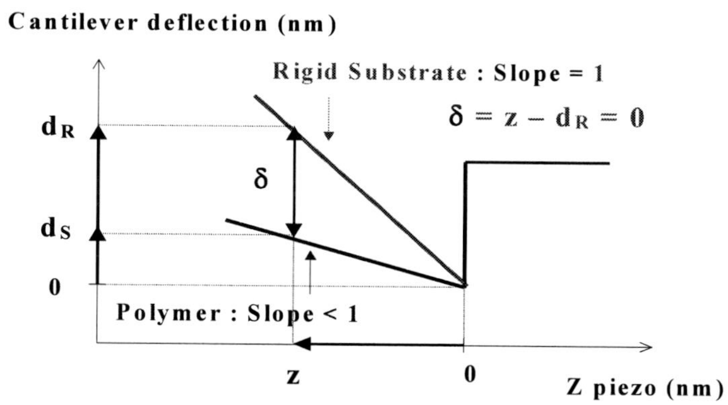

Figure 11. F-I curves obtained from DD curves

B.3. Effect of the Indentation Depth on the Adherence Force Measurements

Figure 12 represents the adherence versus the indentation depth.

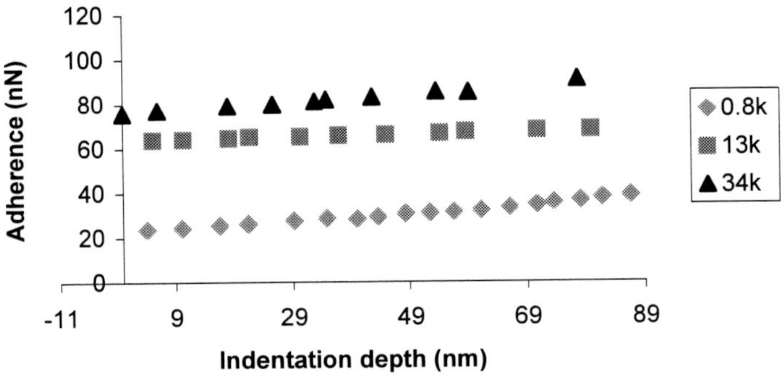

Figure 12: Influence of the indentation depth on the adherence

It shows that adherence linearly increases with the indentation depth. Thus, to compare the adherence forces obtained on different PDMS substrates with different moduli in the same conditions, it is necessary to indent them at the same indentation depth. In our experiments, we have taken care to ensure that we have really indented the polymer. Thus, considering the normal stiffness of the cantilever, it is possible to determine the maximal indentation depth (δ_{max}), using the following relationships:

$$k_{tip}=2\times E^{*}\times\sqrt{(R\,\delta_{max})}$$

k_{tip} is the constant stiffness of the cantilever, E* is the reduced modulus and R is the tip radius. For the hardest substrate, the maximal indentation depth is estimated to be 89 nm. Thus, whatever is the polymer substrate, all our AFM measurements are done at a constant 60 nm indentation depth. Finally, the longitudinal stiffness of the cantilever (k_x), sliding can occur before indenting the polymer. Nevertheless, sliding effect is assumed negligible compared to the total indentation depth.

B.4. DD Curves on Systems of Controlled Surface Chemistry and Controlled Mechanical Properties

The tip-sample interaction force was first measured on chemically modified SAMs obtained on rigid substrates (silicon wafers). Figure 13 (a) shows the jump-off deflection measured on a silicon wafer or on grafted silicon wafers. In our conditions the adhesion forces (which are proportional to the jump-off deflection through the cantilever spring constant k) are sensitive and significant to a chemical modification of the wafer surfaces. In particular, the adherence (or jump-off deflection) increases with the hydrophilicity of the surface ($F_{SiCH3} < F_{Siwafer} < F_{SiNH2}$).

Considering DD curves obtained on a PDMS substrate, a particular attention is taken. All measurements are performed at a constant indentation depth (60nm) of the tip (as discussed previously). Three substrates with different mechanical properties (Figure 13 (c)) were probed. We observed that adhesion force is modulated by varying, either the surface properties or the substrate modulus (Figure 13b). We have realised AFM measurements on cross-linked PDMS substrates with different Young modulus and grafted with the same molecules as for the silicon wafers before. (CH_3 and NH_2 SAMs).

In order to compare the adhesion forces obtained from the different substrates, we have considered the following ratios:

$$\left(\frac{Fadhesion_{PDMSX}}{Fadhesion_{SiX}} \right)$$

($Fadhesion_{PDMS}$ and $Fadhesion_{Si}$ represent the adhesion force measured on respectively PDMS and silicon substrates and X represents the functionality of the grafting).

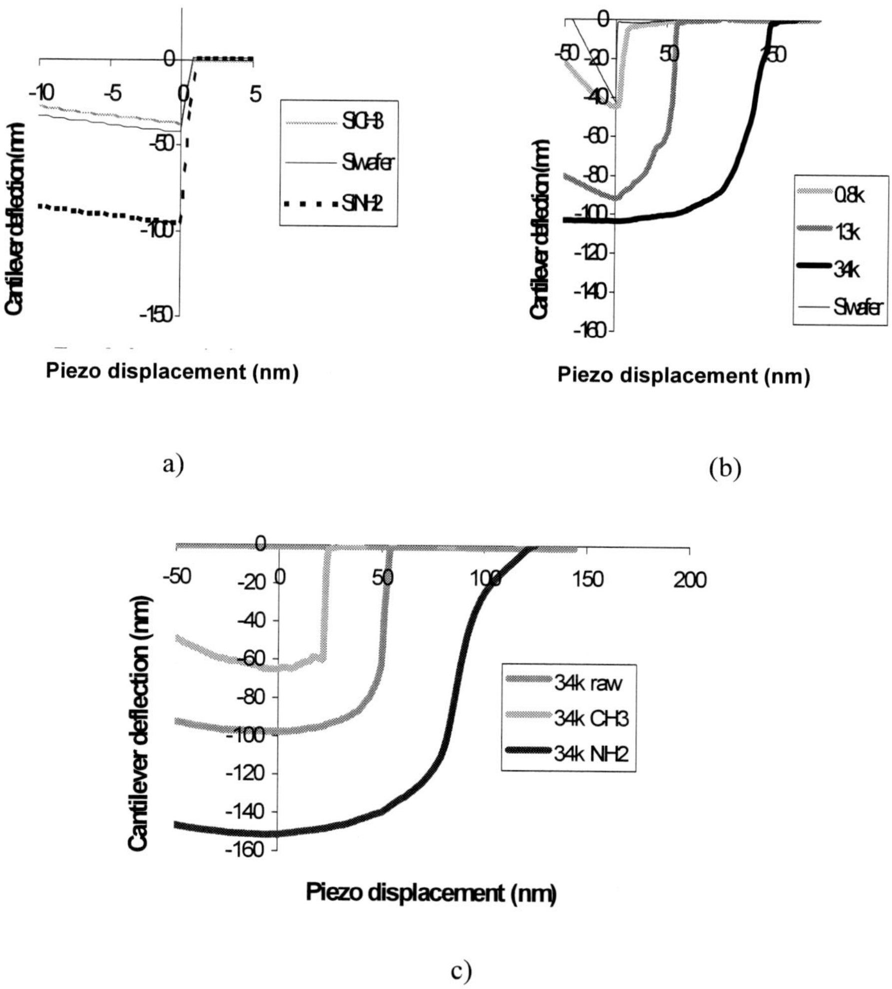

Figure 13: Characteristic experimental force curves (during retraction cycle) obtained from (a) grafted wafers of silicon (raw wafer, CH_3, NH_2), (b) cross-linked PDMS with different mechanical properties, (c) grafted cross-linked PDMS (raw material, CH_3, NH_2)

Table 6 shows that for a given PDMS substrate (for example, 0.8k) or a given Young modulus, the ratios are independent of the surface chemistry, whereas for a given grafting, the ratios are dependent on the mechanical properties of the substrate.

Table 6. Ratios $\left(\frac{FadhX_{PDMS}}{FadhX_{Si}}\right)$ for CH_3, NH_2 and as received substrates and for three PDMS of different mechanical properties (0.8k, 13k, 34k)

Raw	F_{PDMS}/F_{Si}	CH3	F_{PDMS}/F_{Si}	NH2	F_{PDMS}/F_{Si}
0.8k/Si	1.1	$0.8k_{CH3}/Si_{CH3}$	1.0	$0.8k_{NH2}/Si_{NH2}$	1.2
13k/Si	1.8	13kCH3/SiCH3	1.5	13kNH2/SiNH2	1.5
34k/Si	2.0	$34k_{CH3}/Si_{CH3}$	1.9	$34k_{NH2}/Si_{NH2}$	2.2

Thus, for a given substrate, we have the following expression:

$$Fadhesion_{PDMSX} = Fadhesion_{SiX} \times \phi \quad \textbf{Given substrate} \tag{6}$$

(ϕ is a constant dependent on the mechanical properties of the substrate)

However, for a rigid substrate, the DMT theory gives (absolute value of the adhesion force):

$$Fadhesion_{SiX} = 2\pi R\ W_0 \tag{7}$$

(R is the tip radius and W_0 is the thermodynamic work of adhesion)

By replacing (7) in (6), we get:

$$Fadhesion_{PDMSX} = 2\pi R\ W_0 \times \phi \tag{8}$$

By introducing a dimensional constant (k'), we get a separation energy $Gadhesion_{PDMSX}$ that can be expressed as follow:

$$Gadhesion_{PDMSX} = W_0 \times \phi \times 2\pi R \times k'$$

Or

$$Gadhesion_{PDMSX}\ \alpha\ W_0 \times \phi(Mc,T) \tag{9}$$

(G is the separation energy and $\phi(Mc,T)$ is a viscoelastic dissipative function dependent on the network molecular structure (Mc) and on the temperature (T) (the dependence on the temperature has not been yet studied, but as the dissipation of a polymer is temperature dependent, we can do this assumption).

This relationship clearly expresses the mechanical contribution from the chemical one in a force curve measurement with AFM. According to Table 6, ϕ(Mc,T) is determined for each substrate for a given speed and a given temperature, and ϕ is independent of the surface chemistry. Moreover, the experimental ϕ values, reported in Figure 14, are coherent.

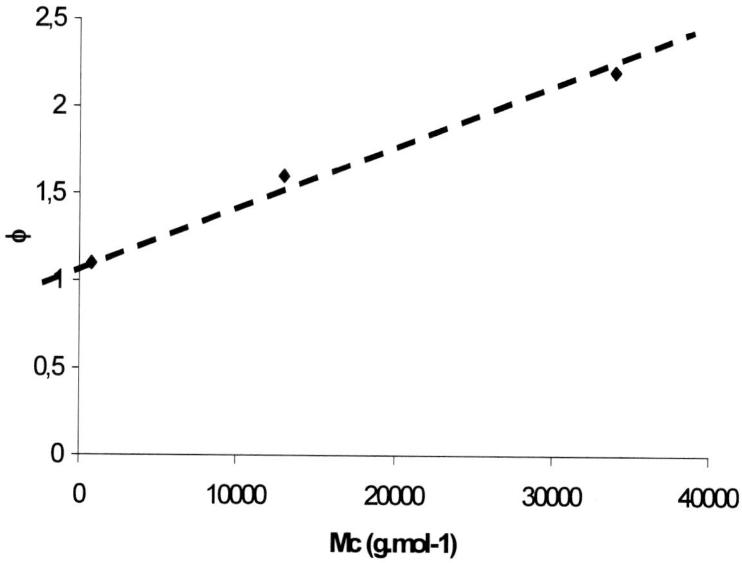

Figure 14. ϕ (viscoelastic dissipative function) dependence with Mc.

Indeed, Φ(Mc,T) cannot theoretically be lower than 1 (the limit value which corresponds to a zero separation speed) and should increase, while the Young modulus decreases (which means, for our substrates, that energy dissipation in the bulk is higher when the network is softer as it has been verified by dynamical mechanical analysis). Moreover, even if the dependance of ϕ on the separation speed v is not dominating in the range of the available rates with the AFM, dynamical mechanical tests showed that dissipation occurs for the 34k sample and that this dissipation is included in the dissipative function ϕ.

5. Conclusions

This study has shown that a rigorous experimental procedure and AFM calibration is necessary to get valuable and reproducible values with the AFM. In particular, comparison of the force curves obtained on two model systems with similar surface chemistry properties but with different mechanical behavior (silicon wafer and cross-linked PDMS substrates) has shown that, adhesion forces can be modulated by varying, either surface chemistry or, mechanical properties of the material. This observation opens interesting perspectives to improve adherence properties.

Finally, comparison of adhesion forces also allowed us to express and quantify the mechanical and chemical surface contributions in a force curve measurement. As a major result, we succeed in proposing a new relationship between the separation energy G_{rupt} and both intermolecular (via a thermodynamic work of adhesion W_0) and mechanical contributions (via Mc, v and T), at the nanoscale:

$$G_{rupt}(v,T,Mc) \propto W_0\left[1 + Mc\phi'(v,T)\right]$$

AKNOWLEDGEMENTS

The authors are grateful to C.Frétigny, H.Haidara and K.Mougin for their fruitful discussions.

REFERENCES

[1] Schultz, J.; Nardin, M. *Adhesion Promotion Techniques*, Marcel Dekker ed.; Marcel Dekker, Inc., 1999.
[2] Cognard, J. *Science et technologie du collage*, Presses Polytechniques et Universitaires Romandes ed., 2000.
[3] Lavielle, L., Schultz, J. & Boutillier. J. in *3ème journées sur l'adhérence*-Société Française du Vide, Lyon 1987.
[4] Gent, A. N. K., H. J. *Rubber Chemistry and Technology* 1990, *63*, 613-623.
[5] Portigliatti, M., Hervet, H. & Leger, L. *C.R.Acad.Sci. série IV* 2000, *1*, 1187-1196.
[6] Gent; Schultz. *J.Adhesion* 1972, *3*, 281.
[7] Andrews, E. H.; Kinloch, A. J. *Proc.Royal Society London* 1973, *A332:385*, 401.
[8] Carre, A.; Schultz, J. *J.of adhesion* 1984, *17*, 135.
[9] Barquins, M., Khandani, B. & Maugis, D. *C.R.Acad.Sci. série II 303, 1517* 1986.
[10] De Gennes, P. G. *Langmuir 12, 4497-4500* 1996.
[11] Reihs, K.; Fuchs-H, e. *Thin solid films* 1995, *264*, 135-140.
[12] Binnig, G.; Rohrer, H.; Gerber, C.; Weibel, E. *Phys.Rev.Lett.* 1982, *49*, 57.
[13] Mougin, K.; Haidara, H.; Castelein, G. *Colloids and Surfaces a Physicochemical and Engineering Aspects* 2001, 193, 231-237.
[14] Chaudhury, M. K.; Whitesides, G. M. *Science* 1992, *255*, 1230-1232.
[15] Vonna, L., 1999.
[16] Porter, M. D.; Bright, T. B.; Alara, D. L.; Chidsey, C. E. D. *J.Am.Chem.Soc.* 1987, *109*, 3559-3568.
[17] J.Wood; R.Sharma. *Langmuir* 1994, *10*, 2307.
[18] Mougin, K.; Mulhouse-France, 2001.
[19] T.Elzein; M.Brogly; J.Schultz. *J.Surf.Interf. anal.* 2003, *35*.
[20] Flory, P. J.; Erman, B. *Macromolecules* 1982, *15*, 800.
[21] Flory, P. *Chem.Rev.* 1944, 35-51.
[22] Girifalco, L. A. G., R. J. *J.Phys.Chem* 1957, *61*, 904.
[23] Cappella, B.; Dietler, G. *Surface science reports* 1999, *34*, 1-104.

[24] Xiao, X.; Qian, L. *Langmuir* 2000, *16*, 8153-8158.
[25] Sader, J. E.; Chon, J. W. M.; Mulvaney, P. *Review of scientific instruments* 1999, *70*, 3967-3969.
[26] Senden, T. J.; Ducker, W. A. *Langmuir* 1994, *10*, 1003-1004.
[27] Sheiko, S. S.; Moller, M.; Reuvekamp, E. M. C. M.; Zandbergen, H. W. *Physical review B Condensed matter* 1993, *48*, 5675-5678.
[28] Gibson, C. T.; Weeks, B. L.; Abell, C.; Rayment, T.; Myhra, S. *Ultramicroscopy* 2003, *97*, 113-118.
[29] Hutter, J. L.; Bechhoefer, J. *Review of scientific instruments* 1993, *64*, 1868-1873.
[30] Gibson, C. T.; Watson, G. S.; Myhra, S. *Nanotechnology Bristol* 1996, *7*, 259-262.
[31] Cleveland, J. P.; Manne, S.; Bocek, D.; Hansma, P. K. *Review of scientific instruments* 1993, *64*, 403-405.
[32] Gibson, C. T.; Weeks, B. L.; Lee, J. R. I.; Abell, C.; Rayment, T. *Review of scientific instruments* 2001, *72*, 2340-2343.
[33] Torii, A.; Sasaki, M.; Hane, K.; Okuma, S. *Measurement science and technology* 1996, *7*, 179-184.
[34] Levy, R.; Maaloum, M. *Nanotechnology Bristol* 2002, *13*, 33-37.
[35] Johnson, K. L.; Kendall, K.; Roberts, A. D. *Proceedings, Royal Society* 1971, *A324*, 301.
[36] Villarrubia, J. S. *Surface science* 1994, *321*, 287-300.
[37] Bemis, J. *Asylum Research Corp.*, Personal communication.
[38] Sarid, D. *Scanning Force Microscopy with applications to electric, magnetic, and atomic forces*, Oxford University Press ed., 1991.
[39] Haugstad, G.; Gladfelter, W. L. *Ultramicroscopy* 1994, *54*, 31-40.
[40] Michel, D.; *Thesis*-Université Bordeaux I (France), 1997.
[41] O.Noel; M.Brogly; G.Castelein; J.Schultz. *MRS Proceedings* 2003-San Francisco-Vol.778-Ed.Warrendale, PA 2003.
[42] Derjaguin, B. V.; Muller, V. M.; Toporov, Y. P. *J.Colloid Interface Sci.* 1975, *53*, 314.
[43] J.Israelachvili. *Intermolecular and surface forces*, 1992.
[44] Riedo, E.; Levy, F.; Brune, H. *Physical review letters* 2002, *88*, 185505.
[45] Fisher, L. R. I., J.N. *Colloids Surf.* 1981, *3*, 303-319.
[46] Noel O., C. G., Brogly M., Schultz J. *Langmuir* 2004, *20*, 2707-2712.
[47] Basire, C.; *Thesis*- Paris VI, 1998.
[48] Pethica, J.; Sutton, A. *J.Vac.Sci.Technol.A* 1988, *6*, 2490.

In: New Frontiers in Polymer Research
Editor: Robert K. Bregg, pp. 99-124

ISBN: 1-59454-956-7

Chapter 5

BIODEGRADABLE PHB/PEG DERIVATIVES AND THEIR DEGRADATION BEHAVIOR

Qiang Zhao and Guoxiang Cheng
School of Materials Science and Engineering, Tianjin University, Tianjin 300072, China
E-mail: gxcheng@tju.edu.cn

ABSTRACT

Poly(3-hydroxybutyrate) (PHB) is a naturally occurring biodegradable polyester produced by many bacteria. In contrast to traditional polymers, the main remarkable characteristics of PHB lie in its biodegradability and biocompatibility. PHB also shows many special properties, such as piezoelectricity, optical activity and so on. However, PHB is a highly crystalline polymer due to its exceptional stereochemical regularity. This gives rise to its low impact strength, narrow processing window and poor hydrophilicity.
Poly(ethylene glycol) (PEG) is a synthetic soluble polymer known to be highly hydrophilic, biocompatible and flexible, and frequently used for modifications of PHB. In the second part of this chapter, four methods were discussed for the preparations of PHB/PEG derivatives. They include: physical blending, graft copolymerization, multiblock copolymerization and interpenetrating network technique. The characterizations of their properties, such as thermal properties, mechanical properties and hydrophilicity were also reviewed.
The degradation mechanism and factors influencing the degradation behavior were described in the final part. PHB is known to be degraded via three kinds of mechanisms: enzymatic degradation, hydrolytic degradation and thermal degradation. Many factors have effects on the process of PHB's degradation and these factors can de divided into two catalogues: exterior factors and material characteristics. The common exterior factors comprise pH, temperature, environmental conditions. Crystallinity, hydrophilicity, tacticity are main material characteristics.
In the final section, the degradation behavior of PHB/PEG blend and PHB/PEG multiblock copolymer were discussed in detail. The degradation of PHB/PEG derivatives proceeded at a higher rate than that of homopolymer PHB. This acceleration is caused by two reasons: the improvement of hydrophilicity within the matrix and the decrease of the crystallinity in the PHB phase.

1. INTRODUTION

Polyhydroxyalkanoates (PHA) are a class of aliphatic thermal plastic polyesters that are biosynthesized by many bacteria as an intracellular carbon and energy storage compound. The simple and most common member of PHA, poly(3-hydroxybutyrate) (PHB) (Fig.1), was discovered in *Bacillus megaterium* by French scientist Lemoigne in 1926.

Currently many organisms synthesize PHB from acetyl-CoA via (*R*)-3-hydroxybutyryl-CoA, but the key enzyme of PHB biosynthesis is *R. eutropha*. The general pathway involved in the biosynthesis of PHB in *R. eutropha* in simplified form is represented in Fig.2. The synthetic process occurs through three steps. First of all, the enzyme β-ketothiolase condenses two molecules of acetyl-CoA to form acetoacetyl-CoA. Then the acetoacetyl-CoA is reduced by the NADH-dependent enzyme (acetoacetyl-CoA reductase) to form (*R*)-3-hydroxybutyryl-CoA. This reaction occurs stereospecifically, which means all resulting 3-hydroxybutyryl-CoAs are in the (*R*)-configuration. In the final step, the monomer is polymerized to form PHB by a synthetase (PHB polymerase) [1-8].

There are 3 different biotechnological processes to produce PHB: fermentative production employing microorganisms, production in transgenic plants and *in vitro* synthesis employing isolated enzymes[9-15].

The main characteristic of PHB is its biodegradability in the natural environment. Besides, PHB shows many special properties, such as biocompatibility, piezoelectricity, optical activity and so on.

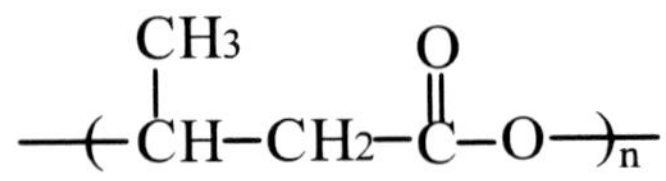

Figure 1. Chemical structure of PHB

Acetyl-CoA —①→ Acetoacetyl-CoA —②→ *R*-3-Hydroxybutyryl-CoA —③→ PHB

①β-ketothiolase ② acetoacetyl-CoA reductase ③ PHB polymerase

Figure 2. PHB-biosynthetic pathway

1.1. Biodegradability

A number of microorganisms have been reported to excrete extracellular enzymes to degrade PHB. They can be degraded in soil, sludge, seawater and so on. Under optimum conditions, the degradation rate is extremely fast. With the aggravation of the environmental pollution, as an environmentally friendly material, PHB has attracted many researchers' attention. The detailed knowledge will be discussed in *3.1* and *3.2*.

1.2. Biocompatibility

Animal tests have shown that PHB is high biocompatible. When subcutaneously implanted into the mice, all samples showed minimal tissue reaction. Hydrolytic degradation of PHB *in vitro* proceeds to the monomer, (*R*)-3-hydroxybutyric acid. This is a normal metabolite in human blood and couldn't cause physiological reactions. In common with acetoacetate and acetone, this acid is one of the three ketone bodies which are produced endogenously by the process known as ketogenesis. *In vitro* tests have also shown that PHB is biocompatible to various cell lines, including osteoblastic, epithelial cell and ovine chondrocytes [16]. Consequently, PHB has become an important candidate for biomedical applications, such as tissue engineering scaffold, drug delivery system, and bone implants [17-30].

1.3. Crystallization Behavior

PHBs are partially crystalline polymers with a high degree of crystallinity in the range of 60-80%. The exceptional purity and stereochemical regularity of PHB make it an ideal model substrate for fundamental studies of polymer morphology, crystallization and nucleation. Single crystals of PHB were first produces by Alper et al. They deduced, from X-ray diffraction data, that PHB has a fiber repeat of 0.59 nm [31]. The detailed crystal structure of PHB was first identified by Okamura and Marchessault. Using X-ray diffraction the unit cell was identified as orthorhombic with the lattice parameters a=0.576nm, b=1.320nm and c=0.596nm (fiber axis). They proposed that PHB existed as a right-handed helix with a two-fold screw axis and that two anti-parallel chain molecules passed through the sub-cell. On the basis of above findings, it was proposed that the space group for PHB was P212121 and that the theoretical crystal density was 1.262g/cm^3 [32].

1.4. Mechanical Properties

The mechanical properties of PHB are comparable to those of isotactic polypropylene. But PHB suffers a limitation in its use: a relatively low impact resistance. This is attributed to the high crystallinity of PHB. When crystallizing from melt, it forms large spherulites. The elongation at break is only about 6%. Table 1 compares the physical properties of PHB with those of PP.

Table 1. Typical physical properties of PHB and PP

Properties	PHB	PP
Melting temperature, ℃	175	176
Glass transition temperature, ℃	15	-10
Density, g/cm^3	1.250	0.905
Flexural modulus, GPa	4.0	1.7
Tensile strength, MPa	40	38
Extension at break, %	6	400
UV resistance	good	bad
Solvent resistance	bad	good

2. Preparation and Characterization of PHB/PEG Derivatives

Even though PHB shows good biodegradability and biocompatibility, it also suffers some inherent flaws due to its structural characteristics. The high crystallinity of PHB gives rise to its low impact strength and poor hydrophilicity. At the same time, PHB has a narrow processing window. It may suffer thermal degradation when heated for a longer time above its melting point. Furthermore, the degradation of PHB, especially the hydrolytic degradation, proceeds relatively slow in comparison with other degradable polymers, such as PCL, PLA, or P(LA-GA). All these limit the application areas of PHB, especially in medical field.

Poly(ethylene glycol) (PEG) is a synthetic soluble polymer known to be highly hydrophilic, biocompatible and flexible. It has been approved by the U.S. Food and Drug Administration (FDA) for internal consumption. Recent attention has been focused on the incorporation of PEG into PHB matrix to improve its properties. The PHB/PEG derivatives can mainly be prepared by two kinds of techniques: physical blending and chemical copolymerization, and the characterizations of their properties are also reviewed in this part.

2.1. PHB/PEO Blends

Blending with other polymers is known to be an effective way to modify the properties of PHB. Up to now, a lot of PHB blend systems have been studied in order to screen materials with ideal characteristics. Our research group have prepared PHB/PEG blends and systematically studied their properties. The results showed that the incorporation of low molecular PEG into PHB matrix can improve the hydrophilicity, biocompatibility and degradation rate by increasing the availability of water within the matrix. On the other hand its mechanical tensile strength and elongation at break decrease [33]. Herein, we select poly(ethylene oxide) with high molecular weight (M_n=20,000) as a second component to blend with PHB. PEO is known with high toughness, and blending of PEO with PHB may improve the poor mechanical properties of PHB without destroying other desirable properties, such as degradability, biocompatibility, and so on.

Blending of PHB with PEO was carried out in chloroform and the films of the blends were obtained by solvent cast technique.

Thermal characterization of polymer blends is a well-known method to determine their miscibility. Differential scanning calorimetry (DSC) was adopted to investigate the thermal properties. From the DSC thermograms (Fig.3), we can see that when the PEO component is more than 10 wt%, both two distinct endothermic peaks can be detected. The higher one represents the fusion of PHB phase while that at lower temperature corresponds to the melting of PEO phase. The apparent melting temperatures of PHB and PEO phase decrease with the increase of the second component (Fig. 3).

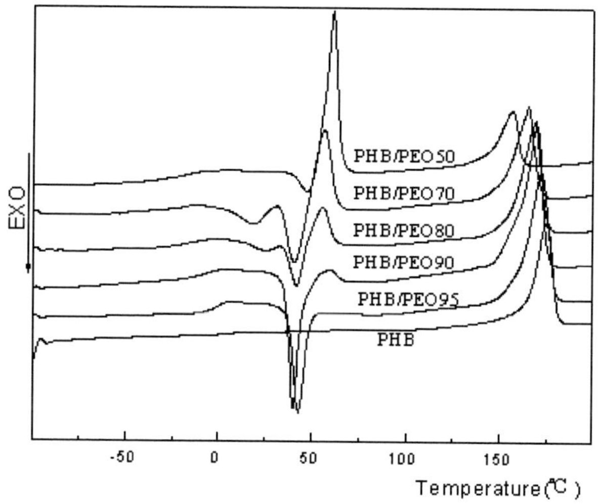

Figure 3. DSC thermograms of PHB/PEO blends (number donates the weight fraction of PHB phase)

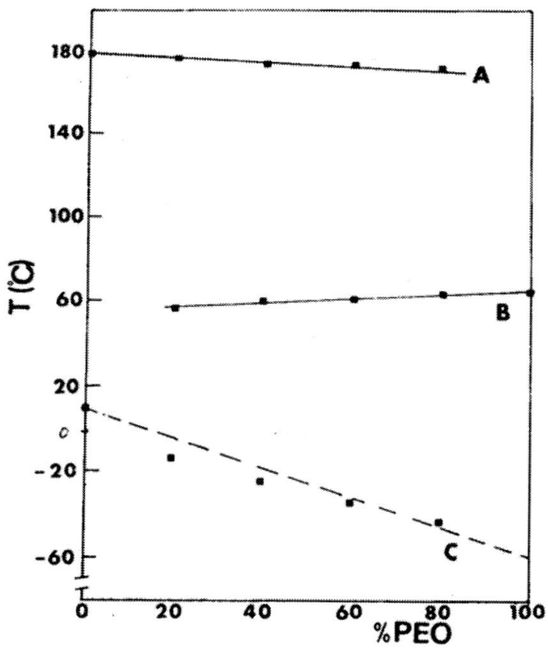

Figure 4. PHB/PEO phase diagrams: (A) T'$_m$ (PHB); (B) T'$_m$ (PEO); (C) ●, T_g (expt),···, T_g (Fox) (ref. [36])

Avella et al. performed detailed studies on thermal and crystallization behavior of PHB/PEO blends[34,36,37]. As far as T_g is concerned, it emerges that PHB/PEO blends are characterized by a single T_g intermediate between those of two component polymers. Moreover, as shown by Fig.3 Tg is composition dependent [34].

For ideal systems, which are miscible and amorphous over the entire composition range, the relationship between the T_g and the composition can be predicted by empirical equations, for instance by the Fox equation [35].

$$1/T_g\,(blend) = W\,(PHB)/T_g\,(PHB) + W(PEO)/T_g\,(PEO)$$

Where T_g donates the glass transition temperature, and W(PHB) and W(PEO) are the corresponding weight fractions.

The relationship between the T_g and the composition was plotted in Fig.4. It can be seen that experimental data fit the theoretical plot obtained by equation quite well. These results strongly suggest that PHB and PEO are compatible in the melt and in the amorphous state.

The PHB-PEO interaction parameter χ_{12}, as calculated by Flory-Huggins equation, is composition-dependent and always has a negative value, indicating strong interactions and miscibility. This finding indicates that some kind of molecular interaction must be present. The special interactions responsible for the PHB-PEO miscibility are likely to be provided by the carbonyl groups of the PHB and the hydrogen of CH_2 groups in PEO.

On isothermal crystallization at temperature above the T_m of PEO, PHB spherulites grow isothermally with a one-phase melt blend (PHB+PEO). After the crystallization of PHB at T_c, the system can be characterized by the presence of a liquid PEO+PHB phase and formed a phase of almost pure liquid PEO. When further cooled to –100℃, the crystallization of PEO phase takes place. From the corresponding DSC thermogram (Fig.5), it emerges that part of the PEO molecules crystallize at a temperature close to that of neat PEO (≈ 40℃) and part at a lower temperature (-20 ~ -30℃).

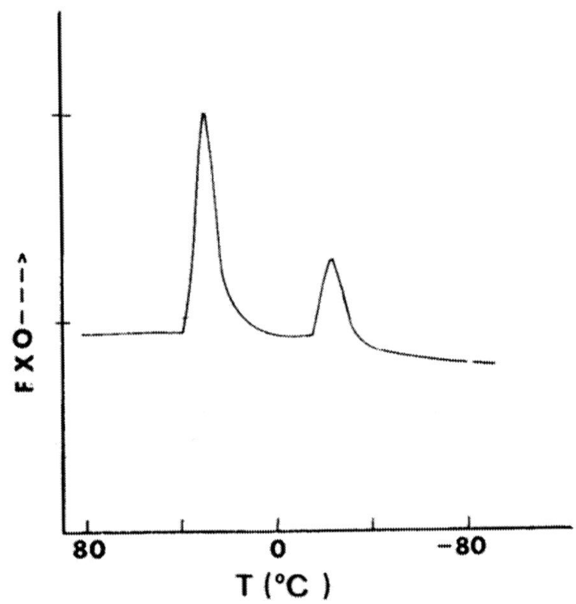

Figure 5. Non-isothermal crystallization DSC thermogram PHB/PEO blend (ref. [36])

The lower exothermic peak must be attributed to the crystallization process of PEO molecules coming from homogeneous PEO+PHB mixture in the interlamellar regions of PHB spherulites. The higher one corresponds to that of the interfibrillar intraspherulitic domains of pure PEO [36].

This fractionated crystallization behavior can occur in two steps at different super-coolings, dependent on the crystallization conditions, thermal history and crystallization rates. These parameters can produce two different modes of nucleation of PEO, homogeneous (at high super-cooling), and heterogeneous (at low super-cooling), and consequently different morphologies [37].

Tensile strength was determined by stress-strain measurements of solvent casting films (Fig.6). The results show that the mechanical properties are strongly dependent on the composition. In general, the Young's modulus and the tensile strength decrease with the increase in PEO content. The addition of PEO results in more flexible and tougher materials with increasing PEO content (Table 2). Blends with a high PHB content (>80%) show brittleness similar to that of homopolymer PHB. Samples with a high PEO content (>50%) show considerable plastic deformations at low stress level. So in a certain range of compositions, improved toughness can be obtained without destroying the tensile strength apparently [38].

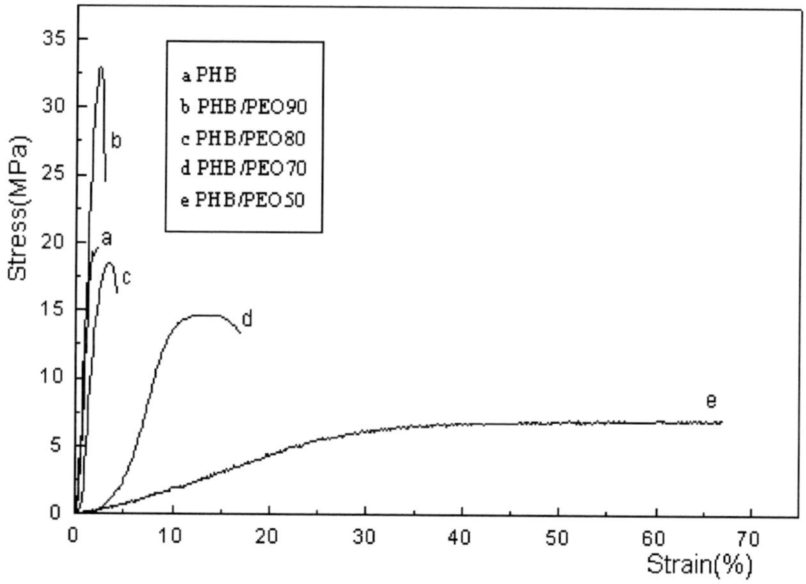

Figure 6. Stress-Strain diagrams of PHB/PEO blends

Table 2. Mechanical properties of PHB and blends, determined from stress-strain diagrams

PHB/PEO	PHB	PHB/PEO90	PHB/PEO80	PHB/PEO70	PHB/EPO50
Tensile strength (MPa)	19.59	32.98	18.25	13.84	6.98
Elongation at break (%)	2.278	2.856	4.195	17.061	67.123

2.2. Semi-Interpenetrating Polymer Networks (Semi-IPNs) of PHB with PEG

Net-PEG is a kind of hydrogel which commonly possesses three-dimensional networks and can retain large quantities of water without dissolution. Although hydrogels are similar in the properties of water-swollen state to natural tissues, they are subject to a decrease in mechanical properties. In order to solve this problem, PHB was combined with net-PEG to produce a PHB/net-PEG semi-IPNs network by Hao and Deng [39].

The synthetic pathway of semi-IPNs is shown in Scheme.1. Hydroxyl end groups of PEG reacted with acryloyl chloride to form PEGM. Then the preparation of PHB/net-PEG semi-IPNs was performed by the radical polymerization of acrylate groups in PEGM under the UV irradiation, owing to their photoactive nature.

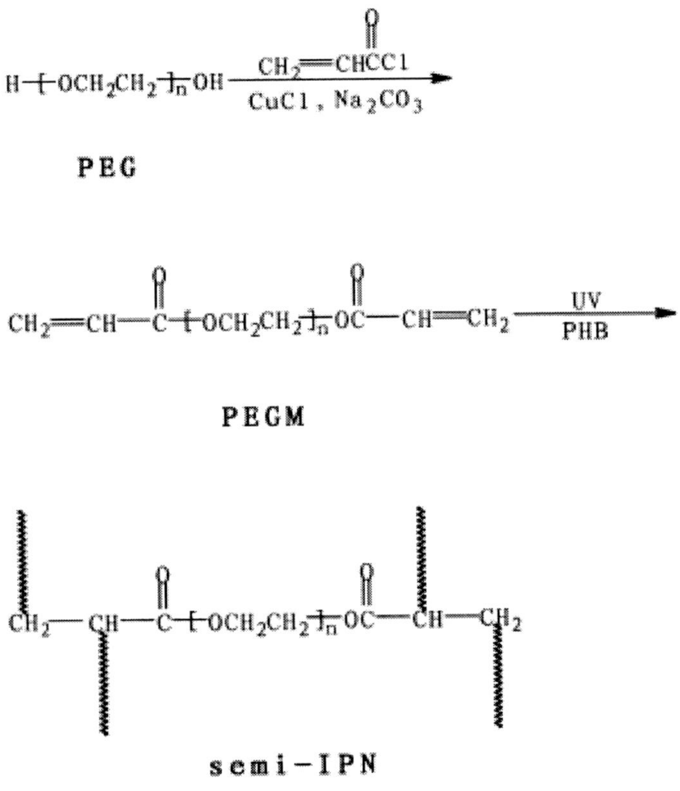

Scheme 1. Synthesis pathway of PHB/net-PEG semi-IPNs (ref. [39])

The swelling behavior of semi-IPN hydrogels was investigated by measuring the equilibrium water content (EWC). The hydrogels swelled rapidly in water and reached equilibrium within 1h. The specific data were plotted against composition in Fig.7. It can be seen that the EWC shows a tendency to decrease with increasing amount of PHB.

Thermal properties were also characterized by DSC. Fig.8 shows the typical DSC thermogram of PHB/net-PEG semi-IPNs as well as that of linear PEGM. The endothermic peak at about 170℃ represents the fusion of PHB phase, whereas the peak below 100℃ is attributed to that of PEG segments. It can be noticed that the melting peak of PEG phase become much broader in semi-IPN after crosslinking than in linear PEGM. In addition, PHB shows double melting peaks. This behavior can be explained in terms of melt/recrystallization process. The lower peak corresponds to the melting of as-formed crystals in the melting hydrogels, while the higher one results from that of more prefect crystals, produced by recrystallization on the heating run.

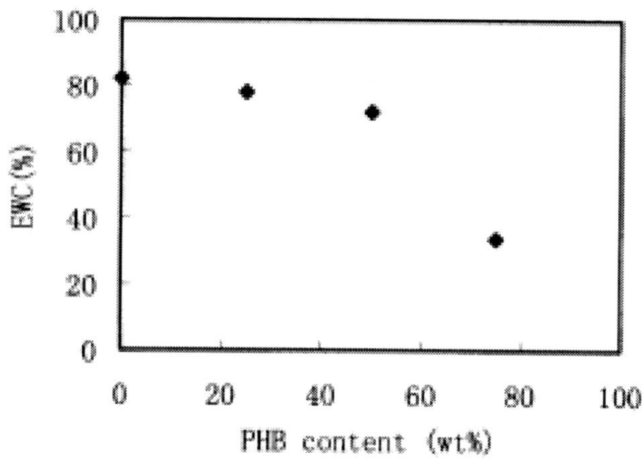

Figure 7. Equilibrium water content (EWC) of PHB/net-PEG semi-IPNs as a function of compositions (ref. [39])

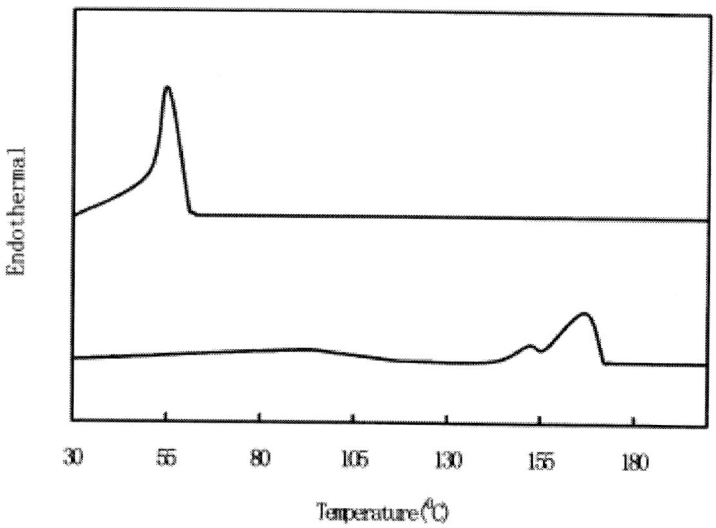

Figure 8. DSC thermograms of : (a) semi-IPN; (b) PEGM (ref. [39])

2.3. PEG-Grafted PHB Copolymers

Graft polymerization is a well-known method for polymer modifications by connecting different types of polymers having the desired properties in the same molecular chain. In our study, graft polymerization of PEG molecule onto PHB backbone was carried out (scheme.2). The mechanism of this reaction is based on radical polymerization. First of all, monomethoxy terminated PEGs were end-capped with acrylate groups to form a polymerizable macromer, PEGM. Then the reaction of PEGM with PHB took place in chloroform solution with two kinds of initiators: chemical initiators (AIBN) and ultraviolet (UV).

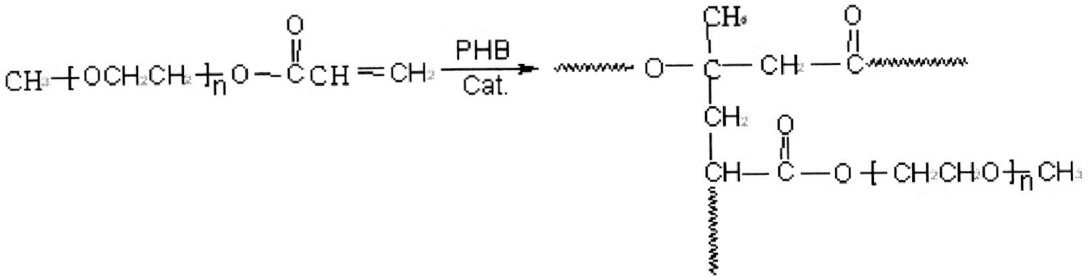

Scheme 2. Synthesis pathway of PHB-*g*-PEG copolymers

The results of thermal analysis are shown in Fig.9. It can be seen that in the heating process, this PHB-*g*-PEG copolymer experienced apparent glass transition. The exothermic and endothermic peaks, corresponding to the crystallization and melting of PHB phase respectively, were also observed. The values of T_g and T_m both decrease with increasing the PEG content. The T_g increases from-20℃ to 0℃ as PEG content decreases from 14 to 2% (Fig.10). The compression of T_m can broaden the processing window to some degree and thus improving the processibility of PHB.

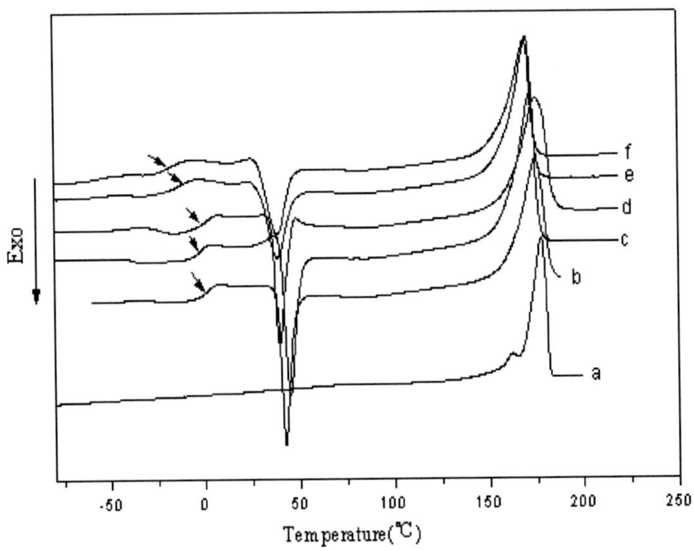

Figure 9. DSC thermograms of PHB-*g*-PEG copolymers

As we all know, PEG is a water-soluble polymer, and the incorporation of it can improve the availability of water within the matrix. Because the PEG molecules are linked to PHB backbone by chemical bonds, they could not be dissolved and liberated into water in the swelling process. The specific data were plotted in Fig.11. As a general rule, the hydrophilicity of the copolymer is gradually increased with an increase in the PEG content.

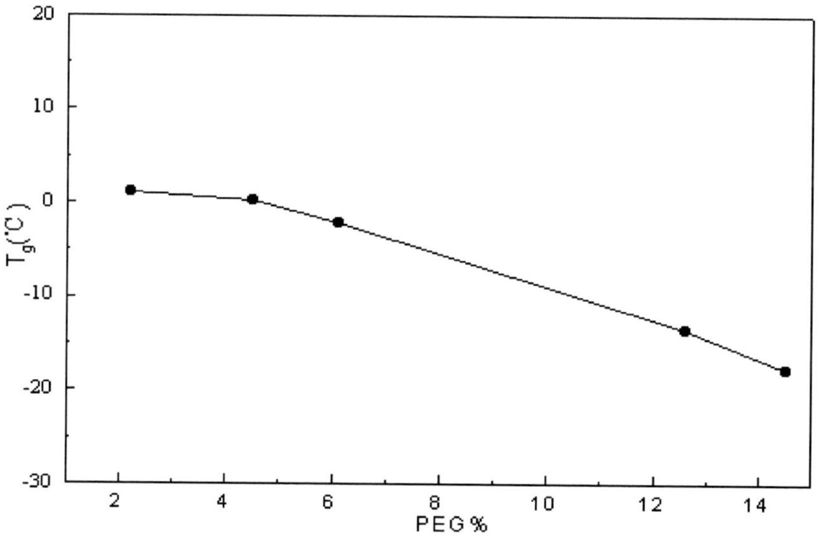

Figure 10. The plot of T_g of PHB phase against PEG content in PHB-*g*-PEG copolymers

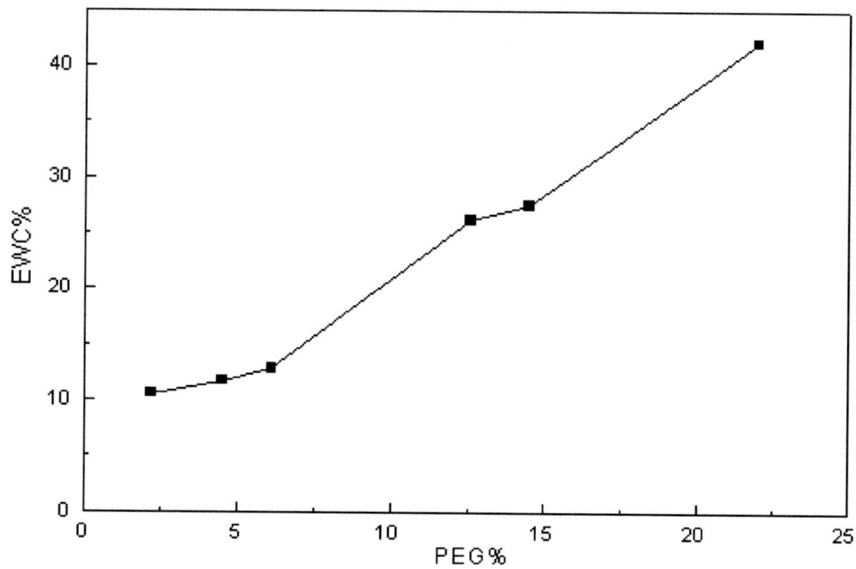

Figure 11. Equilibrium water content (EWC) of PHB-*g*-PEG copolymers as a function of compositions

2.4. PHB and PEG Multiblock Copolymers

As produced in bacteria, PHB has a relatively high molecular weight, which is unsuitable for molecular design of specialty polymers such as block copolymers. In our study, natural-origin PHB of high molecular weight was first degraded by alcohololysis to provide low molecular weight and dihydroxy-terminated prepolymers. Then PHB-diol and PEG were multiblock copolymerized with hexamethylene diisocyanate (HDI) as a coupling agent (Scheme .3).

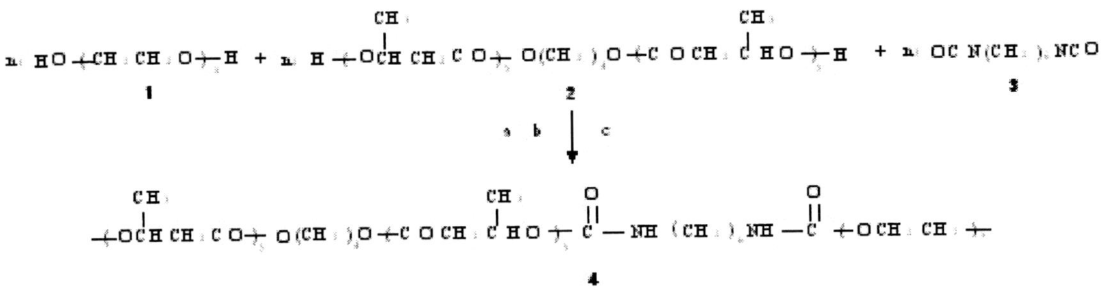

Scheme 3. Synthesis pathway of the PHB/PEG multiblock copolymer a: 75℃; b: 1,2-dichloroethane; c: dibutyltin dilaurate

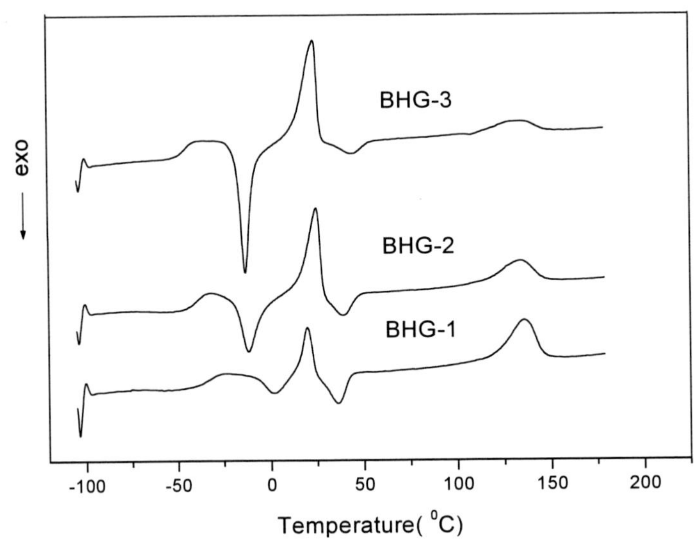

Figure 12. DSC thermograms of PHB/PEG multiblock copolymers

The thermal properties were investigated by DSC (Fig.12). The results show that two components are both able to form crystalline phases. When comparing with PHB precursors (PHB-diol), the PHB phase in the copolymers melts at a lower temperature and the melting temperature (T_m) decreases moderately with the increase of PEG content. The glass transition can be clearly observed in the thermograms. The corresponding transition temperature (T_g) lies between those of the two prepolymers and is much lower than that of PHB.

The tensile properties of copolymers were determined at ambient temperature by stress-strain experiments (Fig.13). This material is characteristic of elastomers with remarkable tensile strength and elongation.

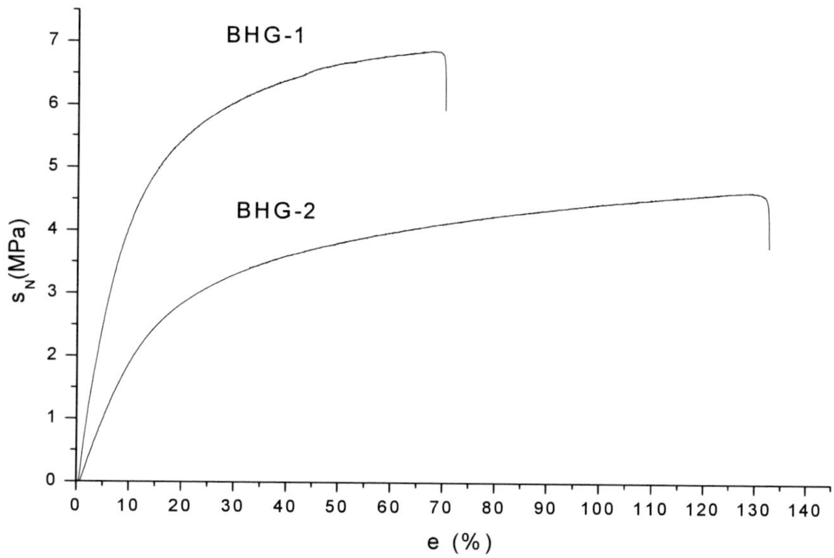

Figure 13. Stress-Strain diagrams of PHB/PEG multiblock copolymers

3. The Degradation Mechanisms of PHB and Degradation Behavior of PHB/PEG Derivatives

3.1. Degradation Mechanisms of PHB

In contrast to traditional plastics, the main remarkable characteristic of PHB is its degradability. PHB is known to have three different kinds of degradations. First of all, it can be hydrolyzed in sterile water, especially in alkaline media conditions. Another is the enzymatic degradation catalyzed by a number of depolymerases. Moreover, it can undergo thermal degradation at temperatures close to its melting point 180℃.

3.1.1. Ezymatic Degradation

One of the attractive features of PHB is its degradation in the natural environments. A number of microorganisms such as bacteria and fungi in the environment excrete PHB depolymerases to degrade the solid PHBs into water-soluble oligomers and monomer, and they utilize the resulting products as nutrients within cells. Aerobic and anaerobic PHB-degrading microorganisms have been isolated from various environments such as soil, activated sludge, compost, lake fresh water, and sea water. In one recent study 295 dominant strains of bacteria were identified which were capable of degrading PHB and P(HB-HV) [40]. These included 105 gramnegative bacteria, 36 *Bacillus* strains, 68 *streptomyces* strains and 86 mold strains.

Although many extracellular enzymes capable of depolymerizing PHB and oligomers derived from the polymer have been found in a few organisms, researches have been concentrated on the depolymerases excreted by *Ralstonia faecalis*. The extracellular depolymerizing enzyme from R. *faecalis* has been isolated and purified by Tanio et al.[41-45].The purified enzyme (MW 47kDa) of R. *faecalis* has a hydrophobic domain (5kDa) as a binding site to adhere to the hydrophobic surface of PHB film, in addition to a catalytic domain as an active site. The presence of binding and catalytic domains in depolymerizing enzymes suggests that the enzymatic degradation on the surface of PHB film takes place via two steps: the first step is adsorption of enzymes on the surface of PHB film via the binding domain, and the second step is hydrolysis of the polymer chain into water-soluble products by the catalytic domain [46]. Fukui et al.[43] reported that the enzyme activity of this depolymerase was found to decrease when treated with trypsin. This loss of activity was attributed to the removal of the hydrophobic site on the depolymerase by trypsin.

The depolymerase converted the PHB to predominately the dimer with a trace of monomer. Quantitation of the (*R*)-3HB liberated and the decrease in ester bonds during the hydrolysis of oligomers suggested that dimer units are released one at a time from the free hydroxyl terminus. However, Hocking et al.[47] reported that *endo* (random) cleavage of PHB chain by the depolymerase occurs together with *exo* (from the chain ends) cleavage.

In addition to PHB depolymerase, R. *faecalis* possesses a quite distinct extracelluar oligomer hydrolase[45,48]. The hydrolase has a high affinity for oligomeric esters, but is inactive with polymers. The oligomer hydrolase appears to act by sequential removal of (*R*)-3HB units from carboxyl terminus

3.1.2. Hydrolytic Degradation

As a normal aliphatic polyester, PHB can also be hydrolyzed in acidic and basic conditions [49]. It is widely believed that not only homogeneous hydrolytic degradation but also heterogeneous hydrolysis of PHB occurs via random scission of the ester linkages throughout the whole polymer matrix. Heterogeneous degradation of PHB has two stages. In the first stage, random scission of the ester linkages reduces the molecular weight and allows a reorganization to a lamellar morphology; weight loss is negligible. In the second stage of the degradation process, an appreciable weight loss of the sample is observed [50-53].

3.1.3. Thermal Degradation

Because of PHB's high crystallinity, it suffers thermal degradation when heated for a long time at the temperature a few degrees higher than its melting point. So very low resistance to thermal degradation seems to be the most serious problem related to the processing of PHB. The mechanism of thermal degradation is sensitive to the investigated temperature range. At moderate temperatures, it is widely accepted that PHB is decomposed through a β-elimination reaction or six-membered ring transition as illustrated in Scheme 4. This mechanism gives almost exclusively unsaturated ends for the products.

Lehrle et al.[54] carried out a detailed study on the mechanism of chain scission by statistical analysis of molecular weight distributions. They found that total thermolysis proceeds exclusively in terms of a random chain scission mechanism. However, in the partial (40%) pyrolysis of PHB, it appeares that there are some kinetically favored scissions occurring near the ends of the molecules.

The thermal degradation of PHB in the presence of two plasticizers (glycerol, glycerol triacetate) has been studied by Janigova et al.[55]. Their results showed that the degradation is substantially accelerated by the presence of glycerol whereas glycerol triacetate is inert. Thermal degradation proceeds rapidly at temperature of 190℃ and higher. But no clearly detectable degradation is observed below 170℃ for virgin PHB as well as PHB containing plasticizers.

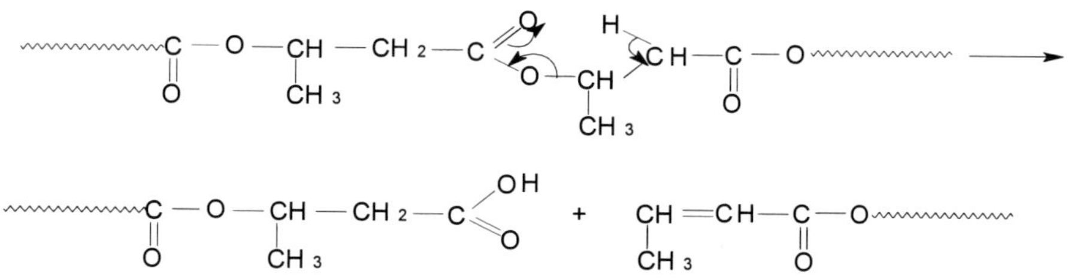

Scheme 4. Thermal degradation of PHB

3.2. Factors Influencing the Degradation Process of PHB

In the process of PHB degradation, there are numerous factors influencing its degradation behavior, such as environmental conditions, temperature, pH, the crystallinity and hydrophilicity of the samples and so on. In general, these factors can be divided into two main catalogues: exterior factors and material characteristics.

3.2.1. Exterior Factors

Environmental Conditions

Although a broad spectrum of bacteria can depolymerize PHB in various environmental conditions, the degradation rates are extremely different. Table 3 indicates typical rates of biodegradation of PHB in soil and various other environments. The biodegradabilities of P(HB-HV) fibers in water decrease in the following order: river freshwater> lake fresh water> sea water. The rate of biodegradation may be dependent on the total number of microorganisms in natural water[56].

PH

The hydrolytic degradation of the polymer is strongly influenced by pH. pH effects must be determined if the material is to be used in the body, because relatively large localized PH changes would alter the degradation profile of the material.

The results of hydrolysis showed that PHB and P(HB-HV) degrade relatively slow at physiological pH (7.0-7.4). The acceleration of hydrolysis can be observed in alkaline conditions [57,58].

Table 3. Biodegradation of PHB in various environments (ref.[50])

Environment Thick packaging	Period for dissolution of 1 mm thick section (weeks)	Average rate of surface erosion (μm/week)	Period of 100% weight loss of 50 μm film (weeks)
Anaerobic sewage	6	100	0.5
Estuarine sediment	40	10	5
Aerobic sewage	60	7	7
Soil at 25℃	75	5	10
Sea water at 25℃	350	1	50

Knowles et al.[58] have reported PHBs degrade by significant different routes at various surrounding medium pH. Natural and acidic solutions produce a more diffuse surface degradation, whereas the alkaline solution appears more aggressive, with specific sites of attack occurring to produce deep points of surface erosion. In addition, the difference in chemical degradation mechanism produced by PH also affects the points on the polymer chain at which the bond scission occurs.

The pH also influences the microbial degradation. The pH dependency of depolymerization has been reported in sewage, with optimum activity occurring between pH7.4 and 8.5[59].

Temperature

The process of degradation is also affected by the medium's temperature, and the rates of hydrolysis increase with the enhancement of the temperature. It has been reported that a change from physiological temperature to 70℃ generally increases the rate some 30 to 100-fold [57,60]. The results of Lauzier et al.[61] indicated that, even under strong acid conditions(HCl, 3.0N), high molecular weight granules are fairly stable at room temperature.

3.2.2. Material Characteristics

Crystallinity

The crystallinity of PHB has a great effect on the rates of its degradation. Amorphous regions are more susceptible to degradation, and PHB is predominately hydrolyzed in the amorphous state [62].It was reported that the rate of enzymatic degradation of bacterial P[(*R*)-3HB] film increased with a decrease in the crystallinity. In addition, the rate of enzymatic degradation of P[(*R*)-3HB] chains in an amorphous was about 20 times higher than the rate for P[(*R*,*S*)-3HB] chains in a crystalline state [63,64].

Abe et al. reported that when chemosynthetic P[(*R*,*S*)-3HB] was blended with bacterial P[(*R*)-3HB], the enzymatic erosion of films was accelerated, and the highest rate of enzymatic hydrolysis was observed at around 50wt% of P[(*R*,*S*)-3HB] content. This rate was about 2-5 times faster than that of P[(*R*)-3HB] film. The acceleration of enzymatic degradation for blends may be caused by the decrease in crystallinity[65]. On the other hand, for amorphous atactic or low crystalline syndiotactic P[(*R*,*S*)-3HB] sample, hydrolysis can hardly take place by PHB depolymerases.

Moreover, a model of enzymatic hydrolysis of PHB films was proposed to explain this phenomenon. In this model, it is suggested that the binding domain of depolymerase adheres selectively to the P[(*R*)-3HB] crystalline phase on the surface of PHB film, and that the catalytic domain predominately hydrolyzes PHB chains in the amorphous phase on the surface and subsequently erodes PHB chains in the crystalline phase [65].

The PHB depolymerase may be liable to adhere to stable crystalline lamellae of P[(*R*)-3HB], while it may hardly bind to mobile PHB chains in the amorphous state. Consequently, atactic P[(*R*,*S*)-3HB] components in the miscible blend films with P[(*R*)-3HB] may be degraded by PHB depolymerase in the presence of crystalline P[(*R*)-3HB], while single amorphous atactic P[(*R*,*S*)-3HB] can hardly be attacked by the enzyme for its lacking of binding support to the enzyme [66,67].

Hydrophilicity

The hydrophilicity of the sample plays an important role in determining its degradability. We have reported that the hydrolysis of PHB/PEG blends and copolymers proceeded faster than that of homopolymer PHB. This is attributed to the enhanced hydrophilicity within the matrix as the introduction of more hydrophilic PEG components [68,69].

Fabrication Technique

The specimen preparation technique is also found to exert a marked effect on the degradation rates. Because the fabrication technique can determine the porosity of the matrix together with the crystallinity of the specimen which are important determinants of polymer degradation. Holland et al.[57] examined the hydrolytic degradation of specimens fabricated by four different preparation techniques. The stability to hydrolytic attack increased in the order: cold compressed discs< solvent cast films< melt pressed discs< injection moulded pieces. In our pervious work, we also found that in the enzymatic hydrolysis the degradation rate of porous PHB scaffold was almost 2 times higher than that of solvent cast PHB films.

Stereochemistry and Tacticity

The tacticity of the PHB has a profound effect on its degradation. In general, high stereochemical regularity leads to a high degree of crystallinity, which is the main determinant of degradations. In addition, the rates of enzymatic hydrolysis of PHB depend on the configuration of 3HB monomer units. The enzyme is incapable of hydrolyzing the sequence of (*S*)-3HB units [65].

3.3. The Degradation Behavior of PHB/PEG Derivatives

These PHB/PEG derivatives are fully degradable. In the hydrolysis, the PHB component can be degraded into monomeric acid and PEG can be dissolved in water, so that no solid remnants will be found.

3.3.1. The Degradation Behavior of PHB/PEG Blends

The enzymatic degradation of PHB and PEG (M_n=6000) blend fims was carried out at 37℃ and pH7.2-7.4 in the aqueous solution of the lysozyme. The samples were fabricated by solvent cast technique. The profile of degradation was investigated by measuring the weight loss of films and the results were plotted in Fig.14. Homopolymer PHB degraded with a relatively low rate in the scale of the experiments, but the degradation rates of blend films were much faster than that of PHB. For the PHB/PEG50 (PHB weight fracation), after enzymatic degradation of 40d the weight loss approximately reached 63%. In general, the degradation rates of the blends are accelerated gradually with an increase in the PEG content.

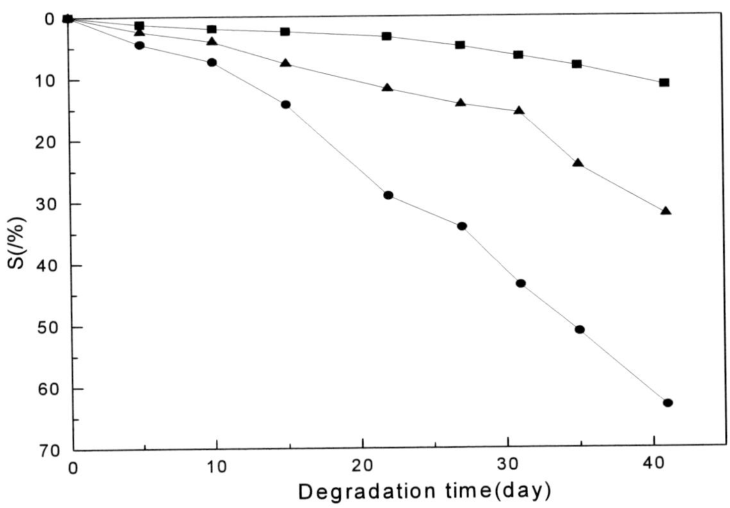

Figure 14. The biodegradation test of PHB/PEG blend films in PBS/lysozyme buffer solution at 37℃ (■) PHB; (▲) PHB80; (●) PHB50

The process of degradation passes through 2 stages: at the initial stage, the weight loss is mainly from the dissolution of PEG phase and PHB component degrades slowly. After the liberation of PEG from the blends, the enzymatic degradation rate of PHB phase accelerates because of the more rapid solvent uptake into the bulk of the matrix which is due to the enlargement of the specific areas.

The reasons for the enhancement of degradation process are as follows:

First of all, PEG is a water-soluble polymer. The incorporation of it can increase the hydrophilicity of the blends. The water contact angle and surface tension of the PHB/PEG blend films are shown in Fig.15. With the increase of PEG content in the blends, the water contact angle decreases, but the surface tension increases. The results indicate that the PHB/PEG blend films become more and more hydrophilic with the increase of PEG content.

Another positive reason for the acceleration of degradation rates is the decrease in crystallinity of PHB. The specific data of crystallinity can be calculated from the peak areas of DSC thermograms, and the relationship between the crystallinity and PHB content in the blends was plotted in Fig.16. The variation tendency reflected in crystallinity is well constant to that in degradation rates.

Meanwhile, the porosity is also an important determinant for the enzymatic degradation. Fig.17 shows the scanning electron micrographs (SEMs) of the surfaces of PHB/PEG blend film and homopolymer PHB film. After degradation, the surface of blend film was apparently blemished by the function of lysozyme and turned porous, while no obvious change took place on the surface of PHB sample. The porous morphology is attributed to the dissolution of PEG componet. With the dissolution of PEGs, more specific areas that can be attacked by enzyme are produced, which in turn accelerates the degradation.

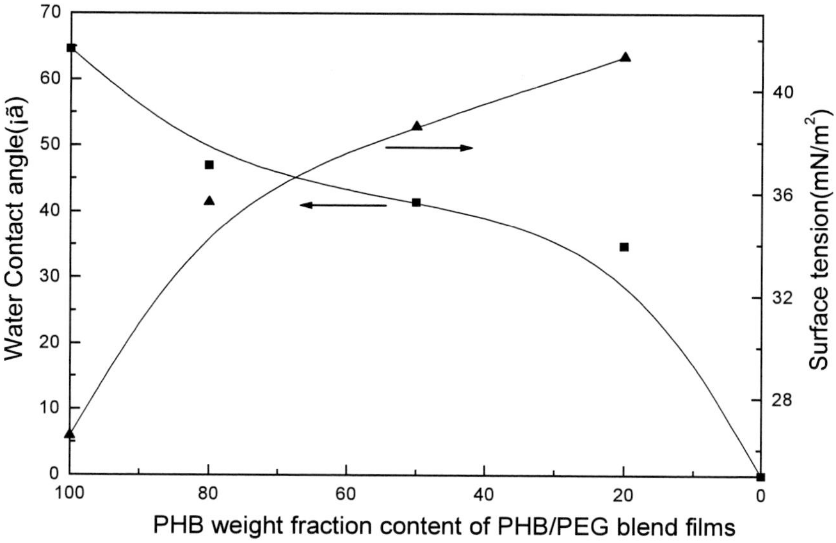

Figure 15. Water contact angle and surface tension of the PHB/PEG blend films

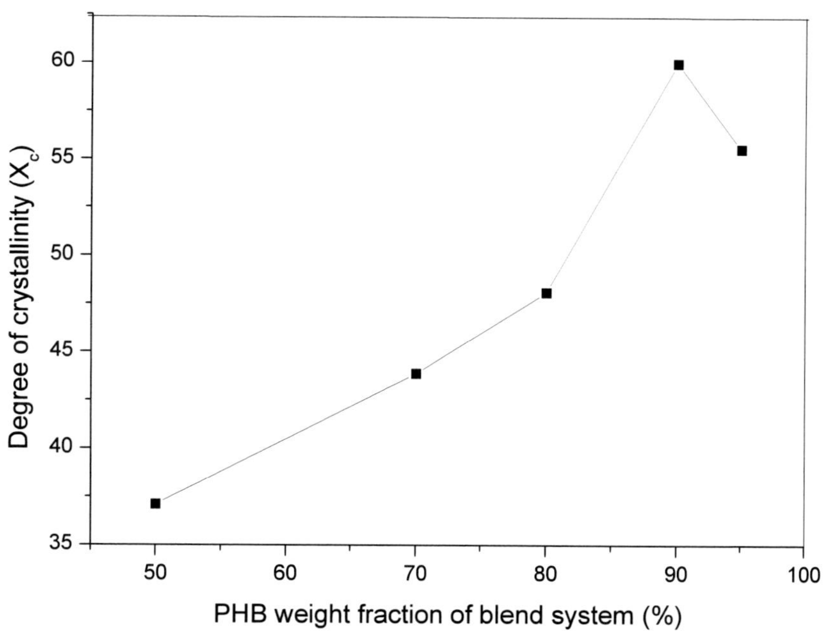

Figure 16. The plot of crystallinity of PHB phase against PHB content in PHB/PEG blends

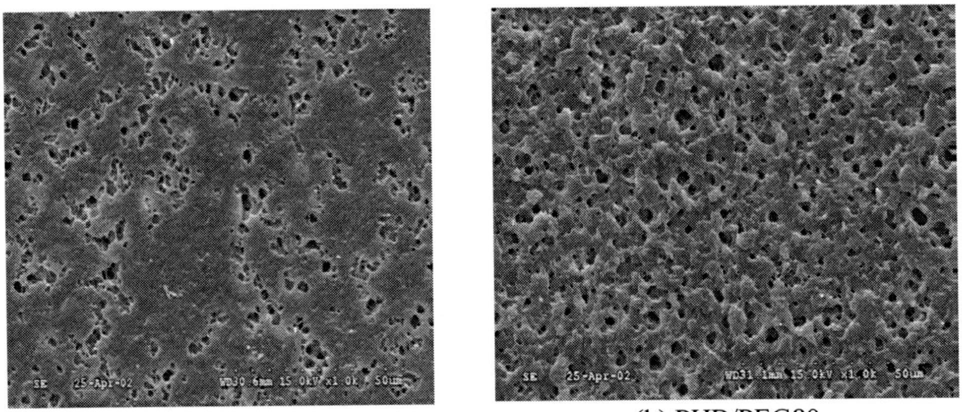

(a) PHB (b) PHB/PEG80

Figure 17. The morphology of the surface of PHB/PEG blend films after degradation in lysozyme/PBS solution for 15 days (a) PHB; (b) PHB/PEG80

3.3.2. The Degradation Behavior of PHB/PEG Multiblock Copolymers

For determining the degradability, samples were cut from dry films fabricated by solvent cast technique and were incubated in PBS buffer solution at 37℃. The degradation process was monitored by weekly recording of the changes in the molecular weight of the copolymers (Fig.18). After 10 weeks, the molecular weight was approximately 35% of the original product. It can be seen that degradation rates of copolymers are much higher than that of homopolymer PHB.

The accelerated degradation rates are caused by two reasons: the improvement of hydrophilicity and the decrease of the crystallinity in the PHB phase.

As shown in Table 4, the PHB phase in the copolymers possesses a much lower degree of crystallinity (X_c) when comparing with PHB precursors (PHB-diol).

On the other hand, due to the introduction of water-soluble PEG, the hydrophilicity of materials was apparently improved (Fig.19). As a general rule, as the PEG component is increased, the EWC increases.

Table 4. Fusional transitions of PHB/PEG multiblock copolymers, determined from DSC

Sample	T_g/℃	PHB-Segment			PEG-Segment		
		T_m/℃	ΔH_m/J/g	X_c/%	T_m/℃	ΔH_m/J/g	X_c/%
BHG-1	-33.3	135.9	43.46	29.8	20.0	17.80	8.2
BHG-2	-39.7	134.1	27.05	18.5	25.0	25.89	12.0
BHG-3	-45.6	134.1	25.54	17.5	23.8	27.34	12.7

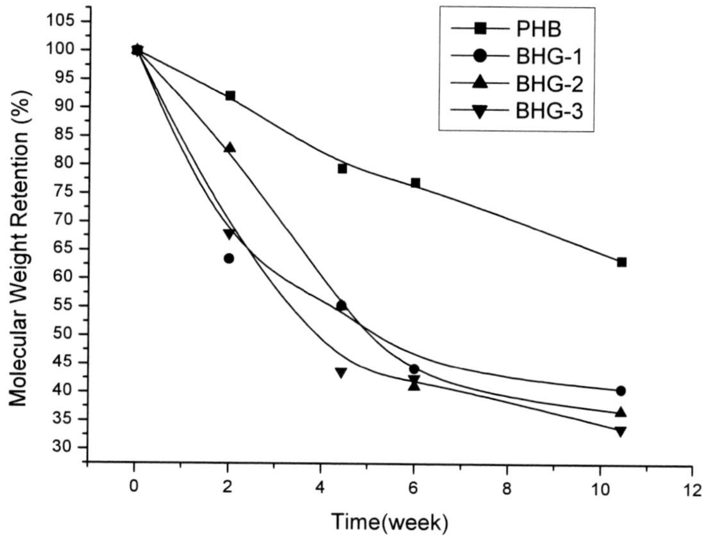

Figure 18. Molecular weight retention of PHB/PEG multiblock copolymers as a function of incubation time at 37°C in phosphate buffered saline

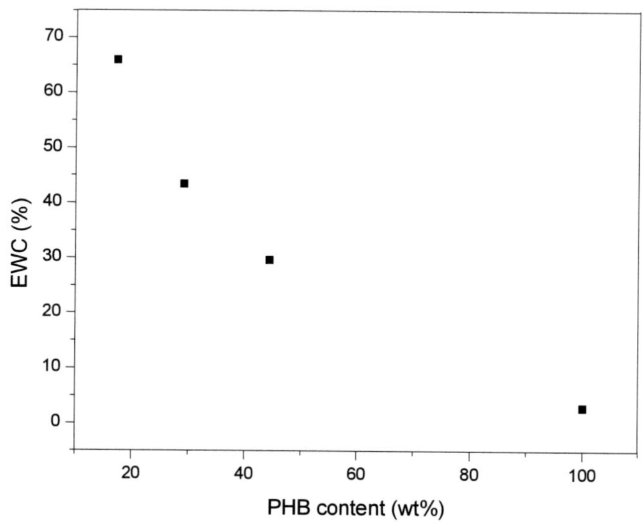

Figure 19. Equilibrium water content (EWC) of PHB/PEG multiblock copolymers as a function of compositions

REFERENCES

[1] Zinn M., Witholt B., Egli T. Occurrence, synthesis and medical application of bacterial polyhydroxyalkanoate. *Advan. Drug. Deliv. Rev.* 2001; 53:5-21.

[2] Fuller R.C. Microbial inclusions with special reference to PHA inclusions and intracellular boundary envelopes, *Int. J. Biol. Macromol.* 1999; 25: 21–29.

[3] Ellar D., Lundgren D.G., Okamura K., Marchessault R.H. Morphology of poly-β-hydroxybutyrate granules. *J. Mol. Biol.* 1968; 35: 489-502.

[4] Ballard D.G.H., Holmes P.A., Senior P.J. Formation of polymers of β-hydroxybutyric acid in bacterial cells and a comparison of the morphology of growth with the formation of polyethylene in the solid state, *Recent Adv. Mech. Synth. Aspects Pol.* 1987 ;215: 293–314.

[5] Klinke S., Ren Q., Witholt B., Kessler B. Production of medium-chain-length poly(3-hydroxyalkanoates) from gluconate by recombinant *Escherichia coli*, *Appl. Environ. Microbiol.* 1999 ;65: 540–548.

[6] Lundgren D.G., Pfister R.M., Merrick J.M. Structure of poly-β-hydroxybutyric acid granules, *J. Gen. Microbiol.* 1964 ;34: 441–446.

[7] Barnard G.N., Sanders J.K.M. The poly-β-hydroxybutyrate granule *in vivo*. A new insight based on NMR spectroscopy of whole cells. *J. Biol. Chem.* 1989; 264:3286-3291.

[8] Bonthrone K.M., Clauss J., Horowitz D.M., Hunter B.K., Sanders J.K.M. The biological and physical chemistry of polyhydroxyalkanoates as seen by NMR spectroscopy. *FEMS Microbiol. Rev.* 1992; 103: 311-316.

[9] Steinbuchel A., Fuchtenbusch B. Bacterial and other biological systems for polyester production. *Trends Biotechnol.* 1998;16: 419-427.

[10] Qi Q., Steinbuchel A., Rehm B.H.A. In vitro synthesis of poly (3-hydroxydecanoate): purification and enzymatic characterization of type II polyhydroxyalkanoate synthases PhaC1 and PhaC2 from *Pseudomonas aeruginosa*, *Appl. Environ. Microbiol.* 2000;54: 37–43.

[11] Kobayashi S., Uyama H. In vitro biosynthesis of polyesters. Babel W., Steinbuchel A. Ed., *Advances in Biochemical Engineering Biotechnology*: Biopolyesters, Springer, Berlin, 2001; Vol. 71, pp. 241–262.

[12] Slater S.C., Voige W.H., Dennis D.E. Cloning and expression in *Escherichia coli* of the *Alcaligenes eutrophus* H16 poly-β-hydroxybutyrate biosynthetic pathway. *J. Bacteriol.* 1988; 170: 4431-4436.

[13] Kidwell J., Valentin H.E., Dennis D. Regulated expression of the *Alcaligenes eutrophus* PHA biosynthesis genes in *Escherichia coli*. *Appl. Environ. Microbiol.* 1995; 61:1391-1398.

[14] John M.E., Keller G. *Proc. Natl. Acad. Sci. U.S.A*. 1993, pp12768-12773.

[15] Hahn J.J, Eschenlauer A.C., Narrol M.H., Somers D.A., Srienc F. Growth kinetics, nutrient uptake, and expression of the Alcaligenes eutrophus poly(β-hydroxybutyrate) synthesis pathway in transgenic maize cell suspension cultures. *Biotechnol. Prog.* 1997; 13 :347-354.

[16] Nebe B., Forster C., Pommerenke H., Fulda G., Behrend D., Bernewski U., Schmitz K.P., Rychly J. Structural alterations of adhesion mediating components in cells cultured on poly-β- hydroxy butyric acid. *Biomaterials*. 2001; 22:2425-2434.

[17] Williams S.F., Martin D.P., Horowitz D.M., Peoples O.P. PHA applications: addressing the price performance issue I. Tissue engineering. *Int. J. Biol. Macromol.* 1999; 25: 111-121.

[18] Wang Y.W., Wu Q., Chen G.Q. Attachment, proliferation and differentiation of osteoblasts on random biopolyester poly(3-hydroxybutyrate-co-3-hydroxyhexanoate) scaffolds. *Biomaterials*.2004; 25:669-675.

[19] Deng Y., Zhao K., Zhang X.F., Hu P., Chen G.Q. Study on the three-dimensional proliferation of rabbit articular cartilage-derived chondrocytes on polyhydroxyalkanoate scaffolds. *Biomaterials*. 2002; 23:4049-4056.

[20] Cai Z.J., Cheng G.X. Novel method to produce poly(3-hydroxybutyrate) scaffolds with controlled multi-pore size. *J. Mater. Sci. Lett.* 2003; 22:153-155.

[21] Koosha F., Muller R.H., Davis S.S. Polyhydroxybutyrate as a drug carrier. *CRC Crit. Rev. Ther. Drug Carrier Systems.* 1989; 6:117-130.

[22] Akhtar S., Pouton C.W., Notarianni L.J. Crystallization behaviour and drug release from bacterial polyhydroxyalkanoates. *Polymer*. 1992; 33:117-126.

[23] Abe H., Doi Y. Controlled release of lastet, an anticancer drug, from poly(3-hydroxybutyrate) microspheres containing acylglycerols. *Macromol. Rep.* 1992; A29:229-235.

[24] Koosham F., Muller R.H., Washington C.W. Production of polyhydroxybutyrate (PHB) nanoparticles for drug targeting. *J. Pharm. Pharmacol.* 1987; 39:136.

[25] Koosha F., Muller R.H. Davis S.S. A continuous flow system for in vitro evaluation of drug loaded biodegradable colloid carriers. *J. Pharm. Pharmacol.* 1988; 40:131.

[26] Akhtar S., Pouton C.W., Notarianni L.J., Gould P.L. A study of the mechanism of drug release from solvent evaporated carrier systems of biodegradable P(HB-HV) polymers. *J. Pharm. Pharmacol.* 1989; 41:5.

[27] Doyle C., Tanner E.T., Bonfield W. *In vitro* and *in vivo* evaluation of polyhydroxybutyrate and of polyhydroxybutyrate reinforced with hydroxyapatite. *Biomaterials*. 1991;12: 841-847.

[28] Knowles J.C., Hastings G.W., Ohta H., Niwa S. Development of a degradable composite for orthopaedic use: in vivo biomechanical and histological evaluation of two bioactive degradable composites based on the polyhydroxybutyrate polymer. *Biomaterials*. 1992; 13: 491-496.

[29] Boeree N. R., Dove J., Cooper J. J., Knowles J., Hastings G.W. Development of a degradable composite for orthopaedic use: mechanical evaluation of an hydroxyapatite-polyhydroxybutyrate composite. *Biomaterials*. 1993; 14: 793-796.

[30] Luklinska Z.B., Bonfield W. Morphology and ultrastructure of the interface between hydroxyapatite-poluhydroxybutyrate composite implant and bone. *J. Mater. Sci.: Mater. Med.* 1997; 8: 379-383.

[31] Alper R., Lundgren D.G., Marchessault R.H., Cote W.A. Properties of poly-β-hydroxybutyrate. 1. General considerations concerning the naturally occurring polymer. *Biopolymers*. 1963;1:545-556.

[32] Okamura K., Marchessault R.H. Physical properties of poly(β -hydroxybutyrate). II. Conformational aspects in solution. Ramachandran G.N. Ed., *Conformation of Biopolymers*, Academic Press, London,1967; Vol.2,pp. 709-720.
[33] Wang L., Cheng G.X., Cai Z.J., Liu X.F., Yao K.D. in *Proceedings of the First International Conference on Biomaterials*, Beijing, 2001, p.153.
[34] Avella M., Martuscelli. Poly-D(-)(3-hydroxybutyrate)/ poly(ethylene oxide) blends: phase diagram, thermal and crystallization behavior. *Polymer*. 1988; 29: 1731-1737.
[35] Verhoogt H, Ramsay B.A., Favis B.D. Polymer blends containing poly(3-hydroxyalkanoate)s. *Polymer*. 1994;35:5155-5169.
[36] Avella M., Martuscelli E., Greco P. Crystallization behavior of poly (ethylene oxide) from poly(3-hydroxybutyrate)/ ploy (ethylene oxide) blends: phase structuring, morphology and thermal behavior. *Polymer*. 1991; 32:1647-1654.
[37] Avella M., Martuscelli E., Raimo M. The fractionated crystallization phenomenon in poly(3-hydroxybutyrate)/ poly (ethylene oxide) blends. *Polymer*. 1993; 34:3234-3240.
[38] Cai Z.J. (2002). *Study on modification of poly(hydroxybutyrate) and construct of scaffold used in tissue engineering*. [Ph.D. Thesis]. Tianjin University.
[39] Hao J.Y., Deng X.M. Semi-interpenetrating networks of bacterial poly (3-hydroxybutyrate) with net-poly(ethylene glycol). *Polymer*. 2001;42:4091-4097.
[40] Mergaert J., Webb A., Anderson C., Wouters A., Swings J. Microbial degradation of poly(3-hydroxybutyrate) and poly(3-hydroxybutyrate-co3-hydroxyvalerate) in soils. *Appl. Environ. Microbiol*. 1993; 59: 3233-3238.
[41] Tanio T., Fukui T., Shirakura Y., Saito T., Tomita K., Kaiho T., Masamune K. An extracellular poly(3-hydroxybutyrate) depolymerase from *Alcaligenes faecalis*. *Eur. J. Biochem*. 1982; 124:71-77.
[42] Shirakura Y., Fukui T., Saito T., Okamoto Y., Narikawa T., Koide K., Tomita K., Takemasa T., Masamune S. Degradation of poly(3-hydroxybutyrate) by poly(3-hydroxybutyrate) depolymerase from *Alcaligenes faecalis* T-1. *Biochem. Biophys. Acta*. 1986; 880: 46-53.
[43] Fukui T., Narikawa T., Miwa K., Shirakura Y., Satio T., Tomita K. Effect of tryptic modification of a bacterial poly(3-hydroxybutyrate) depolymerase on its catalytic activity. *Biochem. Biophys. Acta*. 1988; 952: 164-171.
[44] Korasatko W., Wabnegg B., Braunegg G., Lafferty R.M., Strempfl F. Poly-D-(-)3-hydroxybutyric acid (polyHBA) a biodegradable carrier for long term medication dosage. I . Development of parenteral matrix tablets for long-term administration of pharmaceuticals. *Pharm. Ind*. 1983; 45:525-527.
[45] Anderson A.J., Dawes E.A. Occurrence, Metabolism, Metabolic Role, and Industrial Uses of Bacterial Polyhydroxyalkanoates. *Microbiol. Rev*. 1990;54:450-472.
[46] Mukai K., Yamada K., Doi Y. Kinetics and mechanism of heterogeneous hydrolysis of poly[(*R*)-3-hydroxybutyrate] films by PHA depolymerase. *Int. J. Biol. Macromol*. 1993; 15:361-366.
[47] Hocking P.J., Timmins, M.R., Scherer, T.M., Fuller, R.C., Lenz, R.W. Maechessault, R.H. Enzymatic degradability of isotactic versus syndiotactic poly(β -hydroxybutyrate). *Macromol. Rapid. Commun*. 1994; 15:447-452.
[48] Lotti N., Pizzoli, M., Ceccorulli, G., Scandola, M. Binary blends of microbial poly(3-hydroxybutyrate) with polymethacrylates. *Polymer*. 1993;34:4935-4940.

[49] Yu G., Marchessault R.H. Characterization of low molecular weight poly(β-hydroxybutyrate)s from alkaline and acid hydrolysis. *Polymer*. 2000; 41:1087-1098.
[50] Pouton C.W., Akhtar S. Biosynthetic polyhydroxyalkanoates and their potential in drug delivery. *Advan. Drug. Deliv. Rev.* 1996;18:133-162.
[51] Kanesawa Y., Doi Y. Hydrolytic degradation of microbial poly(3-hydroxybutyrate-*co*-3-hydroxyvalerate) fibers. *Macromol. Rapid. Commun*. 1990;11: 679-682.
[52] Doi Y., Kanesawa Y., Kawaguchi Y., Kunioka M. Hydrolytic degradation of microbial poly(hydroxyalkanoates). *Macromol. Rapid. Commun*. 1989; 10: 227-230.
[53] Doi Y., Kanesawa Y., Kunioka M. Biodegradation of microbial copolyesters: poly(3-hydroxybutyrate-*co*-3-hydroxyvalerate) and poly(3-hydroxybutyrate-*co*-4-hydroxybutyrate). *Macromolecules*. 1990; 23 : 26-31.
[54] Lehrle R., Williams R., French C., Hammond T. Thermolysis and methanolysis of poly(β-hydroxybutyrate) : random scission assessed by statistical analysis of molecular weight distributions. *Macromolecules*. 1995; 28: 4408-4414.
[55] Janigova I., Lacik I., Chodak I. Thermal degradation of plasticized poly(3-hydroxybutyrate) investigated by DSC. *Polym. Degrad. Stab*. 2002; 77: 35-41.
[56] Ohura T., Aoyagi Y., Takagi K., Yoshida Y., Kasuya K., Doi Y. Biodegradation of poly(3-hydroxyalkanoic acids) fibers and isolation of poly(3-hydroxybutyric acid)-degrading microorganisms under aquatic environments. *Polym. Degrad. Stab*. 1999; 63: 23-29.
[57] Holland S.J., Jolly A.M., Yasin M., Tighe B.J. Polymers for biodegradable medical devices II.Hydroxybutyrate-hydroxyvalerate copolymers: hydrolytic degradation studies. *Biomaterials*. 1987; 8:289-295.
[58] Knowles J.C., Hastings G.W. In vitro degradation of a PHB/PHV copolymer and a new technique for monitoring early surface changes. *Biomaterials*. 1991; 12: 210-214.
[59] Briese B.H., Jendrossek D., Schlegel H.G. Degradation of poly(3-hydroxybutyrate-*co*-3-hydroxyvalerate) by aerobic sewage sludge. *FEMS. Microbiol. Lett*. 1994;16:155-160.
[60] Niller N.D., Williams D.F. On the biodegradation of poly-β-hydroxybutyrate (PHB) homopolymer and poly-β-hydroxybutyrate-hydroxyvalerate copolymers. *Biomaterials*. 1987;8:129-137.
[61] Lauzier C., Revol J.F., Debzi E.M., Marchessault R.H. Hydrolytic degradation of isolated poly(β-hydroxybutyrate) granules. *Polymer*. 1994;19:4156-4162.
[62] Abe H., Doi Y., Hori Y., Hagiwara T. Physical properties and enzymatic degradability of copolymers of (R)-3-hydroxybutyric acid and (S, S)-lactide. *Polymer*. 1997;39:59-67.
[63] Kumagai Y., Kanesawa Y., Doi Y. Enzymatic degradation of microbial poly(3-hydroxybutyrate) stereoisomers with different stereoregularities. *Macromol. Chem. Phys*. 1992; 193:53-57.
[64] Doi Y., Kumagai Y., Tanahashi N., Mukai K. Biodegradable polymers and plastics. Vert M., et al. Ed., *Royal society of chemistry*, London, 1992; pp139.
[65] Abe H., Matsubara I., Doi Y. Physical properties and enzymatic degradability of polymer blends of bacterial poly[(R)-3-hydroxybutyrate] and poly[(R,S)-3-hydroxybutyrate] stereoisomers. *Macromolecules*. 1995;28:844-853.

[66] He Y., Shuai X., Cao A., Kasuya K., Doi Y., Inoue Y. Enzymatic biodegradation of synthetic atactic poly(R,S-3-hydroxybutyrate) enhanced by an amorphous nonbiodegradable polymer. *Polym. Degrad. Stab*. 2001;73:193-199.

[67] Scandola M., Focarete M.L., Gazzano M. Crystallinity-induced biodegradation of novel [(R,S)- β -butyrolactone]-*b*-pivalolactone copolymers. *Macromolecules*. 1997;30:7743-7748.

[68] Cheng G.X., Cai Z.J., Wang L. Biocompatibility and biodegradation of poly(hydroxybutyrate)/poly(ethylene glycol) blend films. *J. Mater. Sci.: Mater. Med.* 2003;14:1073-1078.

[69] Zhao Q., Cheng G.X. Preparation of biodegradable poly(3-hydroxybutyrate) and poly(ethylene glycol) multiblock copolymers. *J. Mater. Sci*. 2004;39:3829-3831.

In: New Frontiers in Polymer Research
Editor: Robert K. Bregg, pp. 125-171

ISBN: 1-59454-956-7

Chapter 6

DESIGN OF POLYMER APPENDED CALIXARENES AS ION CARRIERS

Mustafa Yilmaz*[a]*, Shahabuddin Memon*[b]*,
***Mustafa Tabakci*[a] *and Richard A. Bartsch*[c]**

[a]*Department of Chemistry, Selçuk University, 42031 Konya, Turkey; e-mail: myilmaz@selcuk.edu.tr*
[b]*Department of Chemistry, Federal Government Degree College (Men),F-10/4, Islamabad, Pakistan, e-mail: smemon62@yahoo.com*
[c]*Department of Chemistry and Biochemistry, Texas Tech University, Lubbock, TX 79409-1061, USA; e-mail: richard.bartsch@ttu.edu*

ABSTRACT

With a focus directed toward noncovalent interactions, the supramolecular approach is allowing scientists to create quite remarkable molecules and supermolecules. Enhanced knowledge of molecular recognition is being used in an ever increasingly sophisticated manner to produce different classes of compounds, such as crown ethers, cryptands, spherands, cyclophanes, cyclodextrins, cleft-like structures, and other cage-like compounds mainly for complexation of inorganic and organic ions and organic molecules. The design of improved ion-sensing materials has attracted intense attention from chemists in recent years.

Base-catalyzed condensation of *tert*-butylphenol and formaldehyde leads to cyclic oligomers called "calixarenes" or "calix[n]arenes", with n being the number of aromatic rings joined by methylene bridges. These molecules can undergo chemical modifications either by reaction at the hydroxy groups (lower rim) or by replacement of the *tert*-butyl groups (upper rim) by other substituents. Modified calixarenes have been found to be extremely useful in host-guest chemistry due to their tendency to form complexes with cations, anions, and neutral molecules in adsorption, filtration and separation processes, as ionophores for ISFETs (Ion Selective Field Effect Transistors), as catalysts in two-phase reactions, as ion carriers in membrane technology, in nanosize self-assembling systems, which are prospective candidates as materials for microelectronics, mesoscopic devices, *etc.*

Inspired by the increasing use of calixarenes, interest in the design of polymer -appended calixarenes increased as well to make them suitable as new recognition agents with higher effiency. In recent years, various attempts have been undertaken to incorporate calixarenes into different polymers, including self-assembled systems. There are two general strategies for preparing covalently linked, calixarene-based polymers. One is attachment of a calixarene compounds by reaction with a suitably functionalized polymer or oligomer. The other is preparation of a calixarene monomer and its polymerization or copolycondensation with typical vinyl monomers. This survey is intended to provide an overview of calixarene derivatives linked with polymers by conventional covalent bonds.

INTRODUCTION

In recent years, considerable attention has been paid to chemical separation techniques that involve the design and synthesis of new extraction reagents for metal ions. Such attention results from environmental concerns, efforts to conserve energy, and recycling at the industrial level. Supramolecular chemistry has provided solutions in the search for molecular structures that can serve as building blocks for the production of sophisticated molecules by anchoring functional groups oriented in such a way that they delineate a suitable binding site [1,2]. Such macromolecules are widely used in host-guest chemistry for the construction of various receptors for charged species or neutral molecules. Among them, calix[n]arenes with n = 4, 6, and 8 (Figure 1) are a class of supramolecular building blocks or platforms that are readily synthesized by "one-pot" procedures [3]. They are macrocyclic oligomers composed of phenol units joined by methylene bridges. Like crown ethers and cyclodextrins they are attractive inclusion compounds, which can provide three-dimensional structures with almost unlimited derivatization possibilities [4,5]. The highly ordered structures of calixarenes offer not only boundless possibilities for chemical modification, but also makes them extremely useful in the study of molecular recognition and supramolecular processes [6].

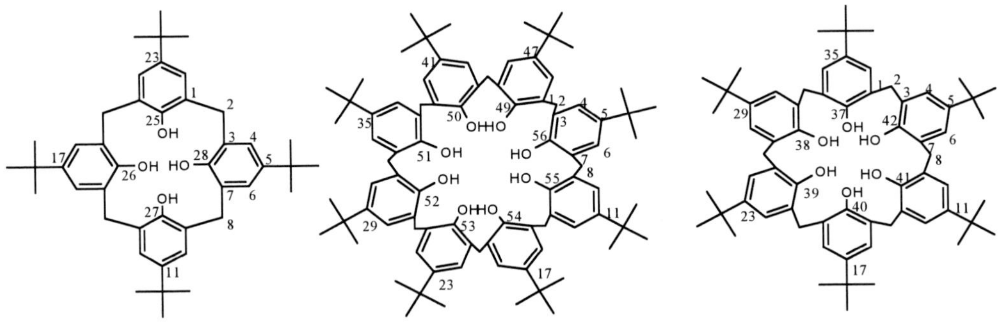

5,11,17,23-tetra-*tert*-butyl-25,26,27,28-tetrahydroxy-calix[4]arene

5,11,17,23,29,35,41,47-octa-*tert*-butyl-49,50,51,52,53,54,55,56-octahydroxy-calix[8]arene

5,11,17,23,29,35-hexa-*tert*-butyl-36,37,38,39,40,41,42-hexahydroxy-calix[6]arene

Figure 1. The Structures of Calix[n]arenes (n = 4, 6 and 8)

One of the most fascinating properties of calix[n]arenes is the variety of conformations that they can assume. Conformation mobility results from quasi-free rotation about the σ bonds of the Ar-CH_2-Ar moieties and formation of cooperative hydrogen bonds [7]. In the case of calix[4]arenes, the relative orientations of the phenol rings are defined according to the nomenclature introduced by Gutsche [8], as cone, partial cone, 1,2-alternate, and 1,3 alternate (Figure 2).

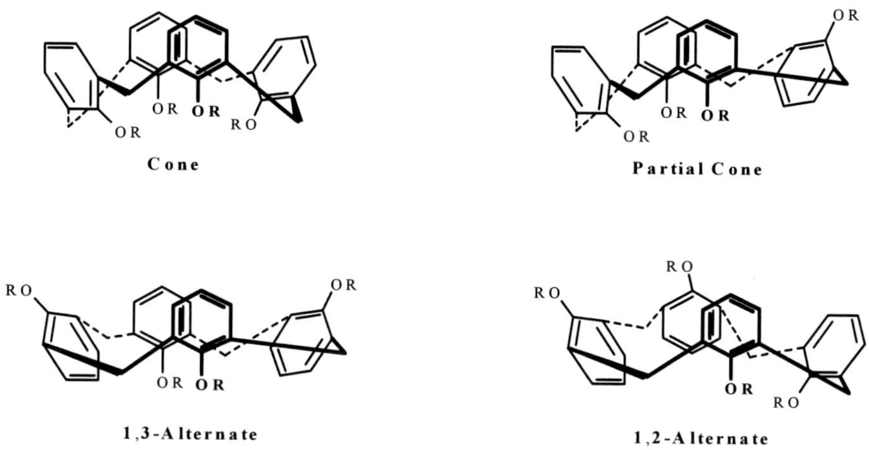

Figure 2. The conformations of calix[4]arenes as defined by Gutsche [8].

The polar lower (narrow) rim comprised of phenolic oxygens has been extensively derivatized to bind and transport ionic species (Figures 3 and 4). The upper (wide) rim is a shallow bowl lined with aromatic rings that can encapsulate small organic molecules or apolar parts of larger compounds. This unique structure of calixarenes has attracted considerable attention in the research area of enzyme mimics because it provides a versatile platform for the construction of organized polytopic ligands. To incorporate appropriate functional groups that are properly positioned on the calixarene framework was a major goal of early calixarene research. A wide range of calixarene derivatives have been reported together with their abilities to form host-guest complexes by trapping organic compounds, small ions, and gases in their torus-like cavities, as well as to mimic enzyme functions [3-11]. The aromatic-rich nature of calixarenes makes them essentially insoluble in water, which enhances their utility as phase transfer agents. For instance, some of them were tested as metal ion extractants and utilized as selective ligands in ion-selective electrodes [12-20].

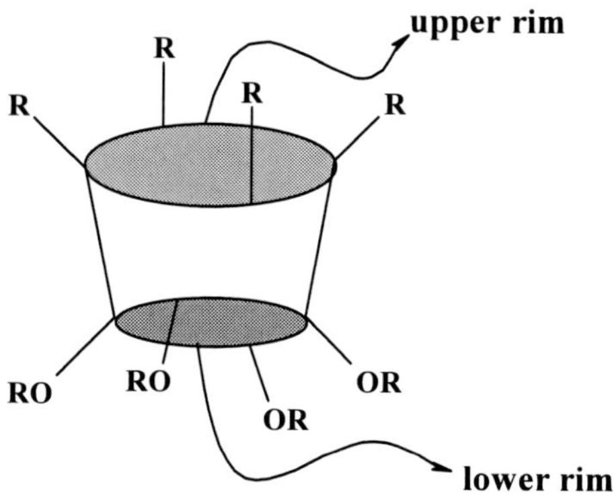

Figure 3. The functionalizable sites 'lower rim and upper rim' of calix[4]arene (R=H).

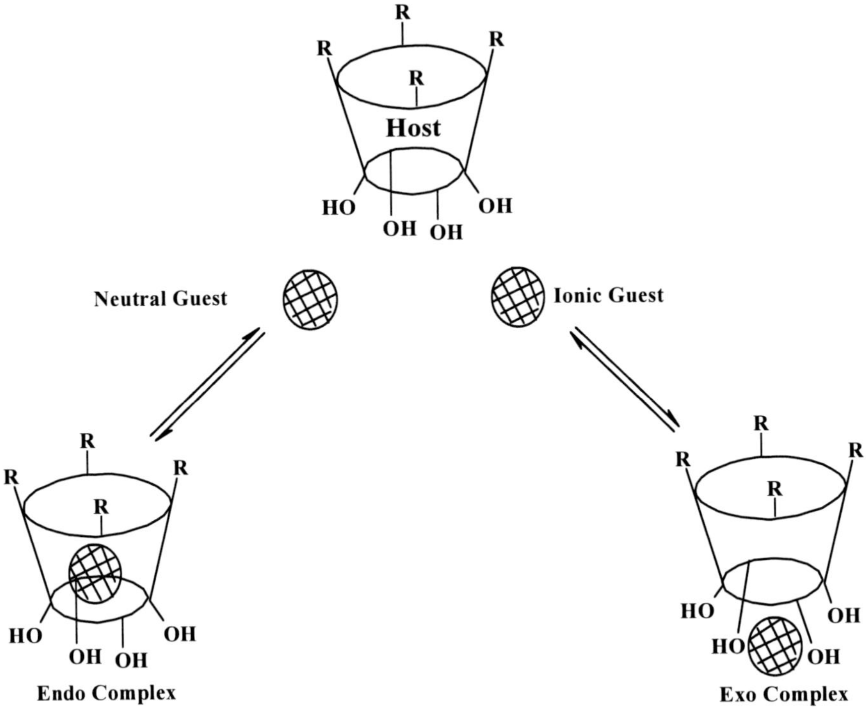

Figure 4. A schematic representation of "endo and exo complexes" of calix[4]arene.

The ability of calixarenes to act as baskets is one of their most intriguing properties, accounting for much of the interest that they have received in recent years and provides them a vast field of applications. Calixarenes can generally form two types of complexes, "endo" or "exo", as shown in Figure 4. Their complexation may be classified further by considering solid-state and solution-state systems [3-6]

In general, two strategies have been adopted to enhance the affinity of calixarenes toward targeted guest species. In one, different ionophoric groups including carbonyl, amide, nitrile and other suitable functionalities have been attached to the calix-platform. In the other, the calixarene units were fixed in a polymeric matrix. Compared to the intense interest in complexation studies of monomeric calixarenes, there are relatively few studies of calixarene-containing polymers, which may be processed into materials suitable for chemical sensor devices [21,22], filtration/extraction membranes [23], and antioxidant additives [24]. Early sensor devices based on calixarenes were prepared by blending technology [25]. Blending methods have several drawbacks including non-uniform ordering of the calixarene units, leaching, and difficulty in reproducing a given set of properties in the blend. Therefore, two strategies have been adopted to prepare covalently linked, calixarene-based polymers. One is attachment of a calixarene derivative by reaction with a suitably functionalized polymer or oligomer. The other is preparation of a calixarene monomer and its polymerization or copolycondensation with typical vinyl monomers. Thus, polyacrylates, polyamides, and polysiloxanes carrying calix[4]arene derivatives as side groups linked to macromolecular chains have been described [26-36].

The preparation of resins containing functionalized calixarenes can be roughly classified into three categories. The first entails the self-polymerization of calixarene compounds containing monomeric functional groups, such as vinyl functions, that take part in the polymerization reaction. The second involves cross-linking of calixarene units with simple groups, such as ether, amide, and methylene groups. The third utilizes immobilization of a functionalized calixarene compound on a polymer matrix.

1. Vinyl Polymers/Copolymers Containing Pendant Calixarenes

In 1987, two patents [37] issued to Hitachi Chemical Co. described polymeric calixarenes that are claimed to have good metal absorbtivity and heat resistance. They were derived from *p*-phenylcalixarenes by acylation with RC(O)Cl (R is $-CH{=}CH_2$, $-C(Me){=}CH_2$, or $-(CH_2)_mCH{=}CH_2$). This introduced unsaturated acyl groups into the 4'-position of the biphenyl ring. Polymerization in the presence of methyl methacrylate was induced by AIBN to give a copolymer with a softening point at 180-185°C.

Blanda and Adou [38] obtained copolymers (**2, 3,** and **4**) (Figure 5) with pendant calixarene systems by copolymerization of monovinyl calixarene derivatives with the typical vinyl monomers of methyl methylacrylate, acrylonitrile, and styrene, respectively. Calix[4]arene monomer **1** was locked in the cone conformation by appending four ethoxyethyl groups on its lower rim.

The upper rim of the calix[4]arene monomer, the site of polymerization, was monofunctionalized with a modified acrylate ester moiety. A seven-atom spacer group was incorporated between the bulky calixarene core and the acrylate moiety to minimize steric interactions, which impeded polymerization. The copolymers were designed to take advantage of the well-established binding interactions of calixarenes with neutral molecules and ions. The copolymers were developed for incorporation into chemical sensor devices, such as ion-specific electrodes, semi-permeable membranes, and quartz microbalances.

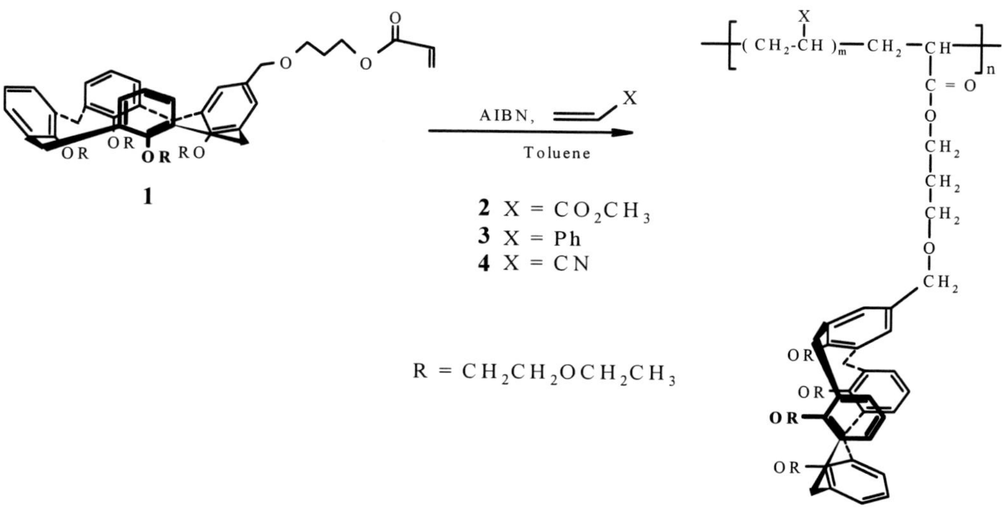

Figure 5. Copolymerization of monovinyl calixarene derivatives with typical vinyl monomers.

Reaction mixtures containing 1:1, 2:1, 3:1, and 10:1 ratios of methyl methylacylate to calixarene monomers were investigated. Oligomeric products with inherent viscosities of 0.06 and 0.11 dL/g, respectively, were obtained from the 1:1 and 2:1 reactant combinations; while higher molecular weight polymers were realized with the 3:1 and 10:1 combinations (Table 1).

Table 1. Physical properties of copolymers

Copolymer	Inherent viscosity (dL/g)[a]	TGA[b] (°C)	Tg (°C)	Average molecular wt.(g/mol)[c]
MA 1:3	0.17	377		37,000
MA 1:10	0.37	376	81	171,000
PS 1:3	0.18	377	-	45,000
PS 1:10	0.30	371	84	53,000
AN 1:3	0.17	341	-	na[d]
AN 1:10	0.28	343	76	na[d]

[a]Measured in chloroform.
[b]10% weight loss in nitrogen.
[c]Measured against polystyrene.
[d]Could be not determined due to insolubility in THF.

The average molecular weights for the two copolymers with methyl methacrylate were found to be 37,000 and 171,000 g/mol, respectively. These are average molecular weights, since there was a high degree of dispersion in each copolymer. The morphology of the polystyrene copolymer was more crystalline than that of methylacrylate copolymers. The average molecular weights for the two copolymers with styrene were 45,000 and 54,000 g/mol, respectively. The copolymers obtained by reacting 1 with 3 and 10 mol equivalents of acrylonitrile were less soluble than those obtained with either methyl methylacrylate or styrene.

It is known that tetraester derivatives of calix[4]arene are sufficiently selective for Na^+ to enable their exploitation in ion-selective electrodes. The efficacy of such Na^+-selective electrodes has been demonstrated [39] in extensive clinical trials for Na^+ analysis in blood. Harris and coworkers [40] synthesized a *p-tert*-butylcalix[4]arene methacrylate from *p-tert*-butylcalix[4]arene and polymerized it under free radical conditions. The resulting polymer **5** (Figure 6) was soluble in chloroform, dichloromethane, and acetone. The molecular weight of the polymer was 6,745. Thus, the average chain length was *ca.* six calixarene units. The polymer formed stable complexes with sodium thiocyanate. Synthesis of calix[4]arene tetraester ionophores **6** and **7** (Figure 7) with polymerizable groups on their upper rims was reported by Parzuchowski and coworkers [28]. Calixarene monomers with one or two methacrylamide groups were copolymerized with methyl methacrylate yielding linear polymers with pendant calixarene groups or cross-linked polymers. Extraction studies with the soluble polymers and with the corresponding pivalamides as model compounds showed no difference in the extraction ability and selectivity. Potentiometric measurements of membrane electrodes prepared with the polymeric calixarene ionophores exhibit high selectivity. The slope of the response curve and low detection limit of membranes containing molecular calixarene ionophores were retained, while the lifetime of the membrane was increased with the polymeric calixarene ionophore.

5

$R = CH_2CO_2C_2H_5$
$X = CH_2CO_2CH_2OCO$

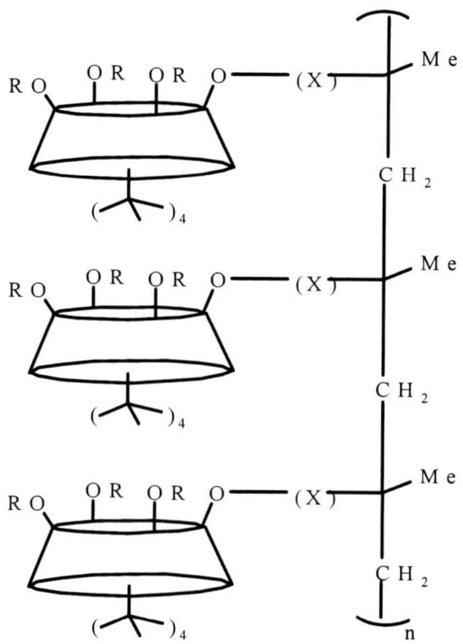

Figure 6. A p-tert-butylcalix[4]arene methacrylate polymer.

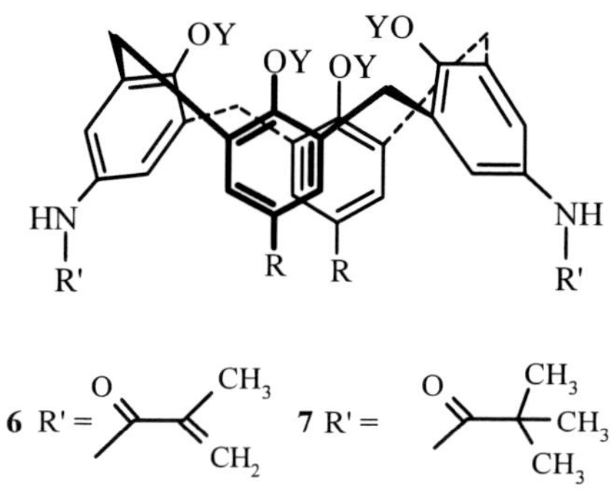

$Y = CH_2COOC_2H_5$ R = *tert*-butyl

Figure 7. Calix[4]arene tetraester ionophores **6** and **7** containing polymerizable groups on their upper rims.

Polymers with pendant calix[4]crown-4 units were synthesized by Memon *et al.* [26] via radical-initiated reactions involving vinylic monomer **8** (5,7,11,23-tetra-*tert*-butyl-25-[2-(acryloxyethoxy)]-27-methoxycalix[4]arene-crown-4) and styrene (Figure 8) . A five-atom spacer was incorporated between the bulky calixarene core and the acrylate moiety to reduce steric interactions that could impede polymerization.

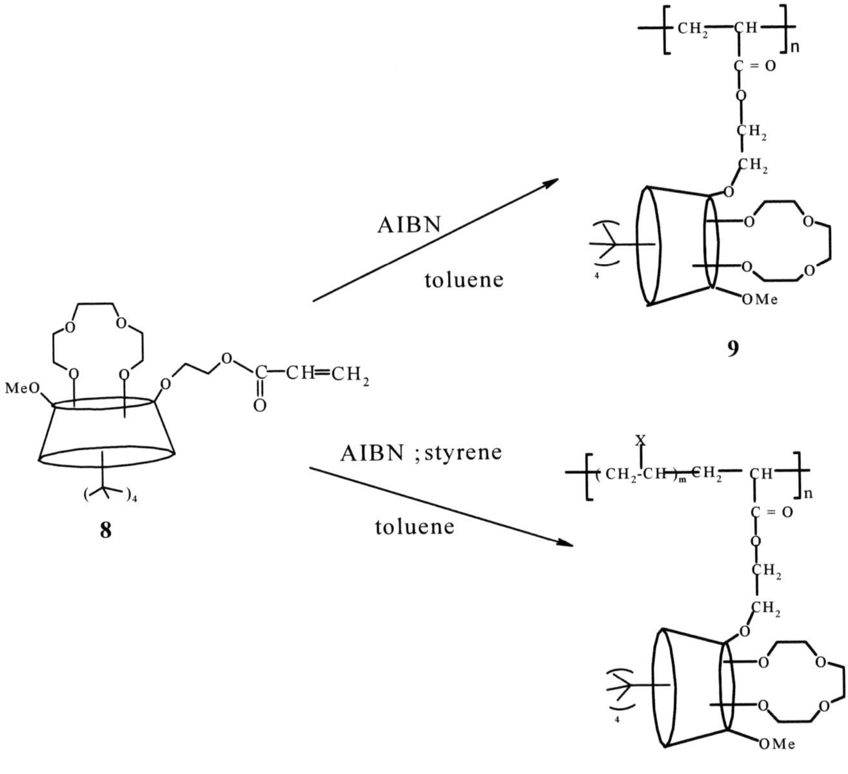

Figure 8. Synthesis of homo- and copolymers with pendant calix[4]crown-4 units.

The average molecular weight of homopolymer **9** was only 4,400 g/mol (pentamer). Presumably the low molecular weight was due to the bulky calixarene moieties with no spacer groups between them. Molecular weights of copolymers **10** and **11** were higher (Table 2). The polymers were found to be efficient, but unselective, extractants for the metal cations examined. It can be concluded that the metal transfer ability of polymers depends upon the conformation of calixarene moiety and the number of calixarene units in the polymer skeleton.

Table 2. Physical properties of white, amorphous calix[4]arene homo- and copolymers 9-11

Polymer	Colour	State	s.p. (°C)	Inherent viscosity (dL/g)[a]	Average molecular wt. (g/mol)[b]
9	White	Amorphous	102	0.033	4,400
10	White	Amorphous	134	0.26	42,000
11	White	Amorphous	138	0.32	60,000

[a]Measured in toluene.

[b]Measured against polystyrene.

The preparation of the calix[4]arene monomer **14** (Figure 9) with one polymerizable group on the lower rim was reported by Tabakci *et al.* [41]. Monomer **14** was copolymerized with styrene (1:5) in sealed vials using degassed toluene as solvent and AIBN as radical initiator. The molecular weight of copolymer **15** was determined by gel permeation chromatography in tetrahydrofuran against polystyrene standards. The chromatogram showed two signals from which the number average molecular weights (M_n) were determined as 3,095 and 6,730 g/mol. The copolymer composition (n/m) of 14 was 1/14. This reveals that the styrene monomer is predominant and suggests that the calixarene units are randomly distributed along the chain. This is a result of steric interactions that occur between bulky calixarene moieties. Extraction studies with this monomer and its polymer showed no selectivity among the metal cations tested.

Anion recognition and sensing is an increasingly important research topic in supramolecular chemistry due to the importance of various anions in biological and environmental locations. For instance, chromate and dichromate anions are important because of their high toxicity [55-57] and their presence in soils and waters [58,59]. For a molecule to be effective as a host it is necessary that its structural features are compatible with those of the guest anions. Chromate and dichromate (CrO_4^{2-} and $Cr_2O_7^{2-}$, respectively) are dianions in which the anion periphery has oxide moieties. These oxygens are potential sites for hydrogen bonding to the host molecule. Recently, efforts have been made to synthesize modified calixarenes that can be used as hosts for simple anions [42-54]. The objective is to synthesize an extractant based on a calixarene framework, which can easily be immobilized in a polymeric matrix and will be suitable for anion extraction.

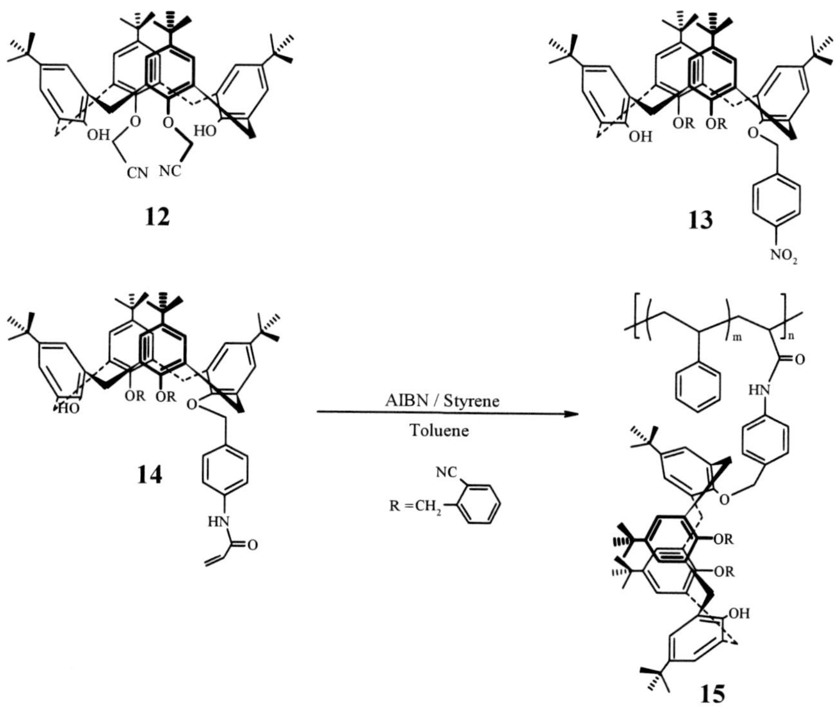

Figure 9. Structures of reference ionophores **12-14** and the synthesis of *p-tert*-butylcalix[4]arene derivative copolymer **15**.

A preliminary evaluation of the binding efficiency of **12**, **13** and **15** was carried out by solvent extraction of $Na_2Cr_2O_7$ from aqueous solutions at different pH values into dichloromethane. The results are shown in Figure 10. There was no transfer of $Na_2Cr_2O_7$ into dichloromethane in the absence of the extractant. Monomer **13** exhibited high binding ability for dichromate anions at low pH; whereas, the dicyanomethyl derivative of *p-tert*-butylcalix[4]arene **12** exhibited poor extraction affinity toward the dichromate anion. The increase in the extraction affinity for **13** may be due to the planar geometry of aromatic rings with a nitrile group and more rigid structural features of the molecule compared to **12.**

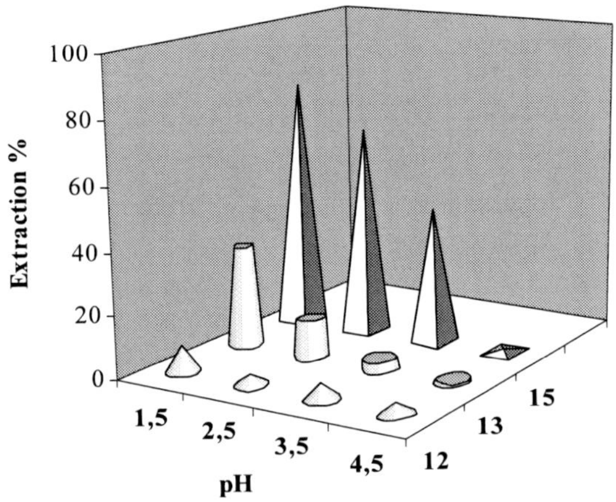

Figure 10. Percent extraction (E%) versus pH for extraction of dichromate from aqueous solutions into dichloromethane by calix[4]arenes **12** and **13** and by calixarene copolymer **15**.

Conversion of monomer **14** into polymer **15** significantly increased the anion extraction ability at pH < 4.0. It was suggested that this results from greater rigidity of the calixarene units in an appropriate structure that aids in the transfer of dichromate anion into the organic phase. For ionophores **13** and **15** lowering the pH of the aqueous phase increased the level of dichromate extraction. This pH dependence was explained by consideration of anion hydration. The importance of ion hydration in liquid-liquid phase extraction has been documented by Hofmeister [60]. In aqueous solutions with low pH, dichromate will be primarily in its protonated form, $HCr_2O_7^-$. This monoanion will have a smaller free energy of hydration than does the dianionic form $Cr_2O_7^{2-}$. As a result, there is a smaller loss in hydration energy as $HCr_2O_7^-$ is transferred from the aqueous phase into the dichloromethane phase. An additional advantage of $HCr_2O_7^-$ over $Cr_2O_7^{2-}$ is that for the former only one Na^+ needs to be co-extracted to maintain charge balance; whereas for $Cr_2O_7^{2-}$, two Na^+ must be extracted, with an additional loss of hydration energy. For these compounds, increased extraction at low pH due to protonation of the nitrile nitrogens to give a dication was discounted. Since the pK_a of CH_3CNH^+ is -4.3 [61], the protonated form is not expected to be present in significant concentrations in aqueous solutions with pH values in the 1.5-4.5 range. The extraction percentages decreased at higher pH. The most significant difference was observed for the case of copolymer **15,** which showed a high affinity toward $HCr_2O_7^-$. Due to the oxidative stability of the nitrile function, copolymer **15** has promise as a candidate for the removal of $HCr_2O_7^-$.

The extraction data for calixarene 13 was analyzed by a classical slope analysis method. Assuming the extraction of an ion-pair $[M^{n+}, A^{n-}]$ by the anion receptor [L] according to equilibrium:

$$m(L)_{org} + nA^{n-}{}_{aq} + nM^{n+} \rightleftharpoons ((L)_{m,} A^{n-}{}_{n}, M^{n+}{}_{n})_{org} \quad (1)$$

the extraction constant, K_{ex}, is defined by

$$K_{ex} = \frac{[((L)_{m,} A^{n-}{}_{n}, M^{n+}{}_{n})]_{org}}{[M^{n+}]^{n} [A^{n-}]^{n}{}_{aq} [L]^{m}{}_{org}} \quad (2)$$

Equation 2 can be rewritten as

$$\log D_A = \log K_{ex} + n \log [L]_{org} \quad (3)$$

where the distribution ratio, D_A, is defined as the ratio of concentrations of the ion-pair $[M^{n+}, A^{n-}]$ in the two phases:

$$D_A = [M^{n+}, A^{n-}]_{org} / [M^{n+}, A^{n-}]_{aq}$$

From Equation 3, it can be seen that a plot of log D_A vs. log [L] will give a straight line whose slope allows the stoichiometry of the extracted species to be determined.

Figure 11 shows the dichromate extracted into dichloromethane by different concentrations of **13**. A linear relationship between log D_A and log [L] was observed with a slope of 1.06 at pH 1.5, revealing that **13** forms a 1:1 complex with dichromate. At this pH, $HCrO_4^-$ is the primary anionic form of Cr(VI) in aqueous solution. This indicates that the ligand **13**, L, is involved in the equilibrium:

$$L_{(org)} + HCr_2O_7^-{}_{(aq)} + Na^+{}_{(aq)} \rightleftharpoons [L, (HCr_2O_7^-), Na^+]_{org}$$

By use of Equation 3 for **13**, log $K_{ex} = 2.60 \pm 0.20$ was calculated.

From spectroscopic data, it was determined that the immobilized dibenzonitrile derivative of *p-tert*-butylcalix[4]arene **15** had a cone conformation in the polymeric backbone. This observation coupled with the good extractant properties of polymer **15** suggests a preorganized extraction complex of the type depicted in Figure 12.

In contrast, monomer **13** is significantly more flexible than copolymer **15**, but exhibited poor extraction ability towards dichromate. Thus, compound **13** does not exhibit enhanced ion-pair extraction of sodium salt at the water-dichloromethane interface. The role of the nitrile functionalities has not been verified, but will be discussed in the next section.

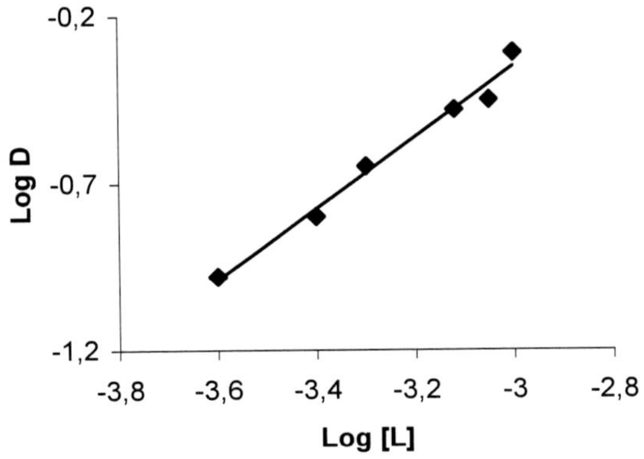

Figure 11. Plot of log D versus log [L] for extraction of dichromate from an aqueous solution into dichloromethane by **13** at 25°C.

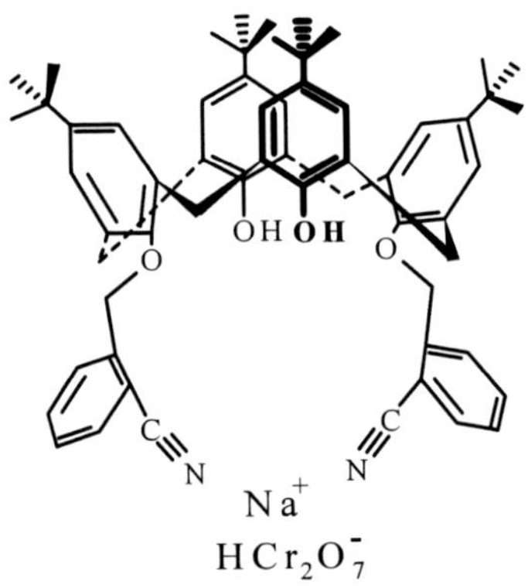

Figure 12. Proposed interactions of *p-tert*-butylcalix[4]arene dibenzonitrile units in polymer **15** with Na^+ and $HCr_2O_7^-$.

The Yilmaz group at Selcuk University has explored the selectity of calixarene polymers with nitrile functionalities towards Hg^{2+} [14, 62]. Polymers **17** and **18** (Figure 13) contain nitrile-functionalized calix[4]arene moieties attached to the polymer backbone. They were synthesized by radical-initiated reactions from monomer **16** [62]. The average molecular weight (M_n) for copolymer **18** was 25,000 g/mol. It was amorphous and exhibited a low Tg with thermal stability up to 380°C. Homopolymer **17** was obtained as an oligomeric product (M_n = 4,300 g/mol).

The extraction properties of the polymers **17** and **18** were studied by liquid-liquid extraction with selected alkali and transition metal picrates, in which the metal ion was Li^+, Na^+, K^+, Cs^+, Ni^{2+}, Co^{2+}, Cu^{2+}, Cd^{2+} and Hg^{2+}, from aqueous solution into dichloromethane (Table 3). As can be seen, both polymers exhibited good selectivity for Hg^{2+}. This can be explained by the fact that the nitrile groups have polarizable π-systems. It is a well known that soft donor group functionalitites (C≡N, C=N, C=O, *etc.*) favor complexation of soft metal cations, such as Hg^{2+} [13, 14, 63].

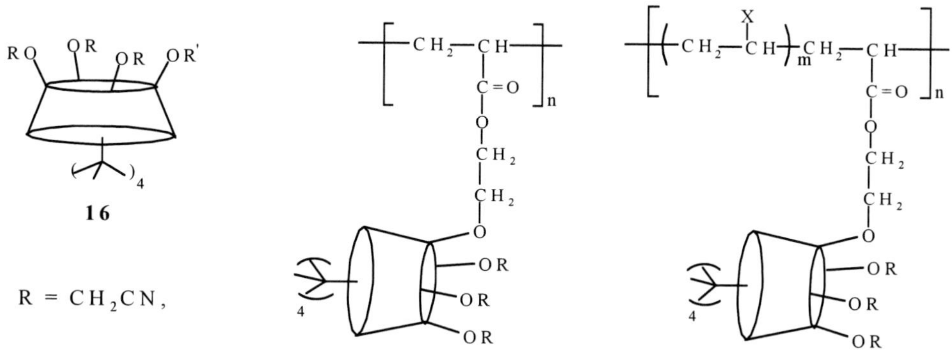

16

R = CH_2CN,

R' = $CH_2CH_2OCOCH{=}CH_2$

17 R = CH_2CN

18 R = CH_2CN, X = Ph

Figure 13. Structures of ionophore 16 and polymers **17** and **18.**

Table 3. Extraction of metal picrates with ligands[a]

	16	17	18
Li^+	7.2	<1.0	<1.0
Na^+	5.6	6.1	<1.0
K^+	7.7	5.0	<1.0
Cs^+	5.7	1.7	<1.0
Cu^{2+}	20	17	7.5
Co^{2+}	20	2.0	<1.0
Cd^{2+}	10	7.1	<1.0
Ni^{2+}	13	8.3	2.1
Hg^{2+}	63	57	34

[a]Aqueous phase, [metal nitrate]= 1 X 10^{-2} M ; [picric acid] = 2.5 X 10^{-5} M ; organic phase, dichloromethane, [ligand] = 1 X 10^{-3} M for 16 and 1 X 10^{-4} mole calixarene unit/g resin for **17** and **18**; extraction at 25 °C, for 1 hour.

2. Condensation Polymerization of Calixarenes with Various Monomers

Synthesis and complexation properties of a series of novel calixcrown telomers **19-22** (Figure 14) were reported by Zong *et al.* [64]. Telomers **19** and **20** were obtained by the condensation reaction of *p-tert*-butylcalix[4]arene and triethyleneglycol ditosylate and decanedioyl chloride, respectively. The calixcrown amide telomer **21** was prepared from the calixcrown carboxylic acid and 1,10-decandiamine and telomer **22** was synthesized by hydrosilylation of *p-tert*-butyldiundecenylcalix[4]crown-5 with dichloromethylsilane and H_2PtCl_6 catalyst to afford the disilylcalix-crown-5. This was polycondensed with silanol-terminated dimethyl silicon oil in a molar ratio of 1:3:3 and then treated with excess chlorotrimethylsilane.

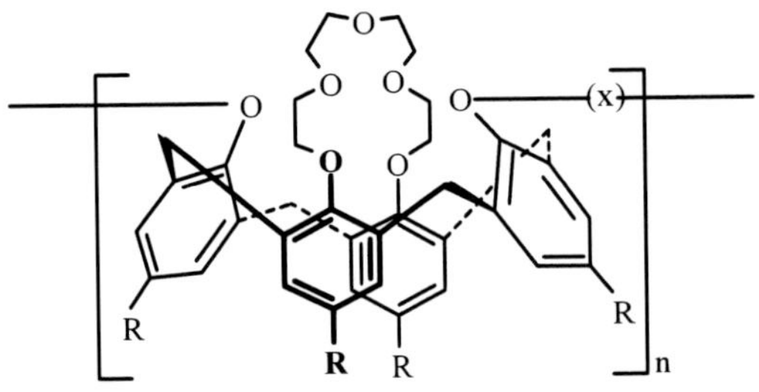

R= *tert*-Butyl

19 X = $CH_2(CH_2OCH_2)_2CH_2$
20 X = $CO(CH_2)_8CO$
21 X = $CH_2CONH(CH_2)_{10}NHCOCH_2$
22 X = $(CH_2)_{11}Si(Me)(Y)O\text{-}(Si(Me)_2O)_m\text{-}Si(Me)(Y)(CH_2)_{11}$
Y = $(OSi(Me)_2)_pOSiMe_3$

Figure 14. Calixcrown-5 telomers **19-22** with different bridging units.

Results from liquid-liquid extraction experiments conducted with the calixcrown telomers and Li^+, Na^+, K^+, and NH_4^+ picrates are summarized in Table 4. Calixcrown ether telomer **19** and the calixcrown amide telomer **21** had much higher phase-transfer efficiencies than the calixcrown ester telomer **20** and the calixcrown siloxane telomer **22**, because the inter-calixcrown bridges provide additional binding sites in the first two telomers. Calixcrown siloxane telomer **22** was a very good stationary phase for gas chromatography due to its low glass transition temperature (a gum at room temperature), excellent chemical and thermal stability, and both hydrophilic and lipophilic cavities, which can interact with various types of compounds.

Blanda and Aduo [65] reported the condensation of calix[4]arene diol or diamine monomers with terephthaloylchloride in dimethylacetamide. The isolated yields of the two white, amorphous copolymers polyester **23** and polyamide **24** were 66 and 50%, respectively. The glass transition temperatures (Tg) for 23 was 141 °C and that of **24** was 100 °C. The thermal stabilities of the copolymers were determined by thermal gravimetric analysis (TGA) in a nitrogen atmosphere. The temperatures at which 10 % weight loss had occurred were 352 and 380 °C for **23** and for **24,** respectively. The inherent viscosities in chloroform at 25 °C were found 0.12 dL/g for **23** and 0.14 dL/g for **24**. Both copolymers were completely soluble in organic solvents and their number average molecular weights (M_n) were 29,000 Daltons for polyester **23** and 52,000 Daltons for polyamide **24**.

Table 4. Percent extraction of picrate salts into chloroform[a] by 19-22 at 25 °C

Cation of picrate salt	**19**	**20**	**21**	**22**
Li^+	3.2	0.4	1.2	0.4
Na^+	7.3	0.7	22	0.4
K^+	33.5	9.2	41	7.6
NH_4^+	13	0.8	1.6	0.5

[a]A 1.00 ml solution of 5.0 mmol/dm³ receptor in chloroform was shaken for 10 minutes with 1.00 ml of a 5.0 mmol/dm³ aqueous picrate salt solution and the percent extraction was determined from the resulting absorbance of the organic layer at 380 nm.

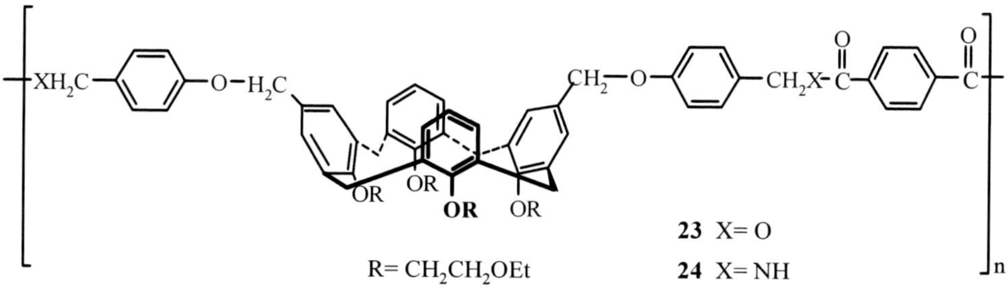

Figure 15. Two copolymers with cone conformations of calix[4]arenes in the polymer backbone.

Modified interfacial polycondensation between Bisphenol A and 1,3-dimethoxy lower rim-substituted calix[4]arene with dibromomethane in the presence of tetrabutylammonium hydrogen sulfate as phase transfer catalyst (PTC) to give copolymers **25-27** (Figure 16) was carried out by Dondoni *et al.* [66]. Polycondensation of the same comonomers with terephthaloyl chloride, but in the absence of the PTC catalyst, afforded the corresponding copolyester **28** (Figure 17). The soluble fraction of this polymeric material in chloroform showed M_w of *ca.* 21,000 and a variable calix[4]arene content depending on the phenolic monomer ratio. Moleculer characterization of the copolyethers **25-27** and copolyester **28** are shown in Table 5.

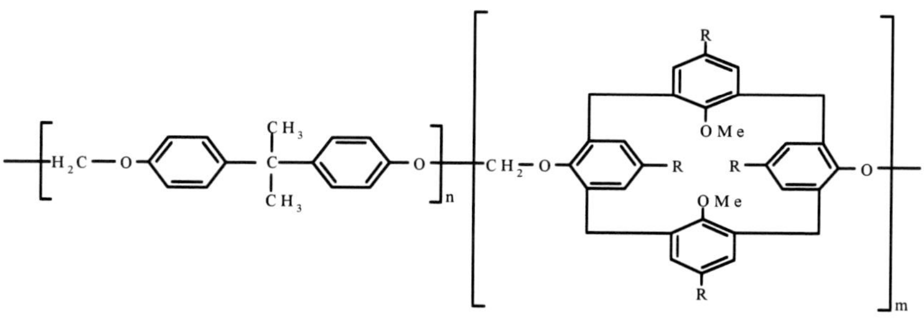

25 m/n= 0.20 R = *tert*-butyl
26 m/n= 0.48 R = H
27 m/n= 0.24 R = H

Figure 16. Products of interfacial polymerization of Bisphenol A, 1,3-dimethoxy lower rim-substituted calix[4]arene, and dibromomethane.

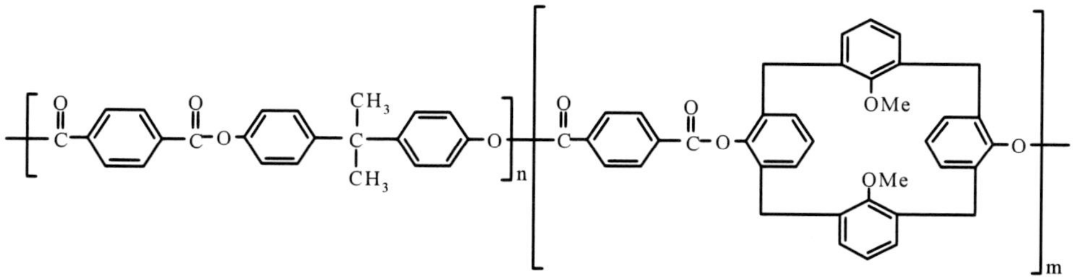

28 m/n= 0.25

Figure 17. Product of interfacial polymerization of Bisphenol A, 1,3-dimethoxy lower rim substituted calix[4]arene derivative, and terephthaloyl chloride.

The results given in Table 5 for copolyester **28** correspond only to a minor soluble fraction (15% by weight) in chloroform, since most of polymeric material was insoluble in this solvent. On the other hand, the DSC curves were obtained using the soluble and insoluble fractions of **28.** This material showed a semi-crystalline behavior with two melting peaks at 235 and 248°C and a T_g value similar to that of the copolyethers. The binding properties of the copolyethers towards Ag^+ was ten-fold higher than those of a model tetraalkoxy calix[4]arene derivative. It was suggested that this increased receptor ability was due to the change of the original cone conformation of the macrocyclic monomer into the partial cone conformation upon incorporation into the polymer. Closely related to this work Memon and coworkers [67,68] synthesized two copolymers **30** and **31** containing pendant calix[4]arene units with nitrile functionalities on their lower rims *via* nucleophilic substitution reactions involving 5,11,17,23-tetra-*tert*-butyl-25,27-bis(cyanomethoxy)-26,28-dihydroxycalix[4]arene (**12**) and 25,27-bis(cyanomethoxy)-26,28-dihydroxycalix-[4]arene **29** with 1,5-dibromo-pentane and Bisphenol A followed by reduction with sodium borohydride in the presence of $CoCl_2{\bullet}6H_2O$ to produce polymers **32** and **33** (Figure 18). Phase transfer studies were performed by liquid-liquid. It was deduced from the observations that both copolymers showed a good phase transfer affinity towards selected alkali, alkaline earth, and transition metal cations unlike their precursor **12.**

Table 5. Macromolecular characterization of copoly-ethers 25-27 and copolyester 28

	25	**26**	**27**	**28**[f)]
R[a)]	1	1	0.5	0.5
M_w[b)]$.10^{-3}$	9.8	24.6	15.0	21.1
M_w/M_n[b)]	3.2	8.1	4.1	14.4
Yield[c)] in %	60	66	55	68
$\frac{T_g}{°C}$[d)]	107	137	109	133
$\frac{T_m}{°C}$[d)]	239	206	245	247
m/n[e)]	0.20	0.48	0.24	0.25

[a]Initial Bisphenol A to calix[4]arene monomer ratio.
[b]GPC molecular weights were determined with polystyrene standards at 30° C.
[c]Weight percentage of the recovered polymer.
[d]Observed peak maxima under second heating by DSC.
[e]Molar percentage of calix[4]arene to Bisphenol A monomers along the chains calculated from ^1H NMR spectroscopic data.
fThe soluble fraction was 15% by weight.

As described previously, a recent development is the discovery that calix[4]arenes with nitrile substituents act as phase transfer extractants for chromium(VI) and that their effectiveness is higher for aqueous solutions that are acidic [68]. In this study, the role of *p-tert*-butyl groups of the calix[4]arene in the extraction of chromium(VI) as $Cr_2O_7^{2-}/HCr_2O_7^-$ anions was explored. The extraction properties of copolymers **30-33** toward $Cr_2O_7^{2-}/HCr_2O_7^-$ anions (Table 6) showed that the calix[4]arene moiety with *p-tert*-butyl groups is a good extractant for transferring the $Cr_2O_7^{2-}/HCr_2O_7^-$ anions from aqueous solutions into a dichloromethane phase.

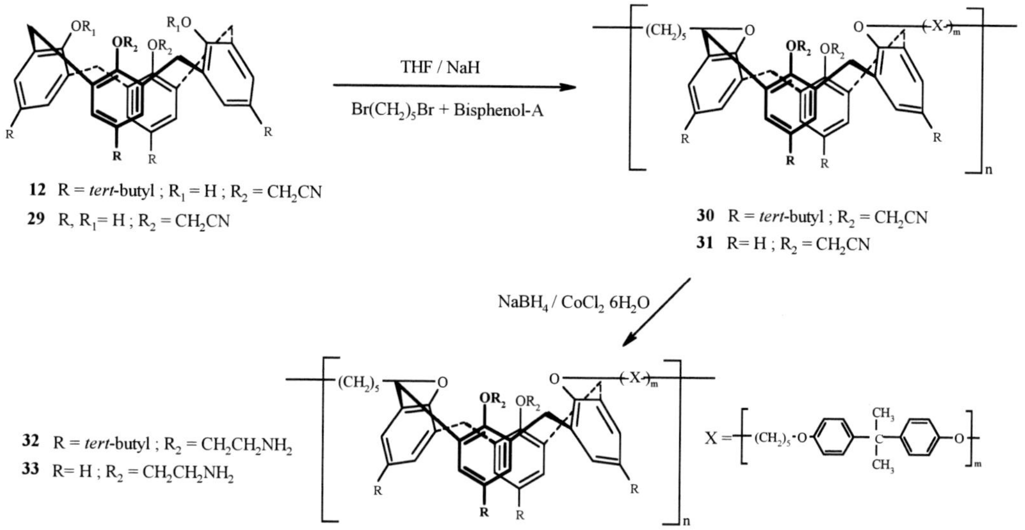

Figure 18. Ionophores **12** and **29** and copolymers **30-33** containing pendant calix[4]arene units with nitrile functionalities on their lower rims.

Table 6. Percent extraction of dichromate[a] by ligands 12 and 30-33

	Dichromate extracted (%)			
pH	1.5	2.5	3.5	4.5
12	8.5	3.2	4.9	3.6
30	76	45	42	40
31	75	38	37	29
32	92	94	62	18
33	68	61	59	58

[a]Aqueous phase, [metal dichromate] = 1 X 10^{-4} M ; organic phase, dichloromethane, [ligand] = 1 X 10^{-3} M or a 1 X 10^{-3} M solution of calix[4]arene unit/g resin for copolymers; extraction at 25 ^{o}C for 1 hour.

Due to the oxidative stability of nitrile functions, polymers **30** and **31** are potential candidates for practical extractants for $Cr_2O_7^{2-}/HCr_2O_7^-$ anions. Better preorganization of the immobilized calix[4]arene receptors **30** and **31**, in which the cooperative effects of both nitrile units are possible, are thought to improve extraction. In contrast, monomer **12** was significantly more flexible than copolymers **29** and **30** and exhibited poor extraction ability towards dichromate anions (Figure 19).

These observations were explained by a model in which the nitrile groups bind the Na^+ counterion and the pH effects are due to hydration effects [68,69]. When a salt is transferred from an aqueous solution an organic phase, both the cation and anion lose their hydration spheres. Size and charge effects are important in the determination of hydration energies [60].

Although both CrO_4^{2-} and $Cr_2O_7^{2-}$ are dianions, $Cr_2O_7^{2-}$ is larger and, therefore, has a smaller hydration energy. As a consequence, transfer of $Cr_2O_7^{2-}$ into an organic phase leads to a smaller loss of hydration energy than is found with CrO_4^{2-}. For $HCr_2O_7^-$, the difference is even more pronounced.

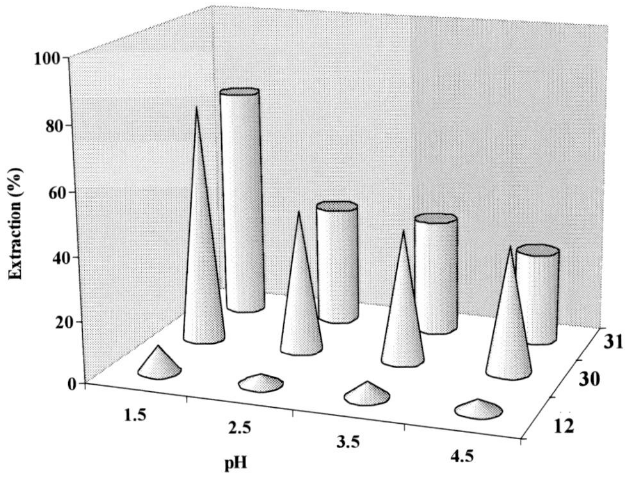

Figure 19. Percent extraction (E%) versus pH for liquid-liquid extraction of dichromate from aqueous solutions into dichloromethane by compounds **12**, **30,** and **31**.

Reduction of the nitrile groups of polymers **30** and **31** into amino groups (**32** and **33**, respectively) had only relative minor effects on the extraction behavior (Table 6). Although the role of the nitrile functionalities has not been verified, comparison with acetonitrile for which the dielectric permittivity is high, suggests the role of the calixarene nitrile may be to solvate Na^+ (Figure 20). In the case of amine groups, which can be protonated, Figure 21 shows the proposed extraction species. In this case, the alkylammonium cation functionalities can hydrogen-bond with $HCr_2O_7^-$.

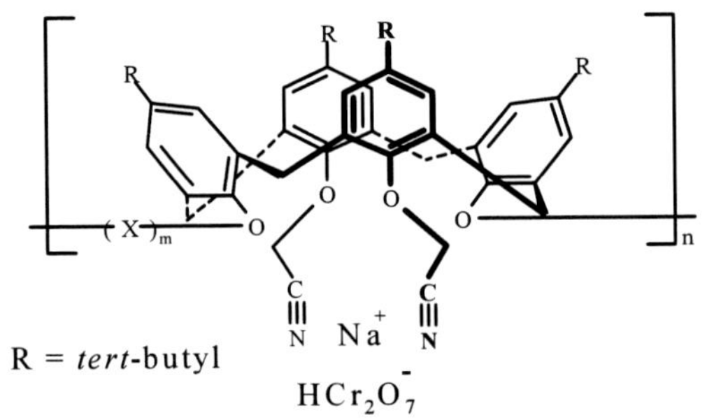

Figure 20. Proposed interactions of *p-tert*-butylcalix[4]arene dinitrile derivative moieties with Na^+ and $HCr_2O_7^-$ ions.

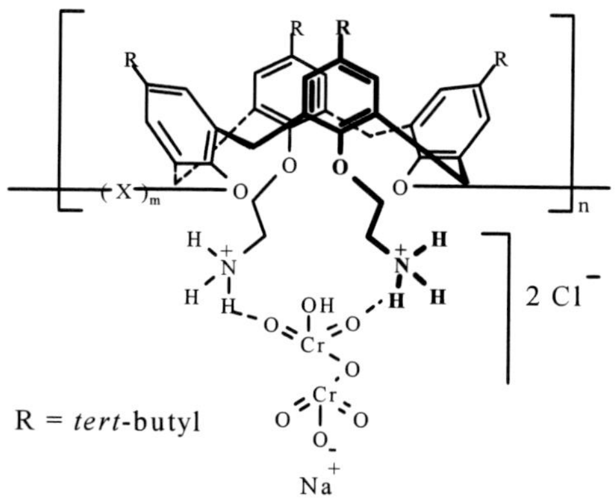

Figure 21. Proposed interactions of *p-tert*-butylcalix[4]arene dialkylamine derivative moieties with Na^+ and $HCr_2O_7^-$ ions.

Four new calix(aza)crown oligomers **36-39** (Figure 22) were synthesized by reacting *p-tert*-butylcalix[4](aza)crowns **34** and **35** with triethylene glycol ditosylate and with 1,5-dibromopentane [70].

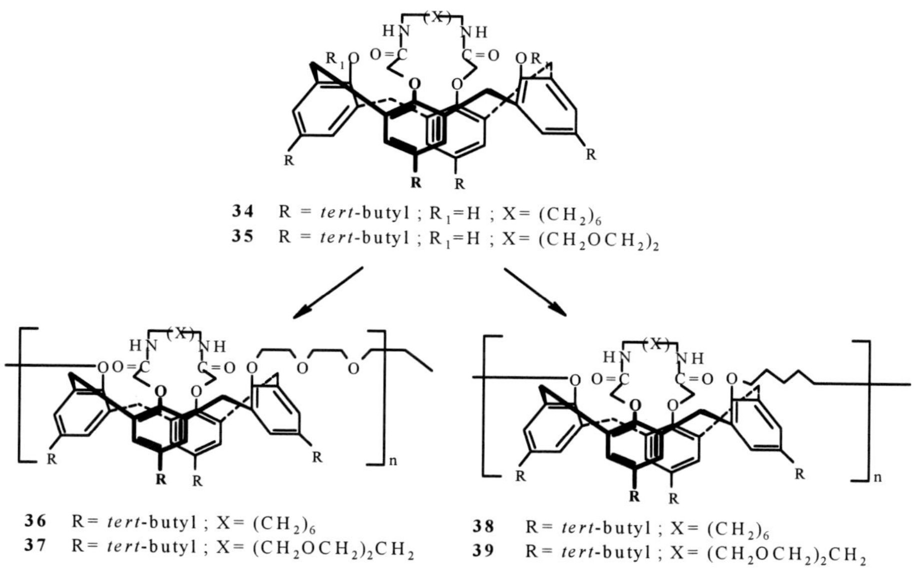

Figure 22. *p-tert*-Butylcalix[4](aza)crowns 34 and 35 and oligomers 36-39.

Their phase transfer behavior (Table 7) showed that oligomers **36** and **38** are good extractants for all of the metal ions studied; whereas, oligomers **37** and **39** exhibited poor extractions of selected alkali metal picrates (Li^+, Na^+, K^+, Cs^+), but high extraction of transition metal picrates (Cu^{2+}, Co^{2+}, Cd^{2+}, Ni^{2+}, Hg^{2+}, Pb^{2+}).

Two different oligocalixcrowns **42** and **43** (Figure 23) were synthesized *via* condensation reactions of *p-tert*-butylcalix[4]areneoxacrown-4 **39** and *p-tert*-butylcalix[4]arenethiacrown-4 **40** with 1,5-dibromopentane [71]. For these oligomerization reactions, it was found that only 3-4 calixcrown units had been linked together through alkyl bridges. Their complexing properties with selected alkali metals and transition metals were reported. Oligomers **42** and **43** were found to exhibit extraction selectivity for Cu^{2+}, Hg^{2+} and Pb^{2+}.

Table 7. Percent Extraction of metal picrates from aqueous solution into dichloromethane[a] by calix[4]arenes 34 and 35 and calixarene polymers 36-39

	34	**35**	**36**	**37**	**38**	**39**
Li^+	98	60	98	12	96	8.2
Na^+	97	63	97	3.6	97	7.8
K^+	97	62	97	5.4	98	8.7
Cs^+	96	58	96	1.5	98	6.5
Cu^{2+}	92	90	97	96	98	98
Co^{2+}	97	89	97	97	98	95
Cd^{2+}	95	88	97	96	98	96
Ni^{2+}	94	84	97	96	96	93
Hg^{2+}	99	92	97	98	98	99
Pb^{2+}	97	92	97	98	96	98

[a]Aqueous phase, [metal nitrate]= 1 X 10^{-2} M ; [picric acid] = 2.5 X 10^{-5} M ; organic phase, dichloromethane, [ligand] = 1 X 10^{-3} M or a 1 X 10^{-3} M solution of calix[4]arene unit/g resin for oligomers; extraction at 25 °C for 1 hour.

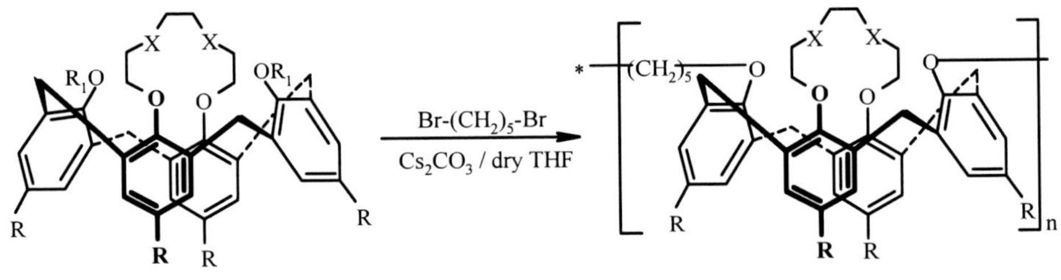

40 R = *tert*-butyl ; R_1 = H, X= O
41 R = *tert*-butyl ; R_1 = H, X= S

42 R = *tert*-butyl ; X= O
43 R = *tert*-butyl ; X= S

Figure 23. *p-tert*-Butylcalix[4]crown monomers **40** and **41** and oligomers **42** and **43**.

Using *p-tert*-butylcalix[4]arene as the starting material, Kim *et al.* [24] prepared two functionalized calix[4]arene monomers through a four-step reaction sequence. New copolyamic acid **44** with calix[4]arene moieties on the polymer backbone was synthesized in N-methyl-2-pyrolidone by polycondensations of 4,4'-oxydiphthalic anhydride (ODPA) with the diaminocalix[4]arene monomers using 4,4'-oxydiphenylene diamine (ODA) as a comonomer.

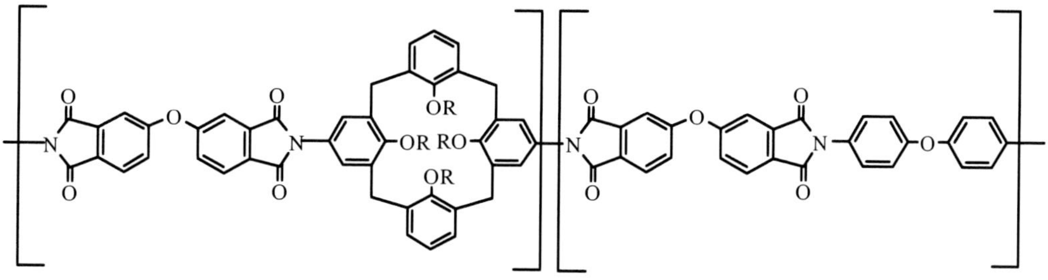

44

Figure 24. A new copolyamic acid **44** with calix[4]arene moieties in the polymer backbone.

These copoly(amic acid)s were soluble in aprotic polar solvents, so they can be processed in various ways. The copoly(amic acid)s precursors were converted thermally into films of the corresponding copolyimides. These copolyimides films were amorphous, but insoluble in common solvents. They were thermally stable up to 366°C. The copolyimides exhibited relatively high TEC's, low Tg's, low refractive indices, low dielectric constants, low optical anisotropies, low dielectric anisotropies, and low water uptake compared to those of conventional ODPA-ODA polyimides. These properties were interpreted as resulting from bulky, cone-like calix[4]arene moieties and their effects on the chain conformation and morphological structure. The copolymers are considered to be potential candidate materials suitable for membranes, antioxidant additives, chemical sensor devices, and microelectronic devices.

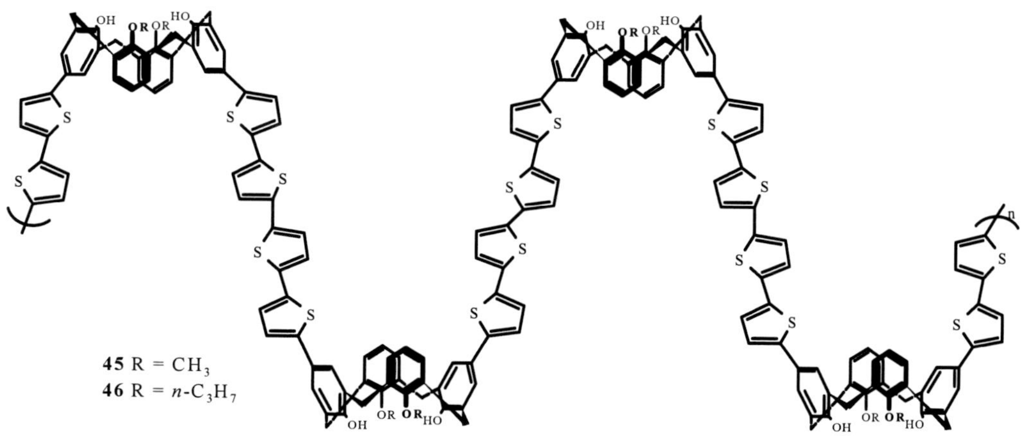

Figure 25. Segmented conjugated polymers **45** and **46** based upon a calix[4]arene scaffold.

Calix[4]arenes undergo dramatic geometric changes involving phenol ring flips between cone, partial cone, and 1,3-alternate conformational isomers [72]. Also they can be constructed to undergo smaller changes, such as pinched cone to pinched cone hinge movements, which vary the distance between two opposing phenol rings [73]. For these reasons, Yu *et al.* [74] sought to develop segmented conjugated polymers **45** and **46** (Figure

25) based upon a calix[4]arene scaffold, wherein calix[4]arene units are flexible and insulating bridges between individual organic electroactive segments. The use of electroactive segments has been demonstrated as a viable route to conducting materials [75]. The cone conformation of the calix[4]arenes creates a zigzag orientation of the polymer segments. These dynamic and porous materials are attractive candidates for sensing [76] and actuating materials [77].

It is well known that Schiff bases are potentially capable of forming stable complexes with many transition metal cations [78]. Thus comparative extraction properties of cyclic Schiff base derivatives of *p-tert*-butylcalix[4]arene **47** and **48** and some of their telomers (**49-51**) toward metal ions were studied [79,80].

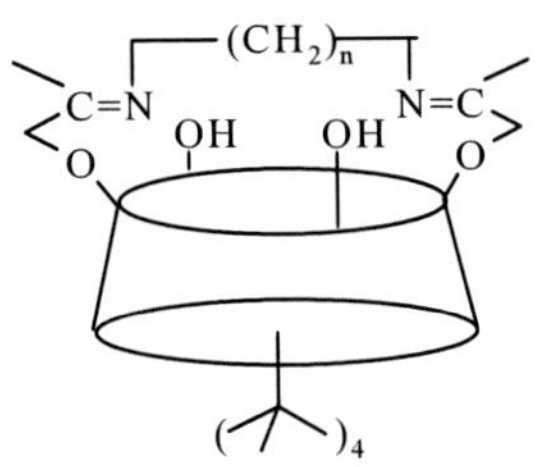

47 n = 0
48 n = 6

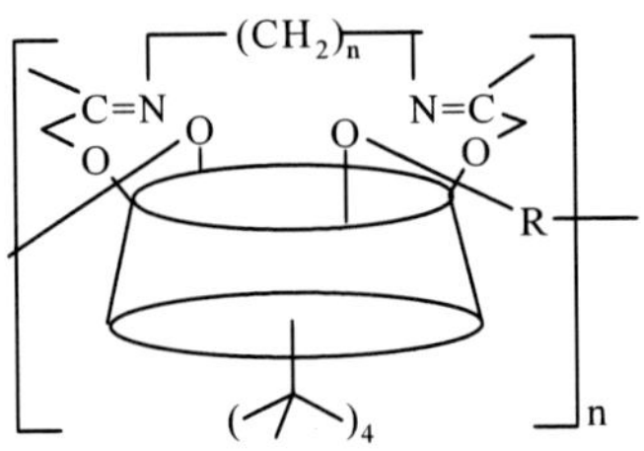

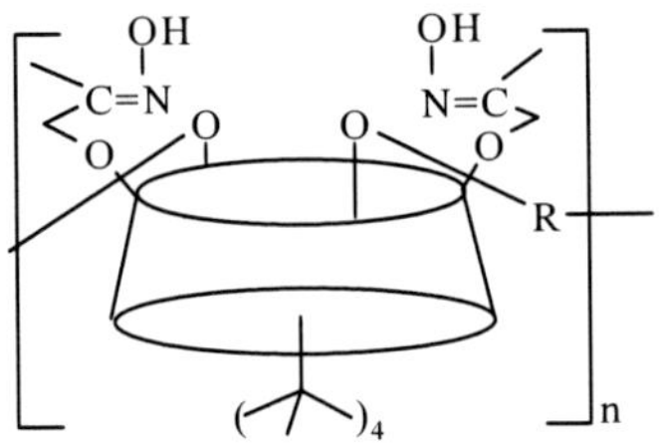

49 n = 0, R= $CH_2(CH_2OCH_2)_2CH_2$ **51** R= $CH_2(CH_2OCH_2)_2CH_2$
50 n = 6, R= $CH_2(CH_2OCH_2)_2CH_2$

Figure 26. Cyclic Schiff base derivatives of *p-tert*-butylcalix[4]arene **46** and **47** and some of their telomers (**49-51**).

The selective extraction ability of the Schiff base-calixarene derivatives in transferring the transition metal cations is due to the presence of soft л-donor systems in the macrocyclic structure (Table 8).

The discovery that calixarenes with nitrile functionalities are very selective for Hg^{2+} has been developed in the work of Memon and Yilmaz [81]. A series of cone conformers of 1,3-dialkylated *p-tert*-butylcalix[4]arenes (**52-55**) and their respective telomers (**56-59**) were synthesized (Figure 27). Complexation behavior was assessed by a liquid-liquid extraction procedure. It was found that among the monomers **55** had very good selectivity for Hg^{2+} (Table 9). Telomer 56 showed greatest affinity toward Cu^{2+}, Hg^{2+} and Pb^{2+}, but the telomer **57** was effective for Hg^{2+} and Pb^{2+} extraction. Telomers **58** and **59** showed little selectivity among transition metal cations (Cu^{2+}, Co^{2+}, Ni^{2+}, Cd^{2+}, Hg^{2+} and Pb^{2+}).

Table 8. Percent extraction of metal picrates from aqueous solutions into chloroform[a] by ligands 47-51

	47	**48**	**49**	**50**	**51**
Li^+	12	30	<1.0	5.2	<1.0
Na^+	10	36	<1.0	10	<1.0
K^+	7.6	39	<1.0	3.9	<1.0
Cs^+	-	79	-	6.0	<1.0
Cu^{2+}	99	80	76	72	18
Co^{2+}	59	59	82	64	7.5
Cd^{2+}	66	60	81	75	7.9
Ni^{2+}	68	59	82	69	4.4
Hg^{2+}	94	95	90	78	78
Pb^{2+}	-	69	-	59	58

[a]Aqueous phase, [metal nitrate]= 1 X 10^{-2} M ; [picric acid] = 2.5 X 10^{-5} M ; organic phase, chloroform, [ligand] = 1 X 10^{-3} M or a 1 X 10^{-3} M solution of calix[4]arene unit/g resin for telomers; extraction at 25°C for 1 hour.

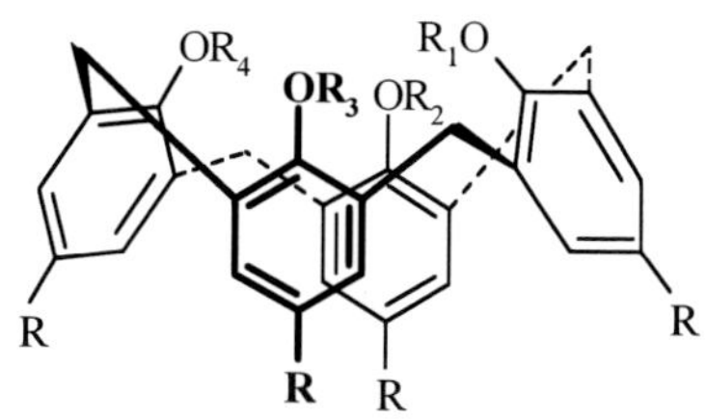

R = *tert*-Butyl

52 $R_1 = R_4 = H$ $R_2 = R_3 = CH_3$

53 $R_1 = R_4 = H$ $R_2 = R_3 = CH_2COCH_3$

54 $R_1 = R_4 = H$ $R_2 = R_3 = CH_2COOC_2H_5$

55 $R_1 = R_4 = H$ $R_2 = R_3 = CH_2CN$

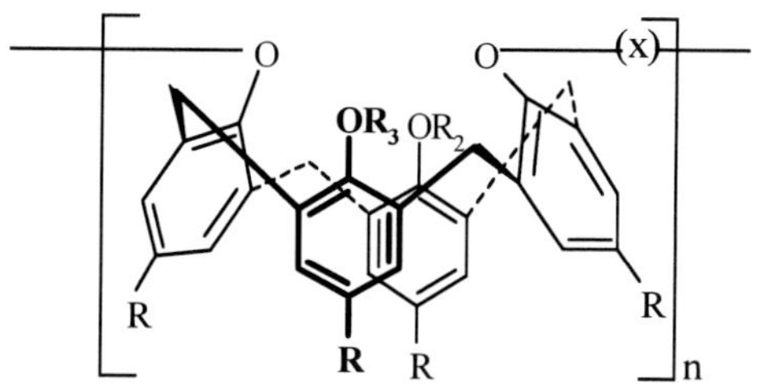

$X = CH_2CH_2OCH_2CH_2OCH_2CH_2$

56 R = *tert*-butyl ; $R_2 = R_3 = CH_3$

57 R = *tert*-butyl ; $R_2 = R_3 = CH_2COCH_3$

58 R = *tert*-butyl ; $R_2 = R_3 = CH_2COOC_2H_5$

59 R = *tert*-butyl ; $R_2 = R_3 = CH_2CN$

Figure 27. Cone conformers of 1,3-dialkylated *p-tert*-butylcalix[4]arenes **52-55** and their respective telomers (**55-58**).

From these observations, it may be concluded that the size of the macro ring alone does not control the complexation phenomenon, but the nature, effective-ness, and aggregation of functional groups are also important factors for metal ion complexation. In the case of telomers, the increased affinity in complexation can be explained by the fact that there is no interference from the remaining two phenolic groups and an important role is played by the ethylene glycol bridges at the water-dichloromethane interface. Moreover, this phenomenon may reflect the "hard and soft acid and base" concept introduced by Pearson and Ho [63]. The complexation sites possess π-bond containing functionalities. Cation-π interactions favor the complexation with the more polarizable transition metal ions, especially Hg^{2+} and Pb^{2+}, which are known as soft metal cations.

Table 9. Percent extraction of metal picrates[a] from aqueous solutions into dichloromethane by ligands 52-59

	52	53	54	55	56	57	58	59
Li^+	<1.0	<1.0	<1.0	<1.0	9.6	<1.0	1.8	12
Na^+	2.5	<1.0	16	<1.0	6.8	<1.0	2.7	8.4
K^+	<1.0	<1.0	4.3	<1.0	5.6	<1.0	3.5	7.5
Cs^+	<1.0	<1.0	2.2	<1.0	2.1	<1.0	3.0	5.7
Cu^{2+}	<1.0	<1.0	51	2.2	28	8.6	32	60
Co^{2+}	<1.0	<1.0	25	1.8	15	5.6	41	54
Cd^{2+}	<1.0	<1.0	24	2.0	11	8.0	56	45
Ni^{2+}	2.8	3.7	37	3.5	15	3.4	45	26
Hg^{2+}	25	29	33	40	28	48	28	64
Pb^{2+}	5.9	11	82	6.6	49	46	76	55

[a]Aqueous phase, [metal nitrate]= 1 X 10^{-2} M ; [picric acid] = 2.5 X 10^{-5} M ; organic phase, dichloromethane, [ligand] = 1 X 10^{-3} M ; extraction at 25°C for 1 hour.

Copolyethers and polyurethanes with lower and upper rim-functionalized calix[4]arene units in the fixed cone conformation were prepared in good yield by Dondoni *et al.* [82], through polycondensation reactions of distal calix[4]arene diols with Bisphenol A/dibromomethane and 2,4-tolylendiisocyanate, respectively (Figures 28 and 29).

Since similar behavior was observed for polyurethane **73** with no calix[4]arene units in the chains (Table 10), it was deduced that the extraction abilities of polyurethanes **68-70**, although very low, were mainly due to the urethane functionalities in the chains, while the calix[4]arene moieties were essentially ineffective. On the other hand, the calix[4]crown ethers **71** and **72** were good ionophores for K^+, Cs^+, and Ag^+. The latter was also effective for Rb^+ extraction. Overall these results demonstrate that the ionophoric properties of calix[4]crown ethers are not substantially modified by insertion of these bis macrocycle structures into a polyurethane chain.

Table 10. Percent extraction of metal picrates from aqueous solutions into chloroform by polyurethanes containing calix[4]arene moieties and a model compound

Host[a]	Cations[b]					
	Li^+	Na^+	K^+	Cs^+	Rb^+	Ag^+
68	-	8.5	4.9	6.0	-	7.9
69	-	7.9	5.1	5.3	-	9.0
70	-	9.5	7.0	5.0	-	6.3
71	14.7	18	98	96	39	85
72	23	64	62	89	79	92
73	-	8.7	<1.0	9.0	-	11.0

[a]The initial host concentration was 2.5 X 10^{-3} M in $CHCl_3$.

[b]Concentration 1 X 10^{-2} M of corresponding nitrate salts in aqueous solution containing 2.5 x 10^{-4} M picric acid.

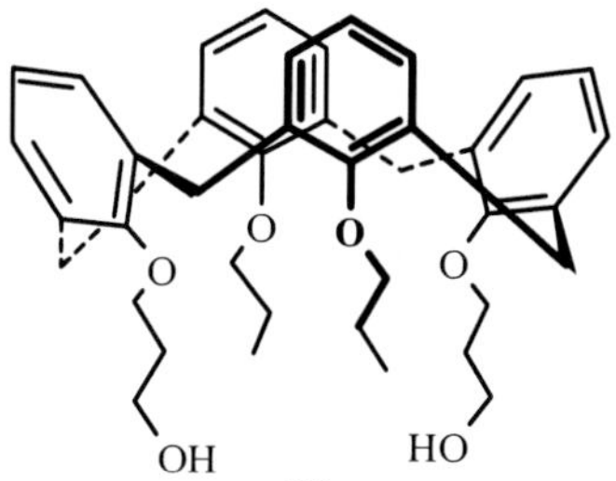

60

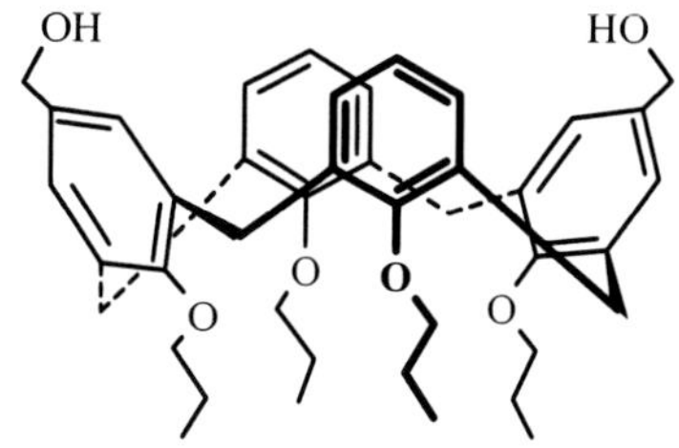

61

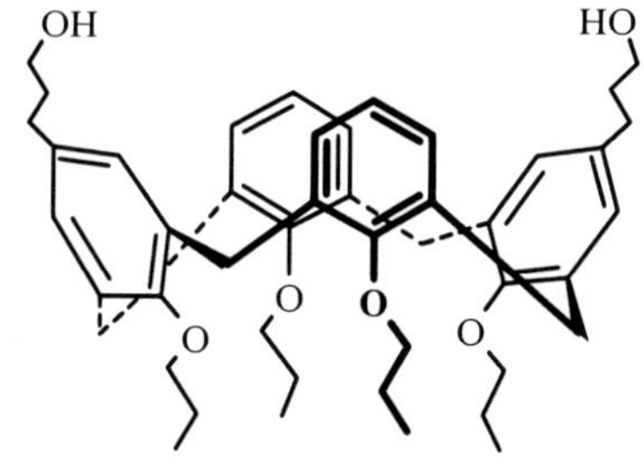

62

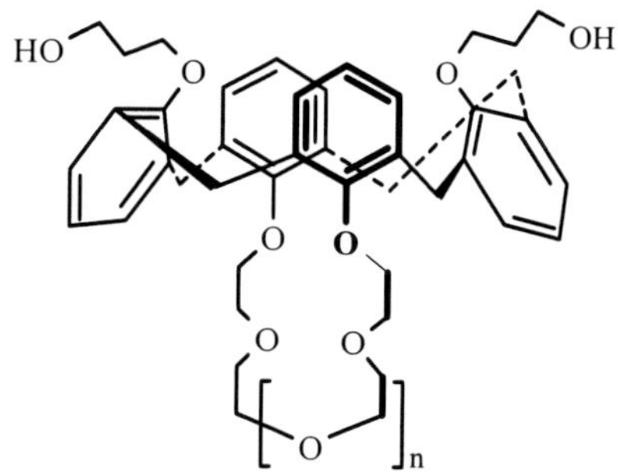

63 n = 1
64 n = 2

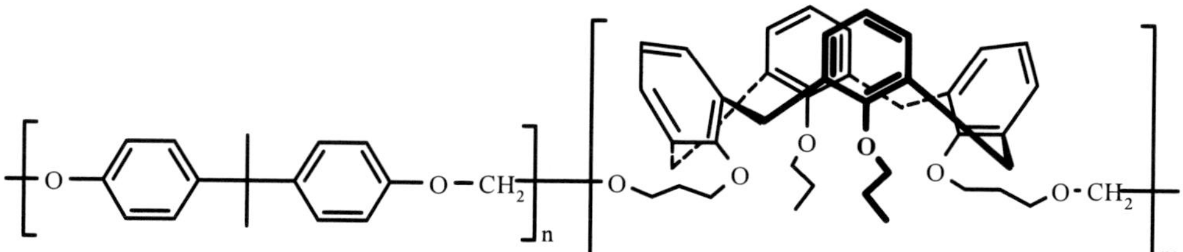

65

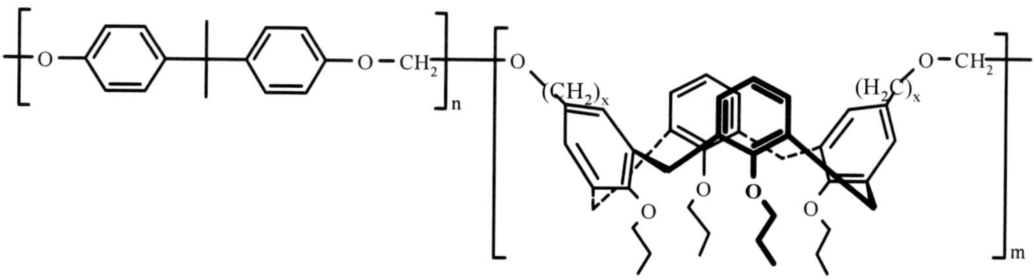

66 x = 1
67 x = 3

Figure 28. Distal calix[4]arene diols and copolyethers containing functionalized calix[4]arene units in a fixed cone conformation.

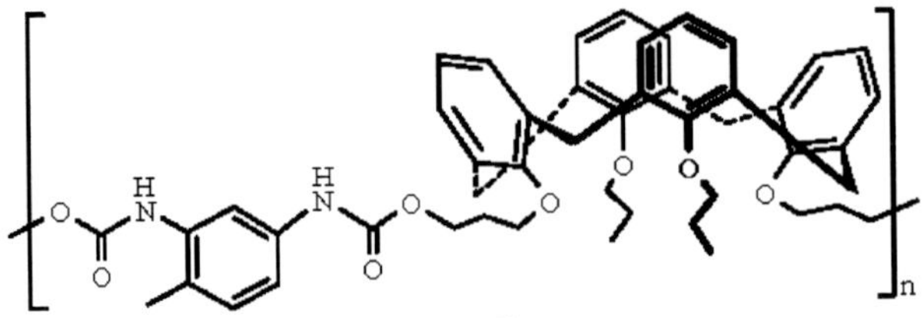

68

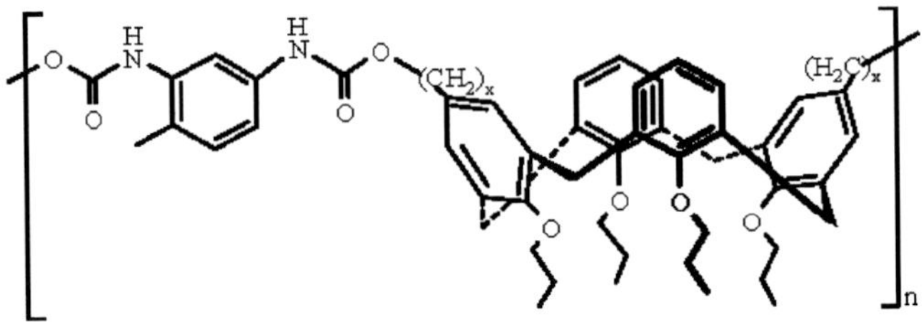

69 x = 1
70 x = 3

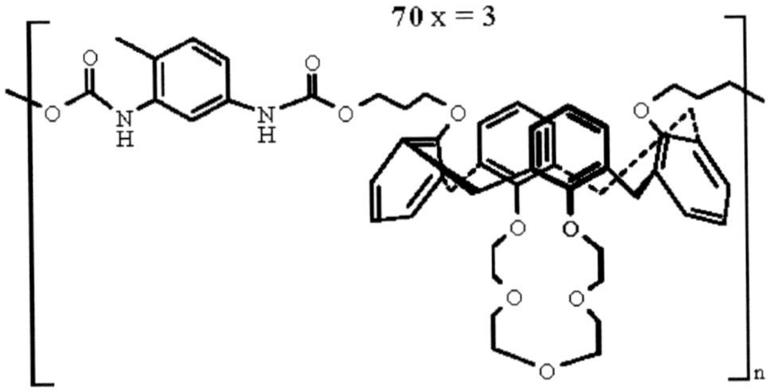

71

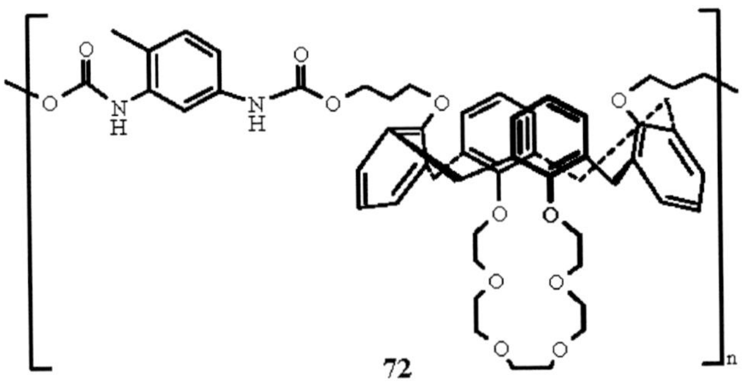

72

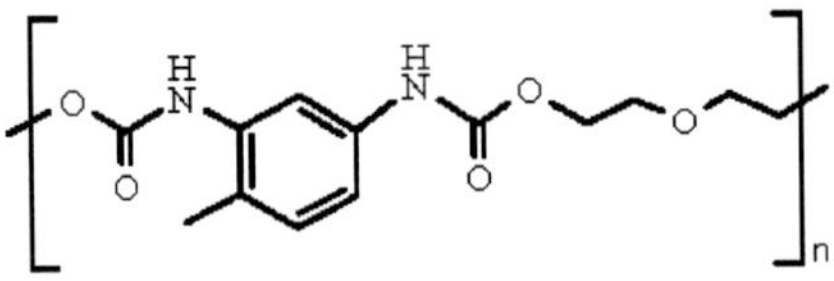

73

Figure 29. Polyurethanes with functionalized calix[4]arene units in a fixed cone conformation and a model polyurethane **73**.

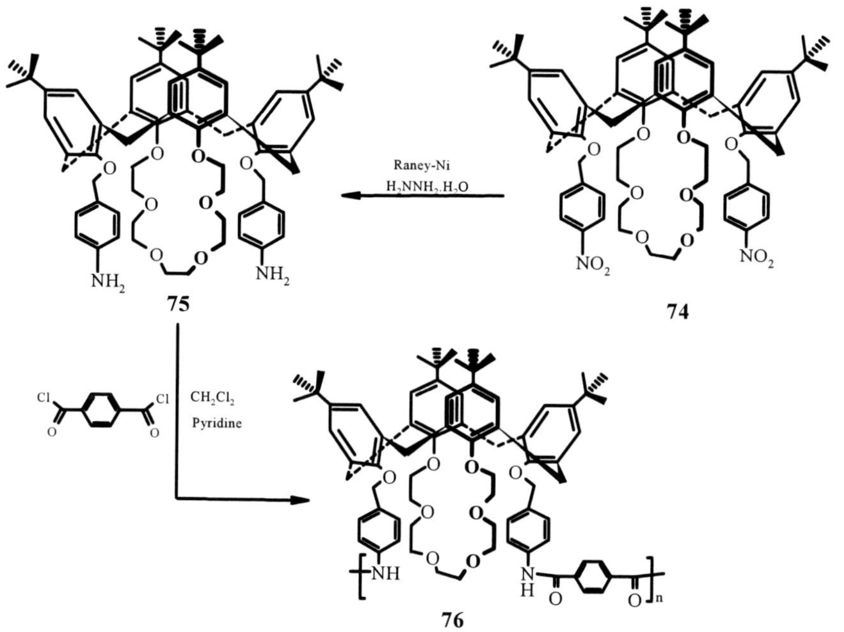

Figure 30. Synthesis of the cone conformer of 5,11,17,23-tetra-*tert*-butyl-25,27-bis(4-aminobenzyl-oxy)calix[4]arene-crown-6 (**75**) containing polymerizable side arms and its oligomeric analogue **76**.

The selective removal of ^{137}Cs from medium-level radioactive waste has been investigated by several research groups [83-85]. High affinity and selectivity towards Cs^+ was reported for some 4-alkylcalix[4]arene-crown-6 ionophores in the 1,3-alternate conformation. On the other hand, calix[4]crown-6 analogues in the cone conformation does not show selectivity for Cs^+ [85]. This could be related to the destabilization of cesium complexes due to the steric effect of the lower-rim alkyl groups that face the polyether ring.

Synthesis of the cone conformer of 5,11,17,23-tetra-*tert*-butyl-25,27-bis(4-aminobenzyloxy)calix[4]arene-crown-6 (**75**) containing polymerizable side arms (Figure 30), along with its oligomeric analogue **76**, was reported by Tabakci *et al.* [86]. Extraction studies performed with oligomer **76** showed no selectivity for Cs^+ (Table 11). Molecular characterization revealed that the cone conformation of calixcrown-6 monomer **75** was maintained in the oligomer.

More interesting and unusual extraction results were observed for the calixcrown-6 cone conformer **74**, which exhibited good extraction ability toward alkali metal cations. Presumably this is due to the influence of the nitrophenyl groups in the side arms, which facilitate binding in a sandwich-like manner from opposite sides of the crown ether part of the molecule. The decrease in the extraction ability in going to **75** may be due to the influence of amino hydrogen atoms forming intramolecular hydrogen bonding between facing oxygen atoms of the crown ether portion of the molecule. Immobilization of monomer **75** by reaction with terephthaloyl dichloride enhanced its extraction ability.

Table 11. Percent extraction of metal picrates into dichloromethane[a] with ligands 74-76

	Picrate Salt Extracted (%)			
Ligand	Li^+	Na^+	K^+	Cs^+
74	39	40	36	39
75	26	24	26	30
76	32	41	25	41

[a]Aqueous phase, [metal picrate] = 2.0 X 10^{-5} M ; organic phase, dichloromethane, [ligand] = 1 X 10^{-3} M or a 1 X 10^{-3} M solution of calix[4]arene unit/g resin for oligomer; extraction at 25 °C for 1 hour.

Barata *et al.* [87] described the synthesis of linear and cross-linked polymers containing hydrophobic and hydrophilic lower rim-substituted derivatives of *p-tert*-butylcalix[4]arenes (Figure 31). The target polymers were prepared either through direct attachment of *p-tert*-butylcalix[4]arene (**77**) or its tripropoxy (**78**) and tetracarboxymethoxy (**80**) derivatives to lightly cross-linked Merrifield's resin (1-2% DVB) or via radical addition copolymerization of 25,26,27-tripropoxy-28-(4-vinyl-benzyloxy)-*p-tert*-butylcalix[4]arene (**81**) with styrene and a cross-linking agent (Figure 32).

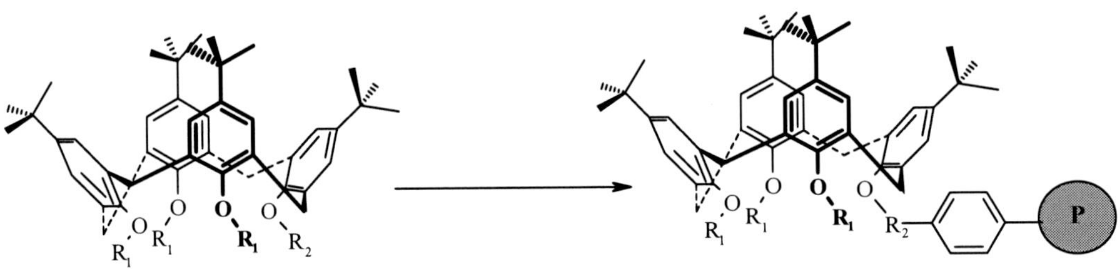

77 $R_1 = R_2 = H$
78 R_1 = Propyl; $R_2 = H$
79 $R_1 = R_2 = CH_2CO_2Et$
80 $R_1 = R_2 = CH_2CO_2H$

81 R_1 = Propyl; $R_2 = CH_2$
82 $R_1 = H$; $R_2 = CH_2$
83 $R_1 = CH_2CO_2Et$; $R_2 = CH_2$
84 $R_1 = CH_2CO_2H$; $R_2 = CH_2$
85 $R_1 = CH_2CO_2Na$; $R_2 = CH_2$
86 $R_1 = CH_2CO_2H$; $R_2 = C(O)OCH_2$

Figure 31. Synthesis of linear and cross-linked polymers containing hydrophobic and hydrophilic lower rim-substituted derivatives of *p-tert*-butylcalix[4]arene.

Calixarene **87** and styrene in various ratios were used in suspension polymerization to produce soluble polymer **88** (Figure 32). Similar levels of conversion were achieved with a calixarene-styrene ratio of 1:10 (58% yield) or 1:20 (51% yield), after reaction for 24 hours. No significant increase in yield was observed with longer times of 63-84 hours. GPC analysis indicated a weight-average molar mass of 41,000 g/mol and a polydispersity (M_w/M_n) of 1.78 for **88** (1:10 ratio). The less calixarene-enriched copolymer **88** (1:20 ratio) yielded a similar polydispersity (1.58), but a lower M_w (19,000 g/mol). An insoluble polymeric matrix **89** was also prepared by suspension polymerization with a feed ratio of 1:20 and five mole percent loading of divinylbenzene.

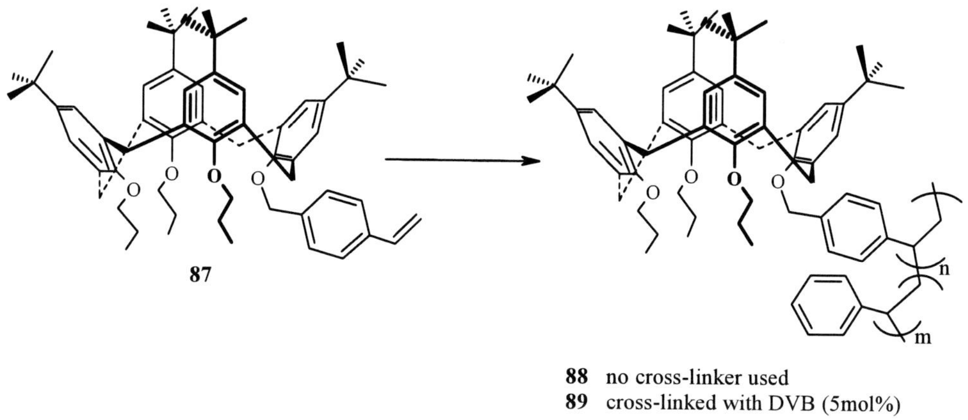

Figure 32. Calixarene monomer **87** and styrene in various feed ratios used in suspension polymerization to produce the soluble polymer **88**.

3. POLYMER-SUPPORTED CALIX[4]ARENE DERIVATIVES

3.1. Organic Polymer-Supported Calixarenes

Polymer-bound calix[6]arene **90** was described in a patent by Shinkai [88], which states that the hexakis(carbethoxymethyl)ether of *p*-sulfonatocalix-[6]arene was partially nitrated, aminated, and fixed on cross-linked chloromethylated polystyrene. This resin absorbed 108 mg of uranium from sea water per 0.10 g of resin in seven days at a flow rate of 30 mL/minute. Another polymer-bound analog was reported by Shinkai *et al.* [29], who treated *p*-(chlorosulfonyl)calix[6]arene with polyethyleneimine to obtain a gel-like product with one calixarene unit for every 15 ethyleneimine units. This polymer showed the same binding power and selectivity for UO_2^{2+} as did the parent *p*-sulfonatocalix[6]arene.

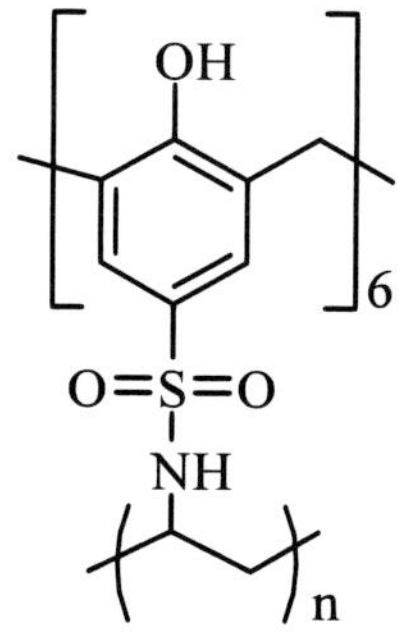

90

Figure 33. Polymeric calixarene 90 which is selective for UO_2^{2+}.

A polymeric compound containing more than one calixarene molecule was synthesized by Deligoz *et al.* [89,90] by reacting an epichlorohydrin oligomer with 25,26,27-tribenzoyloxy-28-hydroxycalix[4]arene (Figure 34). After basic hydrolysis of **91**, the polymeric calixarene **92** was found to be selective for Fe^{3+} in extraction from an aqueous solution containing Cu^{2+}, Ni^{2+}, Co^{2+} and Fe^{3+}. In another study [91], the epichlorohydrin oligomer was reacted with 25,27-diacetonyloxy-5,11,17,23-tetra-*tert*-butyl-26,28-dihydroxycalix[4]arene and then with chloroacetone to obtain an oligomer–supported calix[4]triketone derivative **93**. This oligomeric ionophore was selective for extraction of Li^+ from an aqueous solution containing Li^+, Na^+, K^+, Ag^+, Cd^{2+}, Co^{2+}, Cu^{2+}, and Ni^{2+}.

The epichlorohydrin oligomer and low molecular weight fraction of acryloyl chloride (M_w=3,800 g/mol) were treated with a calix[4]arene tetraester derivative by Deligöz *et al.* [92]. Based on chlorine analysis, the calixarene units did not replace all of the chlorine atoms in the oligomer. The decrease of the chloride content of oligomer **94** was only 50% of that expected for complete substitution.

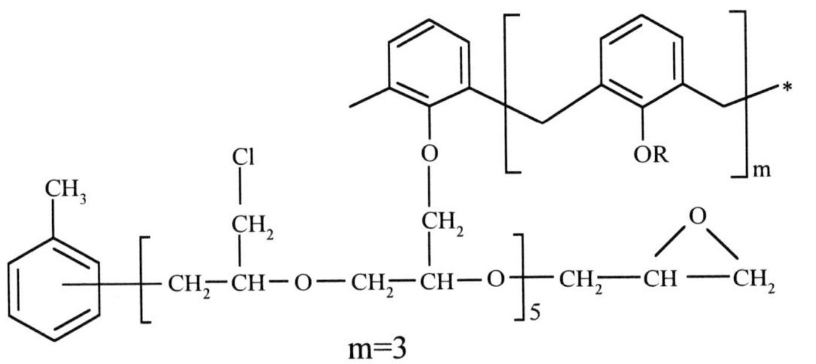

m=3

91 R = COC_6H_5

92 R = H

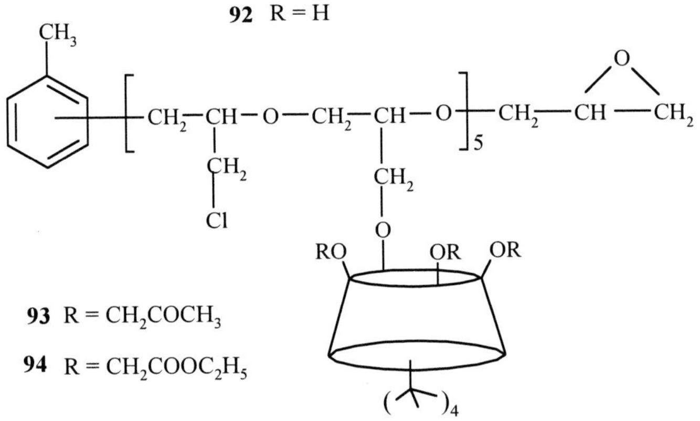

93 R = CH_2COCH_3

94 R = $CH_2COOC_2H_5$

Figure 34. Epichlorohydrin oligomer-supported calixarenes **91-94**.

Calix[4]arene ester derivative oligomer **95** was prepared by reaction of the epichlorohydrin oligomer and calix[4]arene tetraester with boron trifluoride etherate as catalyst. Chlorine analysis of this product indicated that each chlorine of the polymer **95** was replaced by one calixarene unit. Liquid-liquid extraction results showed that both of the compounds were selective for Na^+.

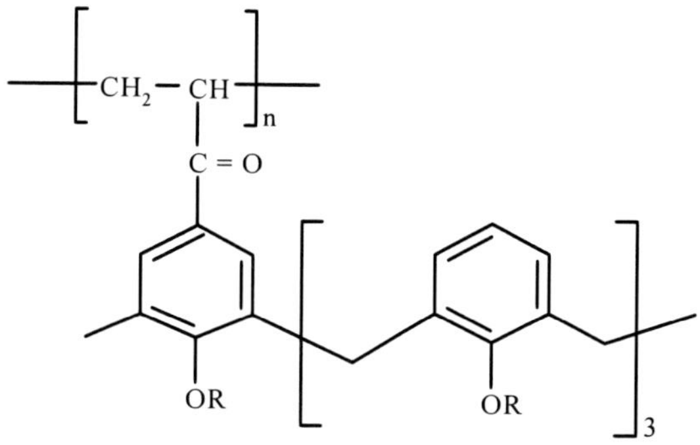

95 R = $CH_2COOC_2H_5$

Figure 35. Oligomer-supported calix[4]arene tetraester 95.

Calix[4]arene-semicarbazone derivative **96** was covalently linked to commercially available Merrifield's resin at the lower rim to obtain polymeric chelating resin **97** (Figure 36). The resin was used for the preconcentration and separation of Ce^{3+}, La^{3+}, Th^{4+}, and U^{4+}. The resin exhibited good separating ability with maximum sorption between pH 2.5-4.5 for Th^{4+} and pH 5.5-7.0 for U^{4+}; whereas La^{3+} and Ce^{3+} were found to have maximum sorption at pH 6.8-8.5 [93].

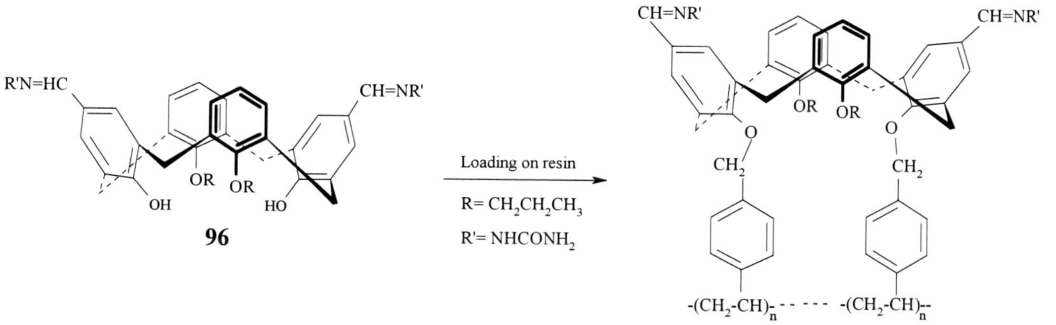

Figure 36. Attachment of calix[4]arene-semicarbazone derivative **96** onto a commercially available Merrifield's resin to obtain polymeric chelating resin **97**.

Studies were performed by Memon and coworkers on monomeric calix[4]arene amide derivative **98** systems that effectively extract chromium(VI) after being attached to a polymer backbone [94]. The polymer chosen for modification was polystyrene. The resulting polymer **100** (Figure 37) was soluble in organic solvents and could, therefore, be used as liquid-liquid extractant.

Polymer **100** was a better extractant for chromium(VI) than its parent monomer 98 (Figure 38). It is possible that the polymer plays a role whereby it folds into conformations that place functional groups from several of the calix[4]arene moieties in the polymer in a preferred conformation where they can associate with the oxoanion.

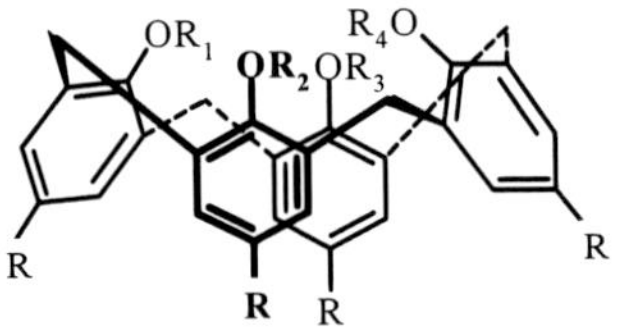

98 R = *tert*-butyl ; R_1,R_4= H
R_2,R_3 = $CH_2CONHCH_2C_6H_5$

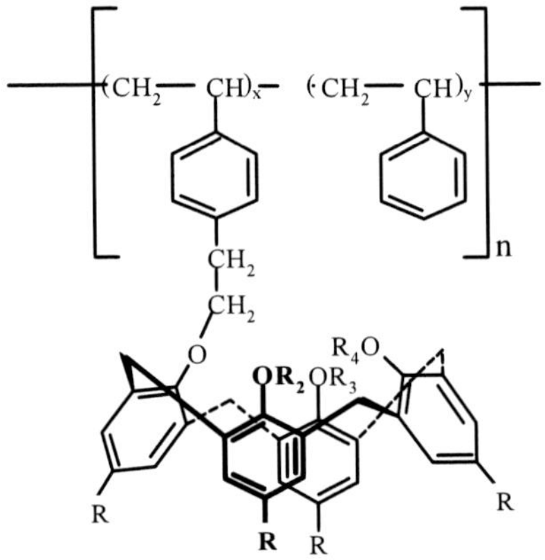

99 R = *tert*-butyl ; R_2 =R_3=R_4= CH_2COOCH_3
100 R = *tert*-butyl ; R_4=H; R_2 =R_3 = $CH_2CONHCH_2C_6H_5$

Figure 37. Attachment of monomeric calix[4]arene amide derivative **98** to a polystyrene backbone.

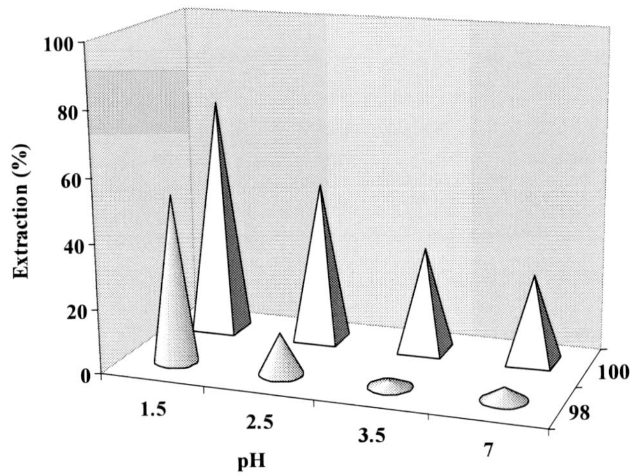

Figure 38. Percent extraction *vs.* pH for liquid-liquid extraction of Cr(VI) by **98** and **100**.

Dextrans are natural, water-soluble carbo-hydrate polymers composed mainly of linear chains of glucose units with α-1,6 linkages. Their bioavailability and biocompatibility explain the interest taken in these structures by industry. They are mainly used in the formulation of cosmetics, pharmaceuticals, imaging agents, and other domains in which they can act as carriers and/or protective agents for other compounds. The presence of secondary alcoholic functions provides these materials with good ability for chemical transformation, allowing cross-linking, such as in Sephadex [95]. The use of dextrans in the field of metal ion complexation was developed for the preparation of new image-enhancing agents, notably with lanthanides [96]. Engrand and Vains [97] prepared the new metal complexing dextran-supported calixarene **102**. The resulting polymer showed an interesting ability to complex Cu^+, giving stable yellowish to light orange solids.

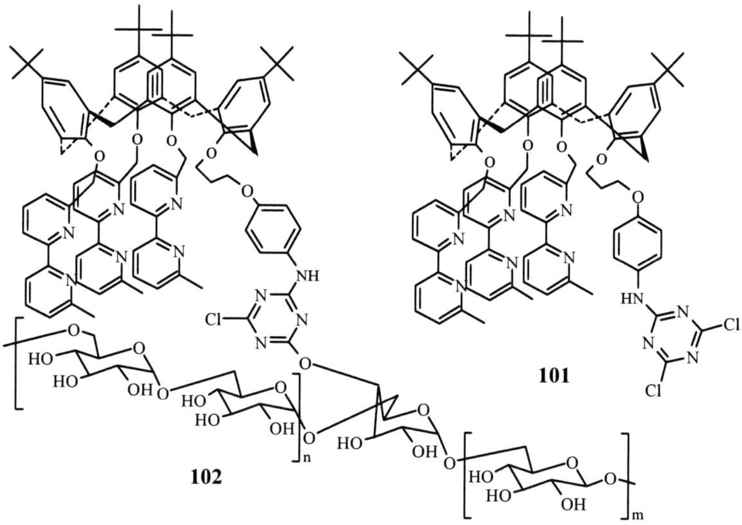

Figure 39. New metal complexing, dextran-supported calixarene **102**.

Chitin is one of the most abundant organic materials that can be easily obtained from nature. Chitin can be readily degraded to chitosan by N-deacetylation to give a nitrogenous polysaccharide composed mainly of β-(1,4)-2-amino-2-deoxy-D-glucopyranose repeating units. Because of its specific structures and properties, chitosan has recently received intense attention for its industrial and medical applications, such as artificial skin and capturing metal ions from waste water [98-101]. Furthermore, the free primary amino groups and hydroxyl groups at the six positions in the pyranose ring of chitosan enable a variety of modifications.

Calixarene-modified chitosans were synthesized by Li *et al.* [102] by reaction of chitosan with 1,3-bis-chloroethoxyethoxy-2,4-dihydroxy-*p*-*tert*-butylcalix[4]arene (Figure 40). Adsorption of Ag^{+}, Cd^{2+}, Cu^{2+}, Hg^{2+}, Ni^{2+}, and Pd^{2+} by polymer **103** was evaluated. The adsorption capacities of **103** for Ag^{+} and Hg^{2+} were much higher than those for Cd^{2+} and Cu^{2+} due to the presence of the calixarene moieties (Table 12).

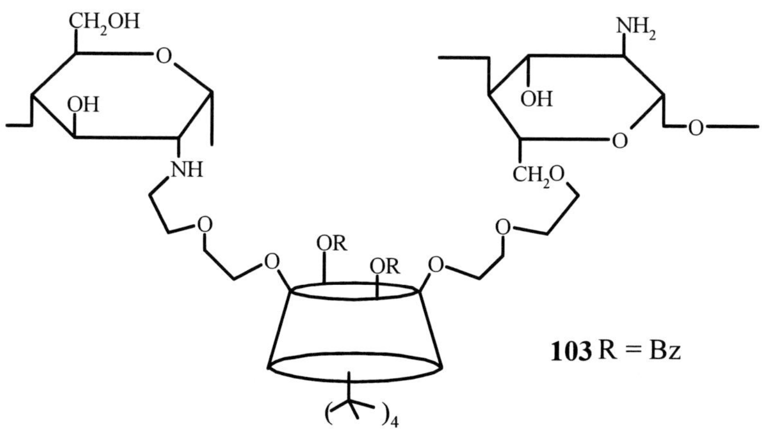

Figure 40. *p*-tert-Butylcalixarene-modified chitosan **103**.

Table 12. Metal ion adsorption ability of calixarene-modified chitosan 103

	pH	Adsorption yield (%)
Cu^{2+}	4.0	60
	5.0	38
	7.0	20
Pd^{2+}	4.0	75
	5.0	76
	7.0	78
Ag^{+}	4.0	46
	5.0	78
	7.0	85
Hg^{2+}	4.0	58
	5.0	63
	7.0	55

Although calixarenes appear to have interesting solvent extraction behavior, they nevertheless possess some serious disadvantages for use as solvent extraction reagents, such as poor solubility in organic solvents, slow extraction rates, and high production costs. In particular, their poor solubility in organic solvents makes them unsuitable for use as extractants on a commercial scale. One way to avoid this drawback and make efficient use of the characteristic properties of calixarenes as ionophores with high recognition ability is to employ them as solid-phase ion-exchange resins. Calix[4]arene carboxylate resin **105** was prepared from calix[4]arene carboxylate **104** on a matrix of polyallylamine to explore the resin's adsorption behavior toward some metal ions including Pb^{2+} [103,104]. The selectivity sequence among the base metal ions on the resin was $Pb^{2+} >> Cu^{2+} >> Zn^{2+} = Ni^{2+} = Co^{2+}$.

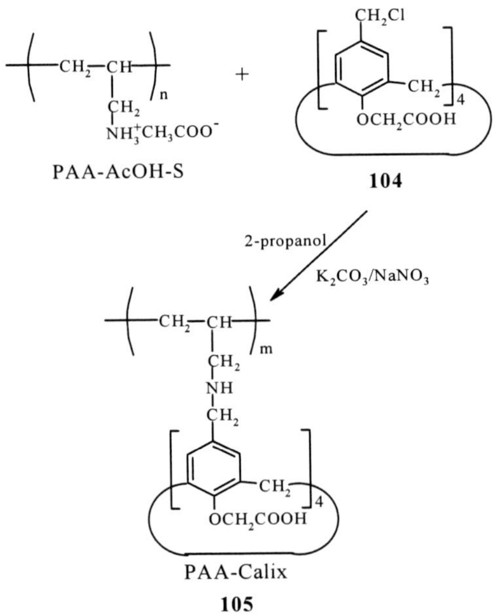

Figure 41. Synthesis of a calix[4]arenecarboxylic acid resin **105** from calix[4]arenecarboxylic acid **104** and a polymer matrix of polyallylamine.

The origin of this selectivity sequence was not attributed to the amino groups in the polymer matrix, but to the functional groups of the introduced calix[4]arenecarboxylic acid units. Although the maximum loading capacity for Pb^{2+} was not very high, the resin showed high Pb^{2+} selectivity and the separation of a trace amount of Pb^{2+} from a large excess of Zn^{2+} was achieved.

Calix[4]arene and phosphorylated calixarene were immobilized onto cross-linked polystyrene supports by Alexandratos and Natesan [31]. Cooperation of the ligating groups induced high selectivity in metal ion complexation by these polymers. The calix[4]arene polymer was highly selective for the removal of Cs^{+}, while the phosphorylated version 106 was specific for the complexation of Fe^{3+} and Pb^{2+}.

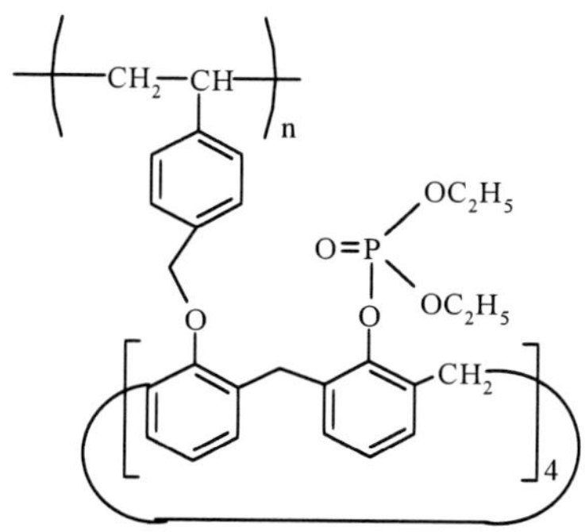

106

Figure 42. A phosphorylated calixarene immobilized on a cross-linked polystyrene support.

3.3. Silica Gel-Bonded Calixarenes

Calix[4]arene tetraethyl ester and calix[6]arene hexaethyl ester were covalently linked to silica gel particles through a hydrosilanization reaction at the upper rim to give **107** and **108** (Figure 48) [34]. When packed into a column with a 1:20 methanol-water solution as the mobile phase, the former was selective for Na^{+}; whereas the latter exhibited selectivity for Cs^{+}, Rb^{+}, and K^{+}.

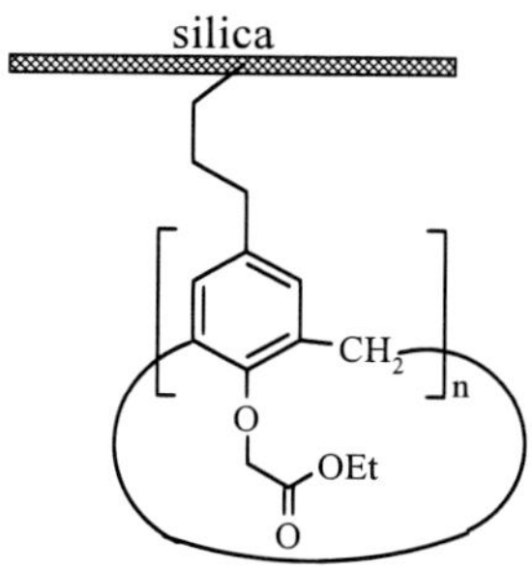

107 n = 4
108 n = 6

Figure 43. Calix[4]arene tetraethyl ester and calix[6]-arene hexaethyl ester covalently linked to silica gel particles.

Chelating calix[4]arenehydroxamates attached to silica gel particles (**109**) (Figure 44) were studied by Hutchinson *et al.* [105] for their uptake of different transition metal ions at varying solution pH. While Pb^{2+} was quantitatively complexed at a pH of 3.5 and Cu^{2+} at pH of 7, the other metal ions (Co^{2+}, Mn^{2+}, Ni^{2+}, and Zn^{2+}) were not removed above 90% even at higher pH values.

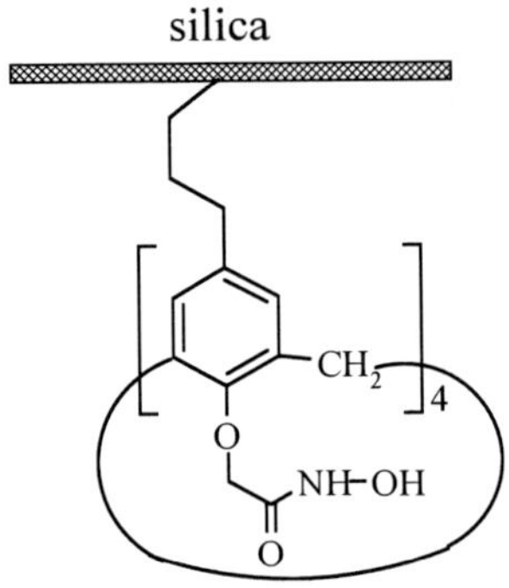

109

Figure 44. Chelating calix[4]arenehydroxamate supported onto silica gel particles.

Silica gel-bonded calix[4]arenetetraethylamide **110** (Figure 45) was evaluated by Glennon *et al.* [30] for its chromatographic selectivity. Retention of Na^+ over other alkali metal cations and of Ca^{2+} over Mg^{2+} was observed.

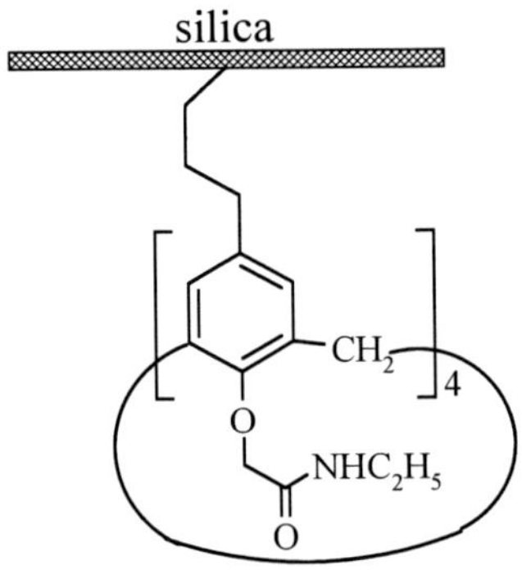

110

Figure 45. Silica-bonded calix[4]arenetetraethylamide.

To separate nucleosides and bases, *p-tert*-butylcalix[6]arene-bonded silica gel **111** stationary phase (Figure 46) was prepared with γ-(ethylenediamino)propyltriethoxylsilane as the coupling agent [106]. The results indicate that this stationary phase behaves as a reversed–phase packing, but its hydrophobicity is much weaker than that of Zorbax C_{18} phase.

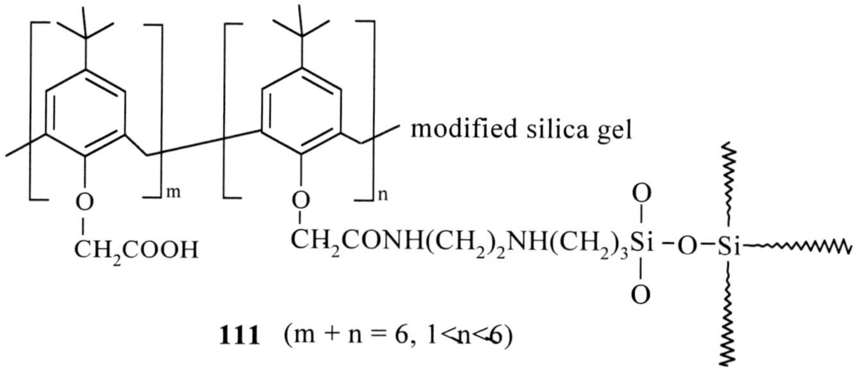

111 (m + n = 6, 1<n<6)

Figure 46. *p-tert*-Butylcalix[6]arene-bonded silica gel.

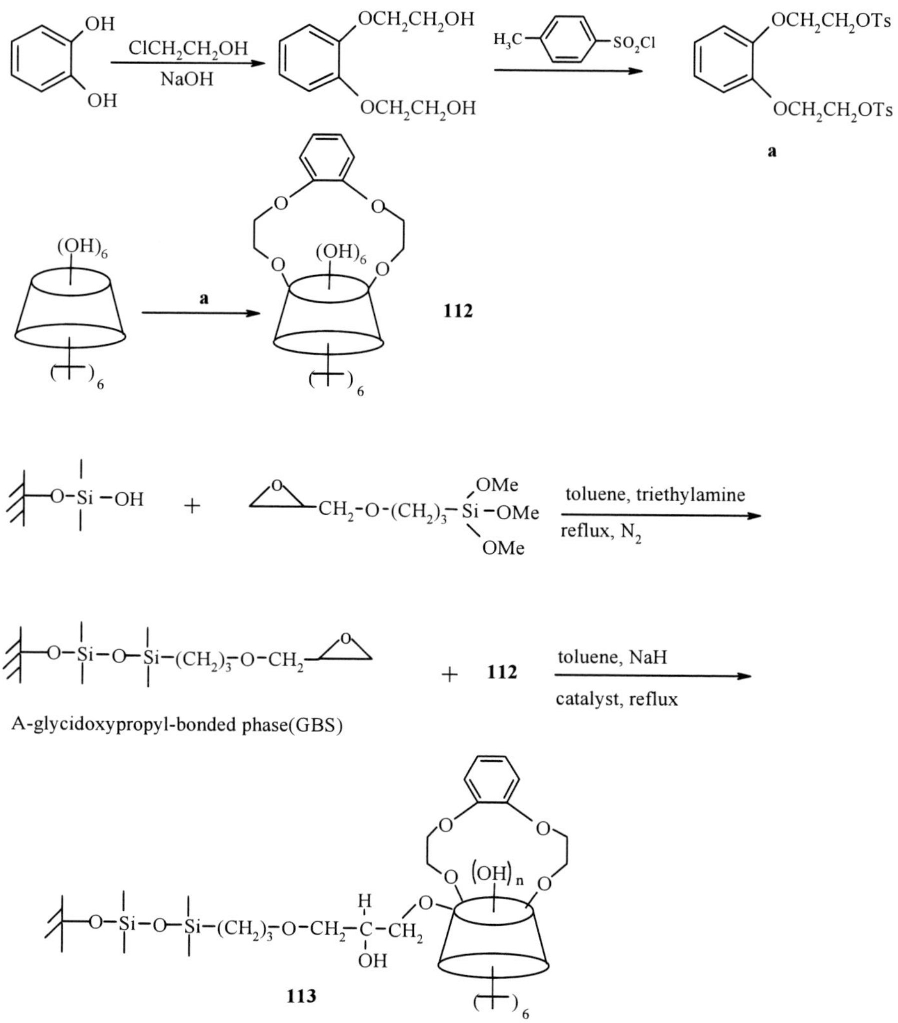

Figure 47. Synthesis of a *p-tert*-butylcalix[6]-1,4-benzocrown-4-bonded silica gel (CR6BS) stationary phase.

For evaluation in high performance liquid chromatography, *p-tert*-butylcalix[6]-1,4-benzocrown-4-bonded silica gel **113** (CR6BS) stationary phase (Figure 47) was prepared by Li *et al.* [27] with 3-glycidoxypropyltrimethoxysilane as the coupling agent The chromatographic performance of the bonded-stationary phase was assessed by use of neutral, acidic, and basic solutes as probes. Based on the chromatographic data, the stationary phase has excellent reserved-phase properties. As a hybrid of calixarene and crown ether, the chromatographic performance of CR6BS is dependent on both the calix[6]arene and the ether bridge moieties. CR6BS exhibited high selectivity in the separation of alkylated aromatics compounds from their parent hydrocarbons.

3.3. Immobilization of Calix[4]Arenes on Silica Surfaces

A monolayer of Na^+-selective fluorescent receptors on glass (**116**) (Figure 48) was prepared [107] by covalent coupling of the bis-isocyanate derivative of the receptor (**115**) to a self-assembled monolayer (SAM) of 3-aminopropyltriethoxysilane (APTES). The optical response of the monolayer to Na^+ was comparable with that of flouroionophore 115 in solution.

The ion receptor site formed by calix[4]arene substituted with four amide moieties and a size and polarity exhibited Na^+ selectivity over the other alkali metal cations [108]. Such ion receptors based on calix[4]arenes have been applied in ion transport [109] and in optical [110] and potentiometric sensors [111].

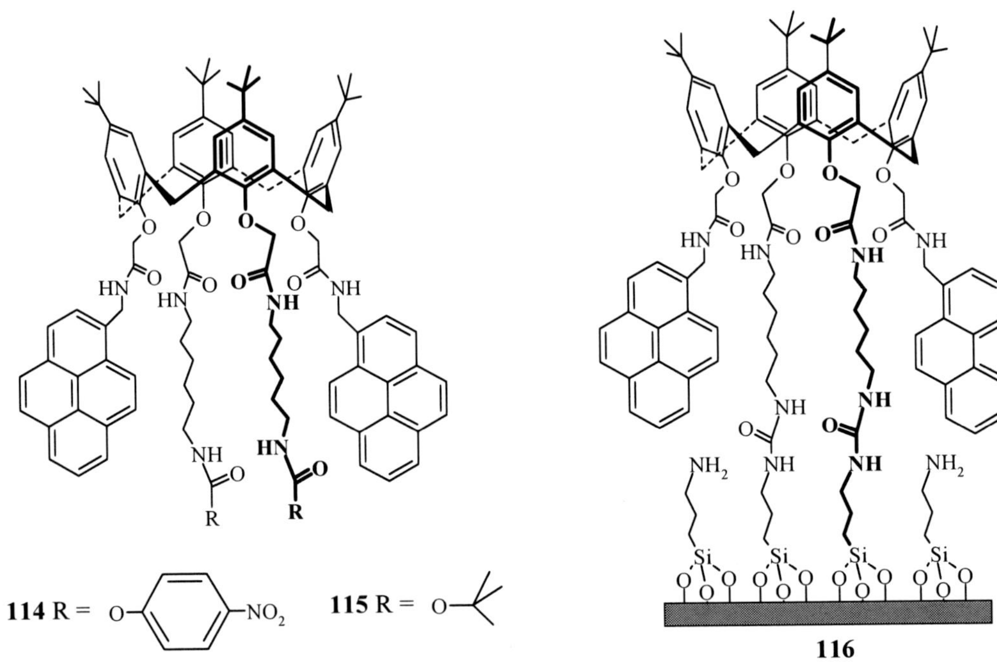

Figure 48. A monolayer of Na^+-selective fluorescent calix[4]arene receptors **116** on glass.

There has been a great deal of interest in the use of calixarenes as designable hosts for the specific adsorption of small molecule guests [10]. The generally poor solubility of calixarenes in protic solvents, such as water and alcohols, has required the use of organic media in studies of adsorption, which results in relatively weak binding for most guests [112]. Calixarene immobilization offers a route to circumvent host solubility limitations. However, this has required rather laborious synthesis of calixarene derivatives that contain reactive functional groups for polymerization on a surface (*e.g.*, groups capable of sol-gel hydrolysis and condensation) [35,113-116].

A novel synthetic method for immobilizing *p-tert*-butylcalixarene on a silica surface (Figure 49) was reported by Katz *et al.* [117]. This method does not require derivatization with flexible ether units. In part because of rigidity the ether linkage between the calixarene and the silica that results from the method, immobilized site densities in excess of 0.2 mmol/g were obtained. To date, this is the highest values reported on a per gram of material basis for calixarene immobilization on silica. Thermal desorption spectroscopy was performed for a sample of **117** onto which phenol was adsorbed from aqueous solution. The data indicate that phenol adsorption onto **117** was reversible.

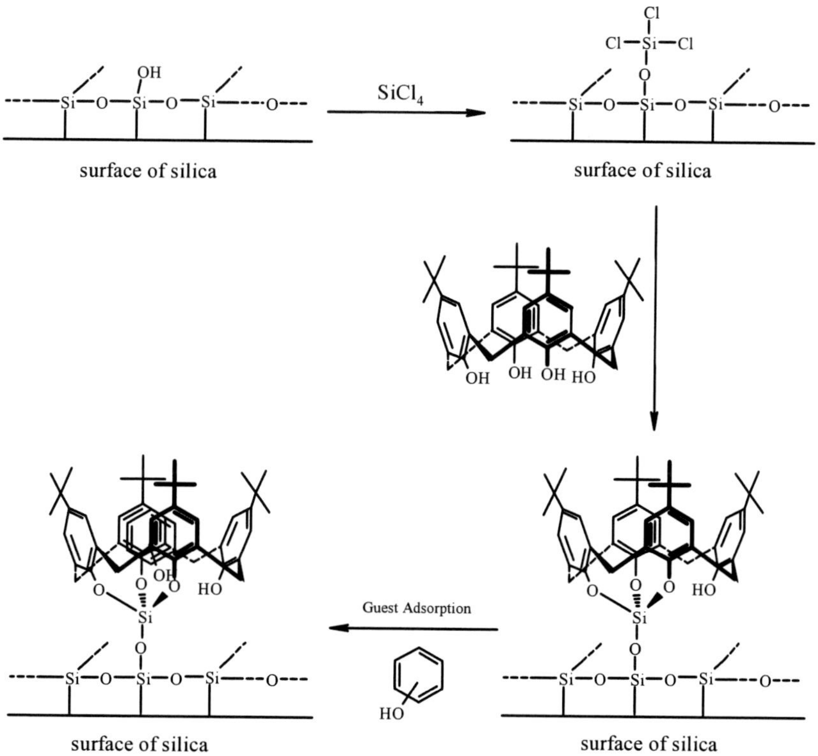

117 **117**

Figure 49. A novel synthetic method for immobilizing a calixarene macrocycle on the surface of silica.

Immobilized calixarenes are expected to find widespread use in fundamental studies, as well as in technological applications, including catalysis [118], adsorption [119], and sensing [120-122].

ACKNOWLEDGEMENT

The authors wish to acknowledge support from the Research Foundation of The Selçuk University (BAP) and the Texas Higher Education Coordinating Board's Technology Development and Transfer Program (RAB).

REFERENCES

[1] Dodziuk, H. Introduction to Supramolecular Chemistry. Dordrecht, The Netherlands: Kluwer Academic Publishers; 2001.

[2] Steed, JW; Atwood, JL. Supramolecular Chemistry. New York: John Wiley & Sons Inc; 2000.

[3] Gutsche, CD. Calixarenes Revisited. Eds.; Cambridge: The Royal Society of Chemistry; 1998.

[4] Asfari, Z; Böhmer, V; Harrowfield, MMcB; Vicens, J. Calixarenes 2001. Dordrecht, The Netherlands: Kluwer; Academic Publishers; 2001.

[5] Yilmaz, M. Solution State Metal Complexes of Calixarenes and Polymeric Calixarenes Handbook of Engineering Polymeric Materials. (Edited by Cheremisinoff, N. P.); New York: Marcel Dekker Inc.; 1997.

[6] Roundhill, DM. Extraction of Metals from Soils and Waters. Dordrecht, The Netherlands: Kluwer Academic Publishers; 2001.

[7] Lhotak, P; Shinkai, S. Calix[n]arenes-Powerful Building-Blocks of Supramolecular Chemistry. *J. Synth. Org. Chem. Jpn.*, 1995, 53, 963-974.

[8] Gutsche, CD; Mukhukrishnan, R. Calixarenes.1. Analysis of the Product Mixtures Produce by the Base-Catalyzed Condensation of Formaldehyde with *p*-Substitued Phenols. *J. Org. Chem.*, 1978, 43, 4905-4906.

[9] McKervey, MA; Schwing-Will, MJ; Arnaud-Neu, F. In Comprehensive Supramolecular Chemistry: (Ed.: G. W. Gokel), Pergamon; 1, 537 (1997).

[10] Böhmer, V. Calixarenes, Macrocycles with (almost) Unlimited Possibilities. *Angew.Chem. Int. Ed. Engl.*, 1995, 34, 713-745.

[11] Ikeda, A; Shinkai, S. Novel Cavity Design Using Calix[n]arene Skeletons toward Molecular Recognition and Metal Binding. *Chem. Rev.*, 1997, 97, 1713-1734.

[12] Grady, T; Cadogan, A; McKittrick, T; Harris, SJ; Diamond, D; McKervey, MA. Sodium-Selective Electrodes Based on Triester Monoacid Derivatives of *p-tert*-Butylcalix-[4]arene. Comparison with Tetraester Calix[4]arene Ionophores. *Anal. Chem. Acta,* 1996, 336, 1-12.

[13] Memon, S; Yilmaz, M. Biscalixarenes: Synthesis and Investigation of the Extraction Behavior of Biscalix[4]arene Derivatives in Two-Phase Extraction System. *Sep. Sci. Technol.*, 2001, 36, 473-486.

[14] Memon, S; Uysal, G; Yilmaz, M. Selective Complexation of Hg^{2+} by Biscalix[4]arene Nitriles. *Sep. Sci. Technol.*, 2000, 35, 1247-1256.

[15] McKittrick, T; Diamond, D; Marrs, DJ; O'Hagan, P; McKervey, MA. Calcium-Selective Electrode Based on A Calix[4]arene Tetraphosphine Oxide. *Talanta*, 1996, 43, 1145-1148.

[16] Erden, S; Demirel, A; Memon, S; Yılmaz, M; Canel, E; Kılıç, E. Using of hydrogen ion-selective poly(vinyl chloride) membrane electrode based on calix[4]arene as thiocyanate ion-selective electrode. *Sensors and Actuators B: Chemical*, 2005, 00, 00.
[17] O'Connor, KM; Svehla, G; Harris, SJ; McKervey, MA. Calixarene-Based Potentiometric Ion-Selective Electrodes for Silver. *Talanta*, 1992, 39, 1549-1554.
[18] Kuruoglu, D; Canel, E; Memon, S; Yilmaz, M; Kilic, E. Hydrogen Ion-selective Poly(vinyl chloride) Membrane Electrode Based on A Calix[4]arene. *Anal. Sci.*, 2003, 19, 217-221.
[19] Demirel, A; Dogan, A; Canel, E; Memon, S; Yilmaz, M; Kiliç, E. Hydrogen Ion-Selective Poly(Vinyl Chloride) Membrane Electrode Based on A *p-tert*-butylcalix[4]arene-Oxacrown-4. *Talanta*, 2004, 62, 123-129.
[20] Knoblauch, S; Falana, OM; Nam, J; Roundhill, DM; Hennig, H; Zeckert, K. Calix[4]arenes with Narrow Rim 2-mercaptoethoxy Substitutents as Potential Precursor Molecules for Metallacages and Sensors. *Inorg. Chim. Acta*, 2000, 300-302, 328-332.
[21] Arrigan, WM; Svehla, G; Harris, SJ; McKervey, MA. Use of Calixarenes as Modifiers of Carbon-Paste Electrodes for Voltammetric Analysis. *Electroanalysis*, 1994, 6, 97-106.
[22] Diamond, D; McKervey, MA. Calixarene-Based Sensing Agents. *Chem. Soc. Rev.*, 1996, 25, 15-24.
[23] Alpoguz, HK; Memon, S; Ersöz, M; Yilmaz, M. Transport of Hg^{2+} through Bulk Liquid Membrane Using Biscalix[4]arene Nitrile Derivative as Carrier: Kinetic Analysis. *New J. Chem.*, 2002, 26, 477-480.
[24] Kim, SI; Shin, TJ; Ree, M; Hwang, GT; Kim, BH; Han, H; Seo, J. Synthesis and Characterization of New Polyimides Containing Calix[4]arenes in the Polymer Backbone. *J. Polym. Sci.: Part A: Polym. Chem.*, 1999, 37, 2013-2026.
[25] Kimura, K; Miura, T; Matsuo, M; Shono, T. Polymeric Membrane Sodium-Selective Electrodes Based on Lipophilic Calix[4]arene Derivatives. *Anal. Chem.*, 1990, 62, 1510-1513.
[26] Memon, S; Uysal, G; Yilmaz, M. Synthesis and Binding Properties of Polymeric Calix[4]crown-4. *React. Funct. Polym.*, 2001, 47, 165-174.
[27] Li, L-S; Da, S-L; Feng, Y-Q; Liu, M. Preparation and Characterization of a *p-tert*-Butylcalix[6]-1,4-benzocrown-4-bonded Silica Gel Stationary Phase for Liquid Chromatography. *J. Chromatogr.*, 2004, A1040, 53-61.
[28] Parzuchowski, P; Malinowska, E; Rokicki, G; Brzozka, Z; Böhmer, V; Arnaud-Neu, F; Souley, V. Calix[4]arene Derived Tetra Ester Receptors Modified at Their Wide-rim by Polymerizable Groups. *New J. Chem.*, 1999, 23, 757-763.
[29] Shinkai, S; Kawaguchi, H; Manabe, O. Selective Adsorbtion of UO_2^{2+} to A Polymer Resin Immobilizing Calixarene-based Uranophiles. *J. Pol. Sci., Part C; Polym. Lett.*, 1988, 26, 391-396.
[30] Glennon, JD; Horne, E; Hall, K; Cocker, G; Kuhn, A; Harris, SJ; McKervey, MA. Silica-Bonded Calixarenes in Chromatography II. Chromatographic Retention of Metal Ions and Amino Acid Ester Hydrochlorides. *J. Chromatogr.*, 1996, 731, 47-55.
[31] Alexandratos, SD; Natesan, S. Synthesis and Ion-Binding Affinities of Calix[4]arenes Immobilized on Cross-Linked Polystyrene. *Macromolecules*, 2001, 34, 206-210.

[32] Memon, S; Uysal, G; Yilmaz, M. Synthesis and Complexation Studies of *p-tert*-Butylcalix[4]crown Telomers. *J. Macromol. Sci.,. Pure Appl. Chem.*, 2001, 38, 173-184.

[33] Engrand, P; Regnouf-de-Vains, J–B. A Bifunctional Calixarene Designed for Immobilization on A Natural Polymer and for Metal Complexation. *Tetrahedron Lett.*, 2002, 43, 8863-8866.

[34] Glennon, JD; Horne, E; O'Connor, K; Kearney, G; Harris, SJ; McKervey, MA. Chromatographic Selectivity of Amino Acid Esters and Alkali Metal Ions on A Silica Bonded Calix[4]arene Tetraester Stationary Phase. *Anal. Proc.*, 1994, 31, 33-35.

[35] Glennon, JD; O'Connor, K; Srijaranal, S; Manley, K; Harris, SJ; McKervey, MA. Enhanced Chromatographic Selectivity for Na^+ Ions on A Calixarene-bonded Silica phase. *Anal. Lett.*, 1993, 26, 153-162.

[36] Deligoz, H. Synthesis of Oligomer and Styrene Polymer-supported Calix[4]arene Derivatives and Selective Extraction of Fe^{3+}. *Supramol. Chem.*, 2003, 15, 317-321

[37] Nakamoto, Y; Ishida, S; Yoshimura, Y. (Hitachi Chemical Co. Ltd.), Jpn. Kokai Tokkyo Koho JP 62,96,505, 1987 (*Chem. Abstr.*, 108:56797f); idem., ibid., 62,96,440, 1987 (*Chem. Abstr.*, 108:6959q).

[38] Blanda, MT; Adou, E. Syntheses and Characterizations of Three Vinyl Copolymers Containing Pendant Calix[4]arenes. *Polymer*, 1998, 39, 3821-3826.

[39] Bakker, E; Telting-Diaz, M. Electrochemical Sensors. *Anal. Chem.*, 2002, 74, 2781-2800.

[40] Harris, SJ; Barrett, G; McKervey, MA. Polymeric Calixarenes. Synthesis, Polymerization and Na^+ Complexation of A Calix[4]arene Methacrylate. *J. Chem. Soc. Chem. Commun.*, 1991, 1224-1225.

[41] Tabakci, M; Memon, S; Sap, B; Roundhill, DM; Yilmaz, M. A Calix[4]arene Derived Dibenzonitrile Receptor Modified at Its 'Lower rim' by A Polymerizable Group. *J. Macromol. Sci. Pure Appl. Chem.*, 2004, 41, 811-825.

[42] Cooper, JB; Drew, MGB; Beer, PD. Alkali Metal Cation Cooperative Anion Recognition by Heteroditopic Bis(calix[4]arene) Rhenium(I) Bipyridyl and Ferrocene Receptor Molecules. *J. Chem. Soc., Dalton Trans.*, 2000, 2721-2728.

[43] Beer, PD; Cooper, JB. Calixarene Based Anion Receptors, *Calixarenes in Action*, pp 111-143, Eds. Mandolini, L; Ungaro, R. London: Imperial College Press, 2000.

[44] Matthews, SE; Beer, PD. Calixarene-based Anion Receptors, *Calixarenes 2001*, pp 421-439 Eds. Z. Asfari, V. Böhmer, J. Harrowfield, J. Vicens, Dordrecht, The Netherlands: Kluwer Academic Publishers, 2001.

[45] Gale, PA. Anion Receptor Chemistry: Highlights from 1999. *Coord. Chem. Rev.*, 2001, 213, 79-128.

[46] Gale, PA. Anion and Ion-pair Receptor Chemistry: Highlights from 2000 and 2001. *Coord. Chem. Rev.*, 2003, 240, 17-55.

[47] Tuntulani, T; Poompradub, S; Thavornyutikarn, P; Jaiboon, N; Ruangpornvisuti, V; Chaichit, N; Asfari, Z; Vicens, J. Aza Crown Ether Calix[4]arenes Containing Cation and Anion Binding Sites: Effects of Metal Ions Towards Anion Binding Ability. *Tetrahedron Lett.*, 2001, 42, 5541-5544.

[48] Aeungmaitrepirom, W; Hagege, A; Asfari, Z; Vicens, J; Leroy, M. Solvent Extraction of Selenate and Chromate Using A Diaminocalix[4]arene. *J. Incl. Phenom. Macrocyclic Chem.*, 2001, 40, 225-229.

[49] Falana, OM; Koch, HF; Roundhill, DM; Lumetta, CJ; Hay, BP. Synthesis and Extraction Studies of 1,2- and 1,3-Disubstituted Butylcalix[4]arene Amides with Oxyions; Geometric and Conformational Effects. *J. Chem. Soc., Chem. Commun.*, 1998, 503-504.

[50] Georgiev, M; Wolf, N; Roundhill, DM. Lower Rim Alkylammonium Substituted Calix[4]arenes as "Proton Switchable" Extractants for Chromate and Dichromate Anions. *Polyhedron*, 1997, 16, 1581-1584.

[51] Wolf, NJ; Georgiev, EM; Yordanov, AT; Whittlesey, BR; Koch, HF; Roundhill, DM. Synthesis and Crystal Structures of Lower Rim Amine and Carbamoyl Substituted Calixarenes as Transfer Agents for Oxyanions between An Aqueous and A Chloroform Phase. *Polyhedron*, 1999, 18, 885-896.

[52] Yordanov, AT; Wolf, NJ; Georgiev, EM; Koch, HF; Falana, OM; Roundhill, DM. Derivatized Calix[4]arenes as Selective Phase Transfer Extractants for Heavy Metal and Oxyion Salts. *Comments Inorg. Chem.* 1999, 20, 163-175.

[53] Memon, S; Yilmaz, M. A Complimentary Study of Calixarene Based Bifunctional Receptors for Alkali or Transition Metal Cations and $Cr_2O_7^{-2}$ Anions. *J. Mol. Struct.*, 2001, 595, 101-109.

[54] Yilmaz, A; Memon, S; Yilmaz, M. Synthesis and Study of Allosteric Effects on Extraction Behavior of Novel Calixarene-Based Dichromate Anion Receptors. *Tetrahedron*, 2002, 58, 7735-7740.

[55] Burrows, D. *Chromium: Metabolism and Toxicity*. Boca Raton, FL: CRC Press, 1983.

[56] De Flora, S; Wetterhahn, KE. Mechanisms of Chromium Metabolism and Genotoxicity. *Life Chem. Rep.*, 1989, 7, 169-244.

[57] Stearns, DM; Kennedy, LJ; Courtney, KD; Giangrande, PH; Phieffer, LS; Wetterhahn, KE. Reduction of Chromium(VI) by Ascorbate Leads to Chromium-DNA Binding and DNA Strand Breaks *in vitro*. *Biochem.*, 1995, 34, 910-919.

[58] Wittbrodt, PR; Palmer, CD. Reduction of Cr(VI) in the Presence of Excessive Soil Fulvic Acid. *Environ. Sci. Technol.* 1995, 29, 255-263.

[59] Roundhill, DM; Koch, HF. Methods and Techniques for the Selective Extraction of Recovery of Oxoanions. *Chem. Soc. Revs* . 2002, 31, 60-67.

[60] Kavallieratos, K; Moyer, BA. Attenuation of Hofmeister Bias in Ion-Pair Extraction by A Disulfonamide Anion Host Used in Strikingly Effective Synergistic Combination with a Calix-Crown Cs^+ Host. *J. Chem. Soc.,. Chem. Commun.*, 2001, 17, 1620-1621.

[61] Perrin, DD. Dissociation Constants of Organic Bases in Aqueous Solution, London: Butterworths, pp 15, 1965.

[62] Uysal, G; Memon, S; Yilmaz, M. Synthesis and Binding Properties of Polymeric Calix[4]arene Nitriles. *React. Funct. Polym.*, 2001, 50, 77-84.

[63] Pearson, RG; Ho, TL. Hard soft acids bases (HSAB) principle and organic chemistry. *Chem. Rev.*, 1975, 75, 1-20.

[64] Zhong, Z-L; Tang, C–P; Wu, C–Y; Chen, Y–Y, Synthesis and Properties of Calixcrown Telomers. *J. Chem. Soc., Chem. Commun.*, 1995, 1737-1738.

[65] Blanda, MT; Adou, E. Syntheses and Characterizations of Two Copolymers Containing Cone Conformations of Calix[4]arenes in the Polymer Backbone. *J. Chem. Soc., Chem. Commun.*, 1998, 139-140.

[66] Dondoni, A; Ghiglione, C; Marra, A; Scoponi, M. Synthesis and Characterization of Bisphenol-A Copolyethers and Copolyesters Carrying Calix[4]arene Units in the Main

Chains and their Binding Properties Towards Silver Cations. *Macromol. Chem. Phys.*, 1999, 200, 77-86.

[67] Memon, S; Oguz, O; Yilmaz, A; Tabakci, M; Yilmaz, M; Ertul, S. Synthesis and Extraction Study of Calix[4]arene Dinitrile Derivatives Incorporated in A Polymeric Backbone with Bisphenol-A. *J. Polym. Environ.*, 2002, 9, 97-101.

[68] Memon, S, Yilmaz, A, Roundhill, DM; Yilmaz, M. Synthesis of Polymeric Calix(4)arene Dinitrile and Diamino-derivatives: Exploration of Their Extraction Properties Towards Dichromate Anion. *J. Macromol. Sci., Pure and Appl. Chem.*, 2004, 41, 433-447.

[69] Memon, S; Roundhill, DM; Yilmaz, M. Remediation and Liquid-Liquid Phase Transfer Extraction of Chromium(VI). *Collect. Czech. Chem. Commun.*, 2004, 69, 1231-1250.

[70] Akkus, GU; Memon, S; Sezgin, M; Yilmaz, M. A Versatile Approach Toward Calix(Aza)-Crown Oligomers: Synthesis and Metal Ion Extraction. *J. Macromol. Sci., Pure Appl. Chem.*, 2003, 40, 95-106.

[71] Akkus, GU; Memon, S; Yilmaz, M. Synthesis of Oligomeric Calix[4]arene-crowns as Novel Ionophores for Alkali and Transition Metals. *Polycycl. Arom. Comp.*, 2002, 22, 1075-1086.

[72] Vicens, J; Böhmer, V. Calixarenes: A Versatile Class of Macrocyclic Compounds. Eds., Dordrecht, The Netherlands: Kluwer Academic Publishers, 1991.

[73] Scheerder, J; Vreekamp, R.H; Engbersen JFJ; Verbom, V; van Duynhoven, JPM; Reinhoudt, DM. The Pinched Cone Conformation of Calix[4]arenes: Noncovalent Rigidification of the Calix[4]arene Skeleton. *J. Org. Chem.*, 1996, 61, 3476-3481.

[74] Yu, H-h; Xu, B; Swager, TM. A Proton-Doped Calix[4]arene-Based Conducting Polymer. *J. Am. Chem. Soc.*, 2003, 125, 1142-1143.

[75] Hong, Y, Miller, LL. An Electrically Conducting Polyester that Has Isolated Quatrathiophene Units in the Main-Chain. *Chem. Mater.*, 1995, 7, 1999-2000.

[76] McQuade, DT, Pullen, AE, Swager, TM. Conjugated Polymer-Based Chemical Sensors. *Chem. Rev.*, 2000, 100, 2537-2574.

[77] Osada, Y; de Rossi, DE. Polymer Sensors and Actuators, Eds., Berlin, Germany: Springer; 2000.

[78] Wieser, C; Dieleman, CB; Matt, D. Calixarene and Resorcinarene Ligands in Transition Metal Chemistry Coord. *Chem. Rev.*, 1997, 165, 93-161.

[79] Yilmaz, A; Memon, S; Yilmaz, M. Synthesis and Binding Properties of Calix[4]arene Telomers. *J. Polym. Sci.: Part A: Polym. Chem.*, 1999, 37, 4351-4355.

[80] Memon, S; Yilmaz, A; Yilmaz, M. Synthesis and Comparative Complexation Studies of Schiff Base Derivatives of *p-tert*-Butylcalix[4]arene Copolymers. *J. Macromol. Sci., Pure Appl. Chem.*, 2000, 37, 865-879.

[81] Memon, S; Yilmaz, M. Synthesis and Complexation Studies of 1,3-Dialkylated *p-tert*-Butylcalix[4]arene Telomers. *React. Funct. Polym.*, 2000, 44, 227-233.

[82] Dondoni, A; Marra, A; Rossi, M; Scoponi, M. Synthesis and Characterization of Calix[4]arene-based Copolyethers and Polyurethanes. Ionophoric Properties and Extraction Abilities Towards Metal Cations of Polymeric Calix[4]arene Urethanes. *Polymer*, 2004, 45, 6195-6206.

[83] Asfari, Z; Lamare, V; Dozol, J-F; Vicens, J. A Tribenzo Modified 1,3- Calix[4]-bis-crown-6, A Highly Selective Receptor for Cesium. *Tetrahedron Lett.*, 1999, 40, 691.

[84] Dozol, J-F; Simon, N; Lamare, V; Rouquette, H; Eymard, S; Tournois, B; De Marc, D; Macias, RM. A Solution for Cesium Removal from High-Salinity Acidic or Alkaline Liquid Waste: The Crown Calix[4]arenes. *Sep. Sci. Technol.*, 1999, 34, 877-909.

[85] Guillon, J; Sonnet, P; Malval, J-P; Massip, S; Gosse, I; Lager, J-P; Lapouyade, R; Rochette, J; Monti, J-P; Jarry C. Synthesis and Cesium Binding Affinity of New 25,27-Bis(Alkyloxy)calix[4]arene-crown-6 Confor-mers in Relation to the Alkyl Pendent Moiety. *Supramolecular Chem.*, 2002, 14, 437-450.

[86] Tabakci, M; Memon, S; Yilmaz, M; Roundhill, DM. Synthesis and Evaluation of Extraction Ability of Calix4-crown-6 Cone Conformer and Its Oligomeric Analogue. *React. Funct. Polym.*, 2004, 58, 27-34.

[87] Barata, PD; Costa, AI; Granja, P; Prata, JV. The Synthesis of Novel Polymer-bound Calix[4]arenes. *React. Funct. Polym.*, 2004, 61, 147-151.

[88] Shinkai, S; Manabe, O; Kondo, Y; Yamamoto, T. (Kanebo Ltd.), Jpn. Kokai Tokkyo Koho, JP 62,136,242, 1987 (*Chem. Abstr.*, 108:64410q).

[89] Deligöz, H.; Yilmaz, M. Selective Extraction of Fe^{3+} by Polymeric Calix[4]arene Tetraesters. *J. Polym. Sci;. Part A: Polym. Chem.*, 1994, 32, 2961-2964.

[90] Deligöz, H.; Yilmaz, M. Synthesis of Polymer-supported Calix[4]arenes and Selective Extraction of Fe^{3+}. *React. Funct. Polym.*, 1996, 31, 81-88.

[91] Yilmaz, M. Synthesis of an Oligomer-supported Calix[4]arene and Selective Extraction of Li^{+}. *React. Funct. Polym.*, 1999, 40, 129-133.

[92] Deligöz, H; Yilmaz, M. Selective Complexation of Na^{+} by Polymeric Calix[4]arene Tetraesters. *J. Polym. Sci;. Part A: Polym. Chem.,* 1995, 33, 2851-2853.

[93] Jain, VK; Handa, A; Pandya, R; Shrivastav, P; Agrawal, YK. Polymer Supported Calix[4]arene-semicarbazone Derivative for Separation and Preconcentration of La(III), Ce(III), Th(IV) and U(VI). *React. Funct. Polym.*, 2002, 51, 101-110.

[94] Memon, S; Akceylan, E; Sap, B; Tabakci, M; Roundhill, DM; Yilmaz, M. Polymer Supported Calix[4]arene Derivatives for the Extraction of Metals and Dichromate Anions. *J. Polym. Environ.*, 2003, 11, 67-74.

[95] Mehvar, R. High-Performance Size-Exclusion Chromatographic Analysis of Dextran-Methylprednisolone Hemisuccinate in Rat Plasma. *J. Chromatogr. B*, 2000, 744, 293-298.

[96] Rongved, P; Fritzell, Th; Strande, P; Klaveness, J. Polysaccharides as Carriers for Magnetic Resonance Imaging Contrast Agents: Synthesis and Stability of a New Amino Acid Linker Derivative. *Carbohydrate Res.*, 1996, 287, 77-89.

[97] Engrand, P; Regnouf-De-Vains Jb. A Bifunctional Calixarene Designed for Immobilisation on a Natural Polymer and for Metal Complexation. *Tetrahedron Lett.* 2002, 43, 8863-8866.

[98] Majeti, NV; Ravi, K. A Review Of Chitin And Chitosan Applications. *React. Funct. Polym.*, 2000, 46, 1-27.

[99] Inoue, K; Yoshizuka, K; Ohto, K. Adsorptive Separation of Some Metal Ions by Complexing Agent Types of Chemically Modified Chitosan. *Anal. Chim. Acta*, 1999, 388, 209-218.

[100] Aly, AS; Jeon, BD; Park, YH. Preparation and Evaluation of the Chitin Derivatives for Wastewater Treatments. *J. Appl. Polym. Sci.*, 1997, 65, 1939-1946.

[101] Peng, CH; Wang, YT; Tang, YR. Synthesis of Crosslinked Chitosan Crown Ethers and Evaluation of These Products as Adsorbents for Metal Ions. *J. Appl. Polym. Sci.*, 1998, 70, 501-506.

[102] Li, H-B; Chen, Y-Y; Liu, S-L. Synthesis, Characterization, and Metal Ions Adsorption Properties of Chitosan-Calixarenes (I). *J. Appl. Polym. Sci.*, 2003, 89, 1139-1144.

[103] Ohto, K; Tanaka, Y; Inoue, K. Adsorptive Separation of Lead and Zinc Ions by Novel Type of Calix[4]arene Carboxylate Resin Immobilized with Polyallylamine. *Chem. Lett.*, 1997, 7, 647-648.

[104] Ohto, K.; Tanaka, Y.; Yano, M.; Shinohara, T.; Murakami, E.; Inoue, K. Selective Adsorption of Lead Ion on Calix[4]arene Carboxylate Resin Supported by Polyallylamine. *Solv. Ext. Ion Exch.*, 2001, 19, 725-741.

[105] Hutchinson, S; Kearney, GA; Horne, E. Solid-Phase Extraction of Metal-ions Using Immobilized Chelating Calixarene Tetrahydroxamates. *Anal. Chim. Acta.*, 1994, 291, 269-275.

[106] Xiao, Y-X; Xiao, X-Z; Feng, Y-Q; Wang, Z-H; Da, S-L. HPLC of Some Nucleosides and Bases on *p-tert*-Butylcalix[6]arene Bonded Silica Gel Stationary Phase. *J. Liq. Chrom. Rel. Technol.*, 2001, 24, 2925-2942.

[107] van der Veen, NJ; Flink, S; Deij, MA; Egberink, RJM; van Veggel, FCJM; Reinhoudt, DN. Monolayer of A Na^+-Selective Flouroionophore on Glass: Connecting the Fields of Monolayers and Optical Detection of Metal Ions. *J. Am. Chem. Soc.*, 2000, 122, 6112-6113.

[108] Brunink, JAJ; Bomer, JG; Verboom, W; Engbersen, JFJ; Reinhoudt, DN. Effects of Anionic Sites on the Selectivity of Sodium-Sensitive Chemfets. *Sens. Actuators B*, 1993, 15, 195-198.

[109] Rudkevich, DM; Mercer-Chalmers, JD; Verboom, W; Ungaro, R; de Jong, F; Reinhoudt, DN. Bifunctional Recognition- Simultaneous Transport of Cations and Anions Through A Supported Liquid Membrane. *J. Am. Chem. Soc.*, 1995, 117, 6124-6125.

[110] van der Veen, NJ; Egberink, RJM; Engbersen, JFJ; van Veggel, FJCM; Reinhoudt, DN. Conformationally Flexible Calix[4]arene Chromoionophores: Optical Transduction of Soft Metal Ion Complexation by Cation-pi Interactions. *Chem. Commun.*, 1999, 681-682.

[111] Cobben, PLHM; Egberink, RJM; Bomer, JG; Bergveld, P; Verboom, W; Reinhoudt, DN. Transduction of Selective Recognition of Heavy-Metal Ions by Chemically Modified Field-Effect Transistors (Chemfets). *J. Am. Chem. Soc.*, 1992, 114, 10573-10582.

[112] Bauer, LJ; Gutsche, CD. Calixarenes. 15. The Formation of Complexes of Calixarenes with Neutral Organic-Molecules In Solution. *J. Am. Chem. Soc.*, 1985, 107, 6063-6069.

[113] Healy, LO; McEnery, MM; McCharty, DG; Harris, SJ; Glennon, JD. Silica-Bonded Calixarenes in Chromatography: Enantioseparations on Molecular Basket Phases for Rapid Chiral LC. *Anal. Lett.*, 1998, 31, 1543-1551.

[114] Friebe, S; Gebauer, S; Krauss, GJ; Geormar, G; Krueger, J. HPLC on Calixarene Bonded Silica-Gels .1. Characterization and Applications of the *p-tert*-Butyl-Calix[4]arene Bonded Material. *J. Chromatogr. Sci.*, 1995, 33, 281-284.

[115] Brindle, R; Albert, K; Harris, SJ; Tröltzsch, C; Horne, E; Glennon, JD. Silica-Bonded Calixarenes in Chromatography.1. Synthesis and Characterization by Solid-State NMR Spectroscopy. *J. Chromatogr. A*, 1996, 731, 41-46.

[116] Nechifor, AM; Philipse, AP; de Jong, F; van Duynhoven, JPM; Egberink, RJM; Reinhoudt, DN. Preparation and Properties of Organic Dispersions of Monodisperse Silica Receptor Colloids Grafted with Calixarene Derivatives or Alkyl Chains. *Langmuir*, 1996, 12, 3844-3854.

[117] Katz, A; Costa, PD; Lam, ACP; Notestein, JM. The First Single-Step Immobilization of Calix[4]arene onto the Surface of Silica. *Chem. Mater.*, 2002, 14, 3364-3368.

[118] Struck, O; van Duynhoven, JPM; Verboom W; Harkema, S; Reinhoudt, DN. Cavity Effect of Calix[4]arenes in Electrophilic Aromatic Substitution Reactions. *Chem. Commun.* 1996, 1517-1518.

[119] Ludwig, R. Calixarenes in Analytical and Separation Chemistry. *Fresenius' J. Anal. Chem.*, 2000, 367, 103-128.

[120] Nabok, AV; Hassan, AK; Ray, AK; Omar, O; Kalchenko, VI. Study of Adsorption of Some Organic Molecules in Calix[4]resorcinolarene LB Films by Surface Plasmon Resonance. *Sens. Actuators B*, 1997, 45, 115-121.

[121] Yang, X; Johnson, S; Shi, J; Holesinger, T; Swanson, B. Polyelectrolyte and Molecular Host Ion Self-Assembly to Multilayer Thin Films: An Approach to Thin Film Chemical Sensors. *Sens. Actuators B*, 1997, 45, 87-92.

[122] Hassan, AK; Ray, AK; Nabok, AV; Davis, F. Spun Films of Novel Calix[4]resorcinarene Derivatives for Benzene Vapour Sensing. *Sens. Actuators B*, 2001, 77, 638-641.

In: New Frontiers in Polymer Research
Editor: Robert K. Bregg, pp. 173-199
ISBN: 1-59454-956-7

Chapter 7

SILICA – ORGANIC POLYMER NANOCOMPOSITES: H-BONDED INTERPOLYMER COMPLEXES AND GRAFTED COPOLYMERS

Vadim V. Annenkov, Elena N. Danilovtseva and Elena A. Filina
Limnological Institute, Siberian Branch of Russian Academy of Science 3,
Ulan-Batorskaya Str., Irkutsk, 664033, Russia, annenkov@lin.irk.ru

ABSTRACT

A set of new organo-inorganic composites stabilized by hydrogen bonds have been obtained by the reaction of silicic acid and tetraethyl orthosilicate with organic polymers: poly(1-vinylpyrrolidone-2), poly(acrylamide), poly(ethylene glycol) and poly(4-vinylpyridine). Considerable influence of the structure of primary silica particles on the composition and morphology of the resulting products have been found: stoichiometry from 1: 2 to 2: 1 and morphology from friable substance of 65±16 nm particles to spherical particles 2.9±1.1 μm in diameter can be obtained in aqueous medium with the same polymer [poly(1-vinylimidazole)]. The use of poly(vinyl butyral) and oxirane-containing polymers as a matrixes in hydrolysis of tetraethyl orthosilicate gives rise to graft copolymer with side polysiloxane chains. Membranes, highly functionalized surfaces for DNA and immuno-analysis, sorbents of heavy metals have been designed on the bases of new nanocomposites.

INTRODUCTION

Silica-based organo/inorganic nanocomposites are widely used in industry, medicine and science as construction materials, sorbents, catalysts, etc. [1-5]. There are several fundamental routes to these materials:

- formation of silica phase from precursors (alkoxysilanes, sodium silicate) in the presence of polymer bearing active groups which can react with inorganic moieties by hydrogen, coordinating or covalent bonds;

- reaction between superfine silica and organic polymer;
- graft polymerization on the specially prepared inorganic phase.

Several years ago interest in the interactions between organic polymers and poly(silicic acid) (PSA) or its precursors had been increased due to study of silicifying bio-organisms such as diatoms, sponges and some higher plants. The most attention is paid now to diatoms because these small unicellular organisms (2-100 μm) give approximately 40% of oceanic primary productivity of organic carbon from CO_2 which is comparable with tropic forests [6]. Diatom algae have unique cell walls constructed from PSA and organic polymers of not completely studied structure. The cell walls consist of highly-ordered micro- and nano-structures (Figure 1, pictures from [7-9]) which are interesting as a renewable and a cheap source of nano-silica. Understanding the nature of the bio-silica structures can open new biomimetic routes to composite nanomaterials. Recently Kröger et al. [10-12] have isolated a set of polypeptides (silaffins) which contain side polyamino chains. Mixing of the silaffins or corresponding polyamines with solution of silicic acid results in spherical submicron particles and other well-ordered patterns. On the other hand, the similar structures were found in the reaction between silicic acid and various bio-polymers: poly-L-histidine [13], poly-L-lysine [14] and other peptides [15-18], so the nature of diatom cell walls is not completely clarified yet.

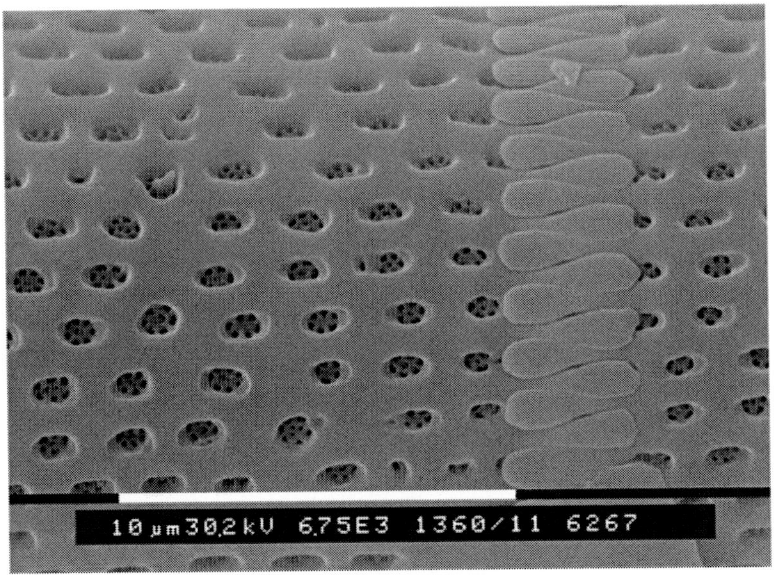

Figure 1. Surface of cell wall of Aulacoseira Baicalensis. Scale bar 10 μm.

H_4SiO_4 is a weak acid (pK_a = 9.18) and it is not ionized at pH < 7 but particles of PSA have a negative charge at neutral pH due to dissociation of some surface silanol moieties [19]. As expected, PSA is active in complexing with basic polymers by hydrogen bonds [13-18, 20-26] and with polycations by ionic bonds [27-33]. These reactions are similar with well-known interpolymer complexes between linear macromolecules. Theory of the interpolymer complexes had been elaborated several decades ago [34-39] and the most important conclusions are the following:

- steric complimentarity of the interacting polymers is important but not absolutely necessary for complex formation;
- total stability of the complex increases exponentially with growing of the length of the interacting segments ("zipping effect"), really there are some critical length above which the reaction becomes irreversible;
- stability of the complex (critical length) depends on the energy of interaction between functional groups, solvation and desolvation energies, hydrophobic interactions and supramolecular effects;
- complexing between hydrophilic functional groups results often in water-insoluble products with specific spherical or fractal morphology.

In the case of PSA or its precursors the situation is more complicated because condensation of silicic acid is a three-dimensional process and the resulting particles are not "dead" macromolecules, they can continue condensation in solution and in the complex with organic polymer. So, synthesis of organo/inorganic nanocomposites from silica precursors include several competitive reactions: condensation of free monomeric units, interaction of the monomers or oligomers with organic polymer and further condensation on the matrix, complexing between relatively large silica particles and organic polymer resulting in stable products. Certainly, chemical structure of the organic macromolecules has an important influence on the structure and morphology of composites but the nature of silica precursors and the condensation conditions can substantially change results of the reaction.

First part of this work is devoted to synthesis of silica nanocomposites in water solutions using poly(1-vinylimidazole) (PVI, molecular mass 20,000) as organic matrix. 1-Vinylimidazole is a commercially available monomer, its polymers were investigated beginning from the sixties, but a number of articles dealing with their basic and complexing properties is not large [40-59]. The basicity of PVI ($pK_{BH^+} = 5.6$) is intermediate between that of stronger bases such as amines and very weak amides and ethers [58]. Polymers of such basicity was not used for synthesis of nanocomposites in aqueous medium yet, so we expected to obtain new data concerning with mechanism of these reaction and to synthesize new materials. Poly(1-vinylpyrrolidone-2) (PVPd, molecular mass 11,000), poly(acrylamide) (PAAm, molecular mass 120,000) and poly(ethylene glycol) (PEG, molecular mass 40,000) were used also in comparison with PVI in the reaction with PSA.

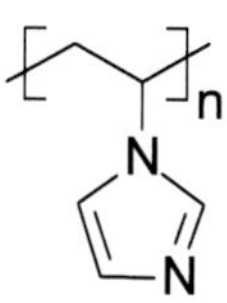

PVI

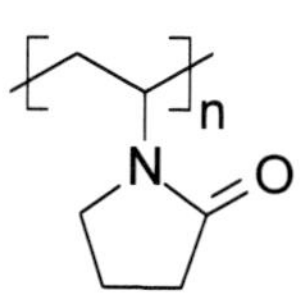

PVPd

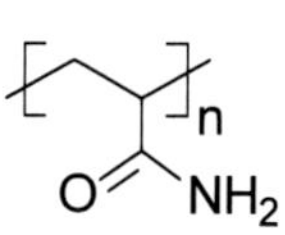

PAAm

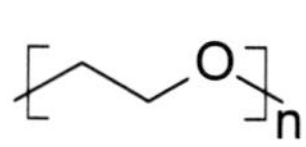

PEG

In the second part of the article we describe a reaction between partially hydrolyzed tetraethyl orthosilicate (TEOS) and such polymers as PVI, PVPd, PAAm and poly(4-vinylpyridine) (PVP, molecular mass 90,000) resulting in H-bonded composites. The use of poly(vinyl butyral) (PVB, molecular mass 70,000-100,000) in this reaction gives rise to graft copolymer with side polysiloxane chains. Some application areas of new materials are discussed in the final part of the work.

RESULTS AND DISCUSSION

Formation of the Composites in Aqueous Media

In the most of experiments devoted to synthesis of silica/organic polymer composites in water a preliminary prepared silicic acid had been mixed with water-soluble polymer [20-22, 27-32]. The solution of silicic acid can be prepared by neutralization of sodium silicate with strong acid or cation exchanger, and organic precursors such as tetramethyl orthosilicate or TEOS were used too. However, it is known [19] that H_4SiO_4 can readily condense into dimers, trimers, tetramers, etc. So, it is very difficult to have a solution of pure silicic acid and as usual we work with a mixture of silica species.

The other way to silica/polymer composites is neutralization of sodium silicate in the mixture with polymer. Here the generated H_4SiO_4 molecules are able to react with polymer right away. Potentiometry titration curves for the solution of Na_2SiO_3 – polymer mixture are similar to pure sodium silicate in the case of very weak polybases such as PVPd, PAAm and PEG. The system with PVI shows (Figure 2) an expected increase of pH after Na_2SiO_3 neutralization because PVI can be protonated in this pH area. The decrease of pH in the basic area in the presence of PVI is similar to the effects found for some polymer-polymer systems with poly(acrylic acid) (PAA) [59, 60]. They had been explained by involving of anions into interaction with basic polymer and by analogy with PVI – polyacid complexes [59, 61] the following triple complex is possible for the first stage of condensation of silicic acid on the PVI matrix:

The addition of HCl to Na_2SiO_3 – polymer solutions gives rise to precipitates for PVI, PVPd and PEG and in the case of PAA or without polymer gelation of the solutions occurs only. Some results on the synthesis of silica/polymer composites are summarized in Table 1. The products with PVI have close to equimolar Si : polymer units ratio near the point of full neutralization of sodium silicate (pH = 5.90-6.65) and the increase of χ up to 2.2 (pH = 4.18) results in depletion of the composite with PVI. The products in the systems with PEG and PVPd have approximately double excess of silica under organic polymer. It seems that in the case of PVI near neutral pH we have preferably condensation of silicic acid on the polymer matrix and the insoluble composite is obtained by further transformations of this 1 : 1 complex. An excess of PVI in the initial mixture results in relatively uniform distribution of silicic acid between PVI chains and the composite has the corresponding composition. The using of weaker polymeric bases (PVPd, PEG) or partial protonation of PVI at low pH decrease the ability of polymer to form 1 : 1 complex, competitive condensation of free silicic acid plays a significant role and the products are enriched with silica.

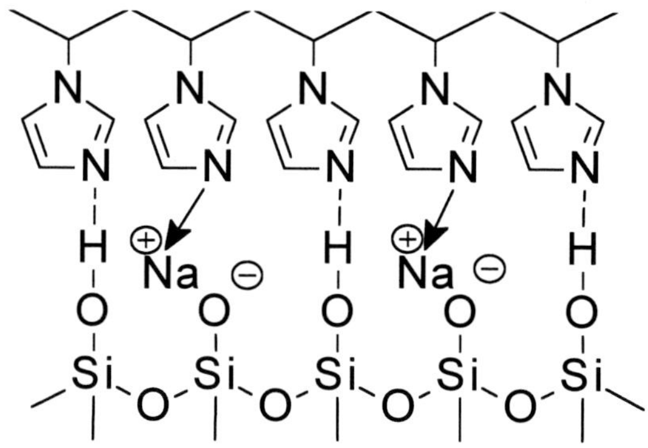

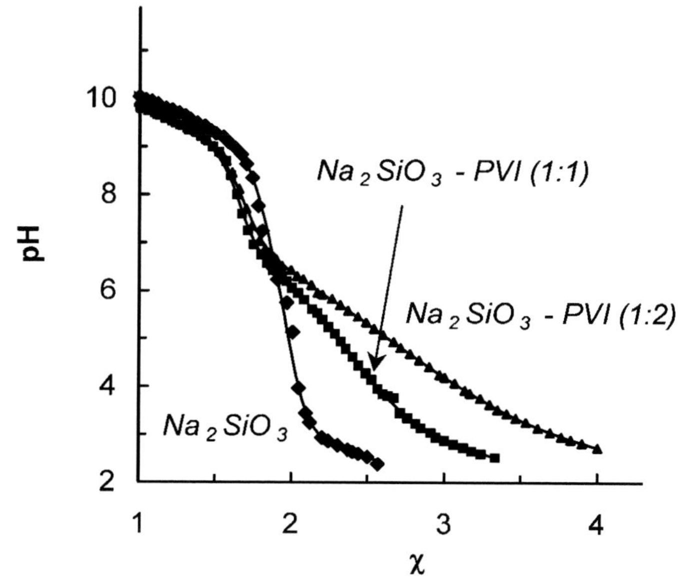

Figure 2. Potentiometry titration data for sodium silicate and its mixtures with PVI. $[Na_2SiO_3] = 0.01$ M, $[HCl] = 0.1$ M.

Table 1. Synthesis of the composites from sodium silicate and water-soluble polymers ($[Na_2SiO_3]$ = [polymer] = 0.05 M, precipitation with 1 M HCl, $\chi = [HCl]/[Na_2SiO_3]$)

Polymer	χ	pH	Precipitate composition, %			Yield, %		Si : polymer units in composite
			N	Si	C	Calculated on polymer	Calculated on Si	
PVI	1.96	6.65	13.4	18.3	29.7	69.1	94.0	1.36
	2.06	6.60	13.8	17.3	29.2	53.8	67.3	1.25
	2.10	5.90	15.0	15.7	32.7	82.8	86.1	1.04
	2.20	4.18	12.1	20.6	24.6	30.6	52.1	1.70
	1.96*	6.77	17.9	11.0	37.8	78.1	95.3	0.61
PEG-40000	1.96	5.72	-	30.2	15.6	50.7	84.2	1.66
PVPd	1.97	5.82	5.4	22.9	25.8	32.9	69.8	2.12

*double excess of PVI

The bonding between silica and organic polymers in the obtained composites is realized by hydrogen bonds which is evident from FTIR spectra (Figure 3): the bands of the polymers in composites are slightly shifted relatively free substances. The band of Si-OH (960 cm^{-1}, [33, 62]) is more intensive in the presence of PVI or PVPd which is also connected with hydrogen bonding in the composites.

The formation of inorganic phase in the presence of organic polymer usual gives rise to submicron particles containing inorganic core coated with the polymer [63-65]. Size of these particles depends on the strength of organic-inorganic interactions: after some radius of the growing inorganic particle the interaction with polymer becomes irreversible and further growth becomes impossible. In the case of silica composites typical size is < 100 nm [20], which corresponds to the size of building blocks of the bio-silica in diatom algae [66]. SEM data for PVI-silica composite (Figure 4) point to formation of large spherical particles 2.9±1.1 μm in diameter.

Alternatively, reaction of PVI with H_4SiO_4 from sodium silicate and cation exchanger results in friable substance which consists of 65±16 nm particles (Figure 5). Thus, the following two routes towards silica/polymer composites can be proposed depending on polymer nature and reaction conditions:

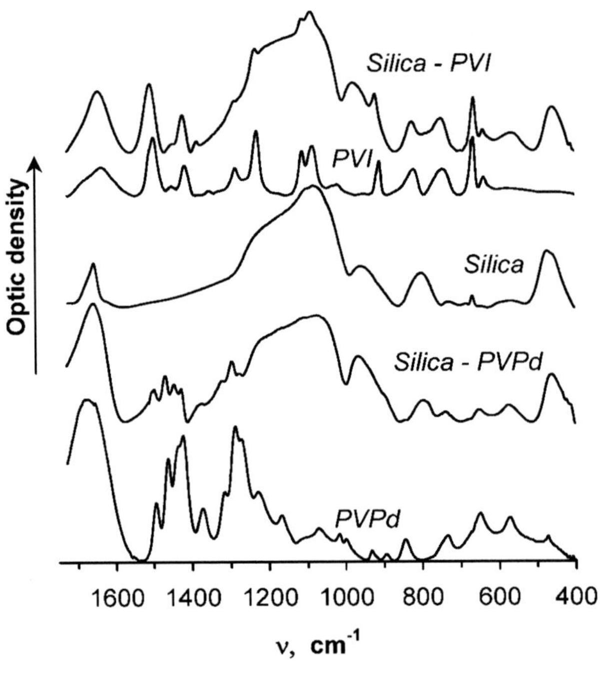

Figure 3. FTIR spectra of organo/inorganic composites, films on KBr substrate.

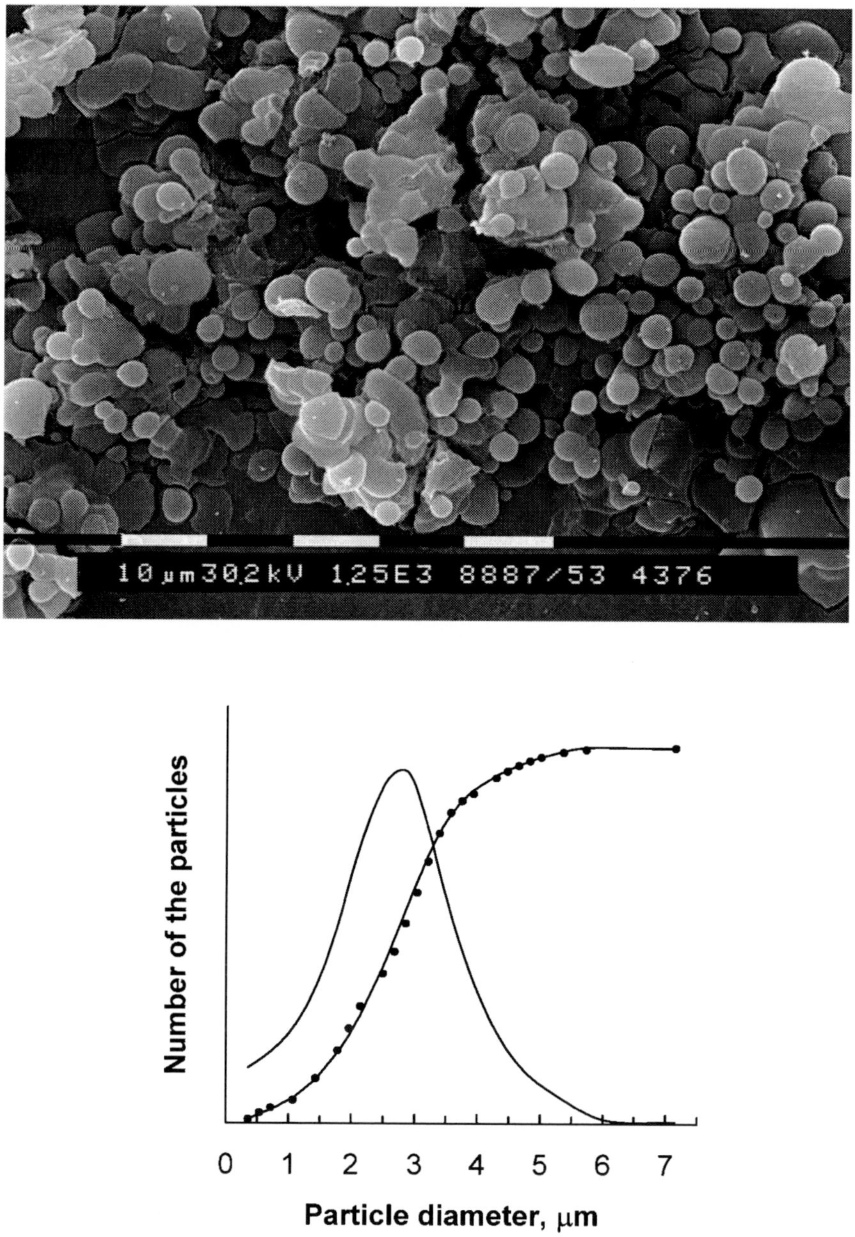

Figure 4. SEM data for PVI-silica composite obtained by neutralization of Na_2SiO_3 – PVI mixture with HCl. Scale bar 10 µm.

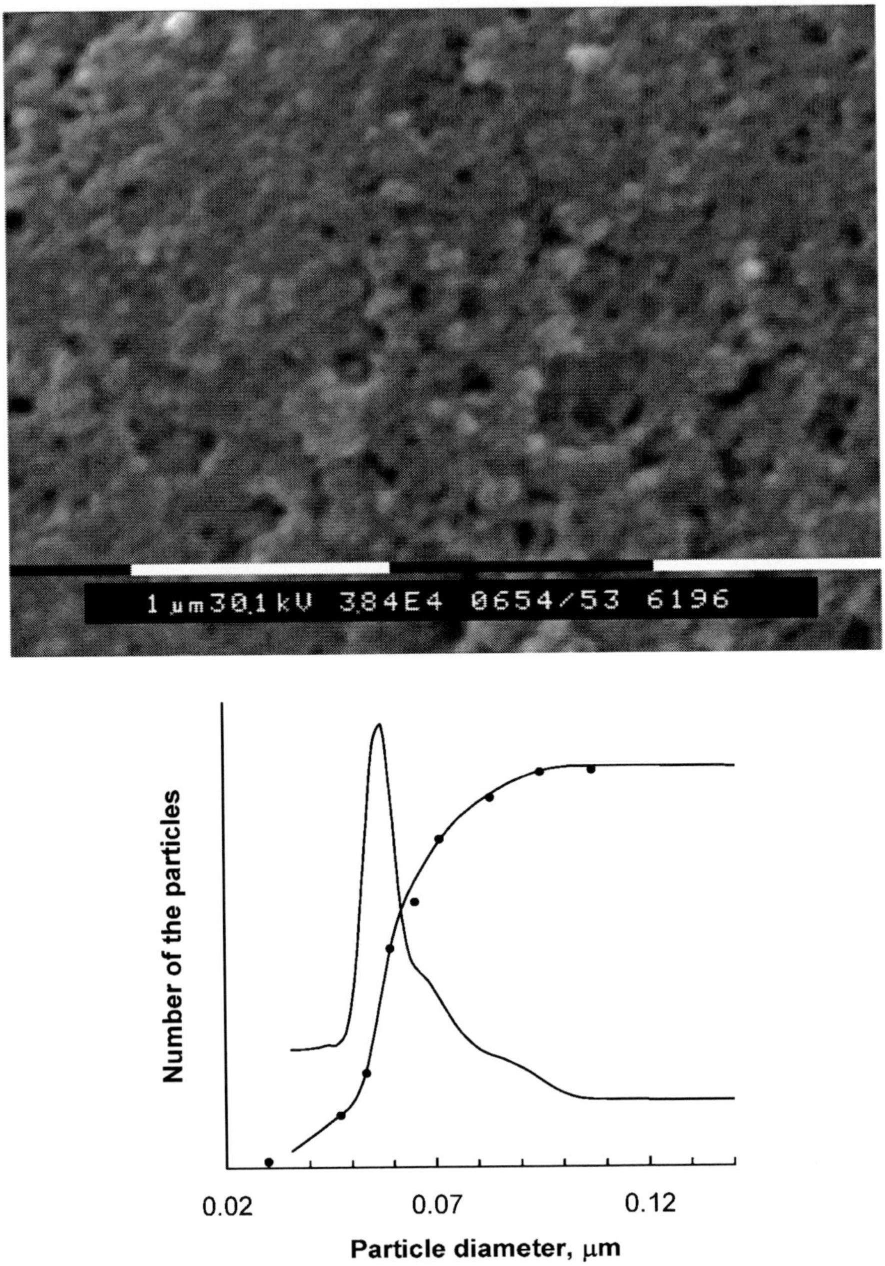

Figure 5. SEM data for PVI-silica composite obtained by the reaction of PVI with H_4SiO_4 from sodium silicate and cation exchanger. Scale bar 1 µm.

A. Condensation of silicic acid prevails over its interactions with organic polymer or we use preliminarily prepared (and already partially condensed) silicic acid. Organic polymer reacts with growing silica particles and accelerate their growth and precipitation by compensation of negative charge on the silica surface. This scheme is typical for very weak polybases, such as PEG [20] which can form hydrogen bonds with silica decreasing acidity of silanol moieties. On the contrary, polymeric amines are significantly protonated at pH below 7 and so they are able to form ionic bonds with ionized silica only. The examples of such systems are peptides [13-18] and polymeric amines from diatoms [10-12]. In any case, the process finishes with formation of stable submicron particles. According to [42] average size of PVI macromolecules from this work is 10 nm, so only several PVI chains can be present in 65 nm particles from Figure 5.

B. Condensation of silicic acid proceeds as matrix reaction along polymeric chain. This is possible if silicic acid or its short oligomers are able to react with the matrix. PVI is the appropriate polymer here because of the intermediate level of basicity: it is slightly protonated (< 10%) at pH > 6 but its basicity is high enough to give strong hydrogen bonds. Under the neutralization of Na_2SiO_3-PVI mixture with HCl the interaction of PVI with silicon species begins at high pH (Figure 1) and near $\chi = 2$ we must have some kind of stoichiometric complex PVI – PSA. So, morphology of the resulted insoluble particles depends on aggregation of this interpolymer complex. The intermediate situation occurs when Na_2SiO_3-PVI mixture is neutralized with cation exchanger: 100-200 nm particles of irregular shape were obtained (Figure 6), they contain 1: 2 PVI: Si units. Evidently, the lack of Na^+ ions renders impossible formation of the mentioned above PSA-Na^+-PVI complexes at pH > 7. So, condensation of H_4SiO_4 units and interaction with PVI proceed simultaneously giving rise to irregular, intergrown particles.

We also have used the ability of PVI to control silica condensation in the reaction of cross-linked PVI with H_4SiO_4 from sodium silicate and cation exchanger. Transparent PVI hydrogel became white in silicic acid solution and its swelling degree drops in 2-3 times. SEM data (Figure 7) show smooth surface of the product: in contrast with soluble PVI the rigid structure of the gel prevents formation of spherical particles, at least of the particles > 20-30 nm (microscope resolution). These observations might be useful for understanding of bio-silica structures because their morphology (spherical or smooth) can be connected with rigidity of the matrix biopolymer.

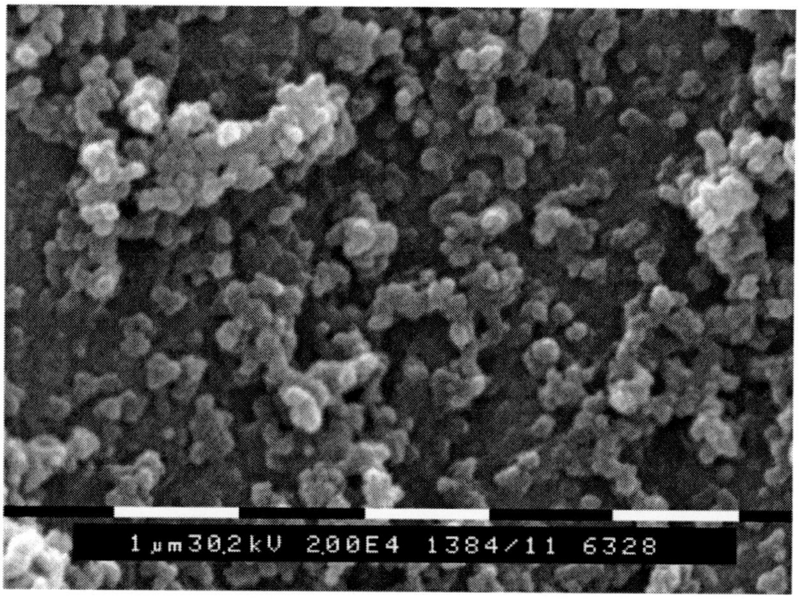

Figure 6. SEM data for PVI-silica composite obtained by neutralization of Na_2SiO_3 – PVI mixture with cation exchanger. Scale bar 1 μm.

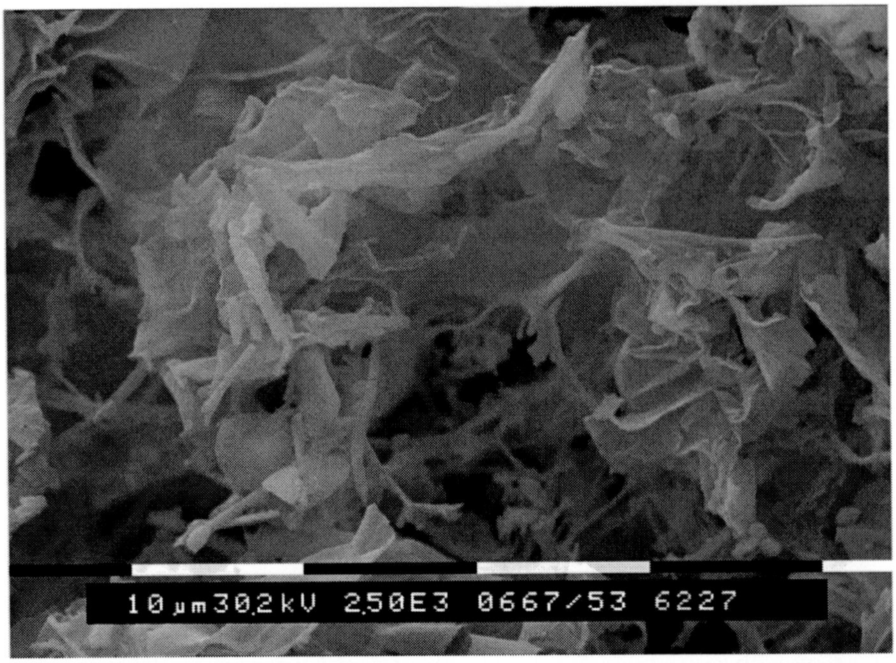

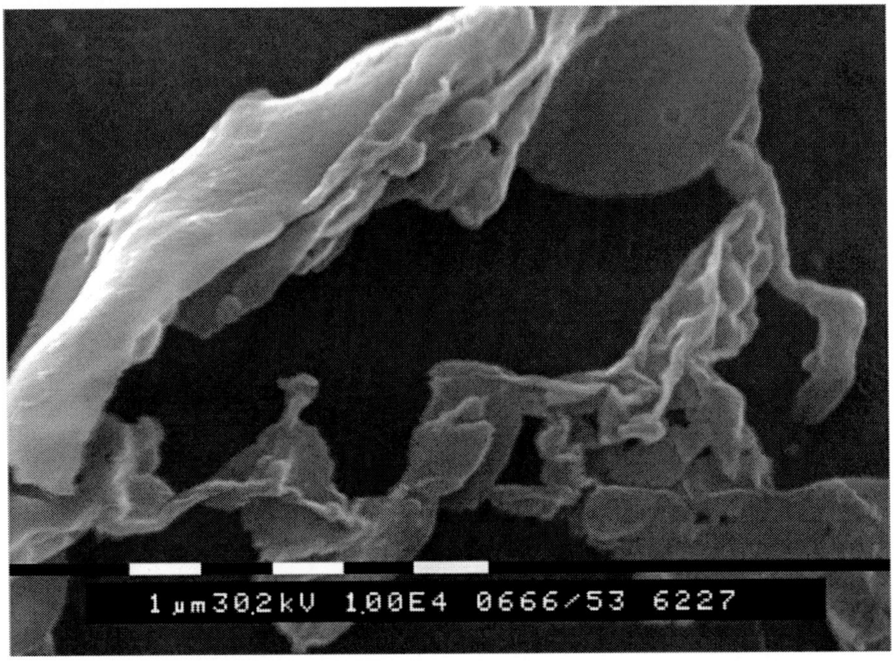

Figure 7 continued on next page

Figure 7. SEM images of the composite structures obtained by treatment of cross-linked PVI with solution of silicic acid. Scale bar 10 μm (top) and 1 μm (bottom).

TEOS-Based Composites

Acid-catalyzed hydrolysis of TEOS in the presence of organic polymer is a general route to silica composites. We have studied four polymeric bases of various basicity (PVI, PVP, PVPd and PEG) in this reaction in comparable conditions [13% TEOS in ethanol, TEOS: H_2O = 1: 3, 0.01 M HCl, TEOS: polymer = 1: (0.25-1) by mass, 1 h boiling]. The reaction with PVI, PVP and PVPd results in homogeneous solutions stable for several days, in the case of PEG-3000 gelation of the solution was observed due to cross-linking of the composite by terminal hydroxyls of PEG [64]. Solid composite films can be easily obtained on the polyethylene surface by evaporating of the solutions.

FTIR spectra of PVI composite are similar with pure PVI spectrum (Figure 8) which is connected with relatively low absorption of silica bonds. This had been confirmed by model spectrum of PVI-silica mixture which do not contain Si-O-Si band at 1070 cm^{-1}. From the other hand, in the spectra of composite we can see absorption at 978 cm^{-1} (Si-OH). Increase of these band comparing with model spectrum and shift comparing with poly-TEOS (949 cm^{-1}) are the confirmations of stabilization of PVI composite by hydrogen bonds with silanol moieties. The same change of silanol band was observed in the system with PVP (Figure 8). In the case of PVPd solid composite is significantly soluble in ethanol (up to 70%) and FTIR spectra of the soluble and insoluble parts are very similar (Figure 8). Evidently, the interaction between polysiloxane chains and PVPd is strong enough to prevent cross-linking by Si-OH condensation but insufficient for stabilization of the composite.

The other way to polysiloxane composites is the interaction between poly-TEOS and polybases due to silanol moieties in poly-TEOS. The addition of ethanol solutions of PVI, PVP and PVPd to solution of hydrolyzed TEOS results in precipitate and gelation of the solution. FTIR-spectra of the composite precipitates and dried gels show characteristic shift of heterocyclic or carbonyl bands (Table 2) and Si-OH band (949 cm^{-1}) shifts to 965-975 cm^{-1} which is the evidence of hydrogen bonding in these materials.

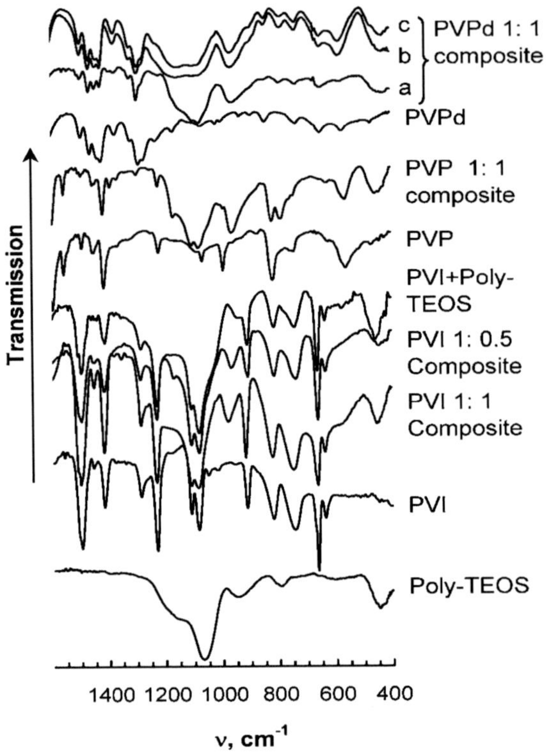

Figure 8. FTIR spectra (KBr pellets) of polysiloxanes from TEOS (Poly-TEOS), PVI, Poly-TEOS – PVI 1: 1 and 1: 0.5 composites, 1: 1 model mixture of Poly-TEOS with PVI, PVP and its composite, PVPd and Poly-TEOS – PVPd composite (**a** – solid composite, **b** – composite after ethanol washing, **c** – washed out products).

Table 2. Shifts of the characteristic bands in FTIR spectra of poly-TEOS – polybase composites

Polybase	Shift of bands in FTIR, cm^{-1}
PVI	912 → 918; 1416 → 1420; 1497 → 1508
	912 → 920; 1416 → 1420; 1497 → 1509
PVP	1597 → 1603; 1415 → 1420
	1597 → 1604; 1415 → 1421
PVPd	1668 → 1653; 1493 → 1496; 1460 → 1466; 1436 → 1444

Composition of the obtained precipitates (Table 3) is poor dependent on the initial poly-TEOS – polybase ratio: two imidazole units of PVI correspond to one silicon atom and for PVP and PVPd we have 1-2 silicon atoms per polymeric unit. In any case, conversion of poly-TEOS to the precipitate is far from 100% and increases from PVI to PVPd with decreasing of basic strength of the polymer. Evidently, we have a competition between two reactions here: precipitation of insoluble interpolymeric complex and formation of three-dimensional H-bounded net:

Table 3. Interaction of poly-TEOS with polymeric bases ([TEOS] = 29.6 %, H_2O/TEOS = 5)

Polymer	[TEOS]: [polymer]	Precipitate composition, %				Conversion into precipitate, %		Si : polymer units in precipitate
		N	Si	C	H	Polymer	poly-TEOS	
PVI	1 : 0.2	19.1	9.8	42.2	6.8	86.2	8.8	0.51
	1 : 0.3	17.6	11.1	37.8	5.9	82.7	16.3	0.63
	1 : 0.65	19.9	11.3	44.5	6.9	25.5	9.4	0.56
	1 : 1.3	19.9	10.1	43.5	6.7	13.1	8.4	0.51
PVP	1 : 0.17	4.3	17.3	43.9	5.2	58.6	19.7	2.00
	1 : 0.3	6.6	13.4	48.4	5.7	81.0	25.4	1.01
	1 : 0.64	7.1	13.9	48.2	5.2	54.2	34.3	0.98
	1 : 1.17	5.9	15.8	46.2	4.8	35.8	56.8	1.34
PVPd	1 : 0.17	4.1	19.5	27.7	5.3	55.5	22.9	2.37
	1 : 0.3	6.6	17.2	34.4	5.8	85.8	36.0	1.29
	1 : 0.63	8.6	18.0	35.0	6.1	85.2	56.5	1.04
	1 : 1.3	4.5	16.3	32.6	6.2	21.4	50.7	1.81

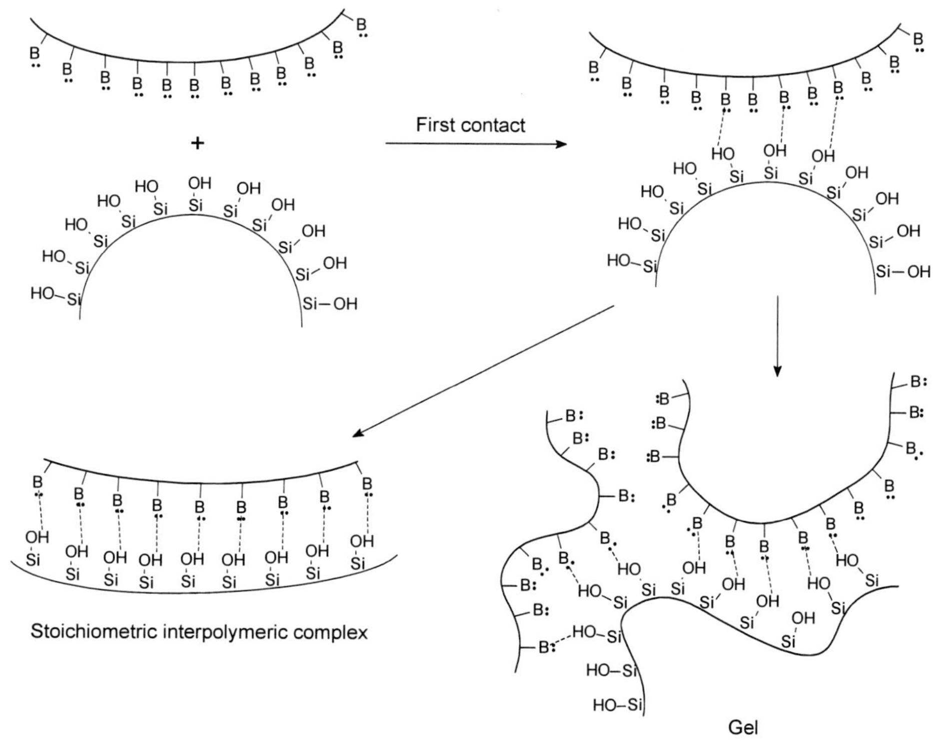

The ratio between these two routes of the reaction depends on the ability of the system to attain the equilibrium state: strength of the hydrogen bonds, chains flexibility, etc. PVI is the strongest base among the studied polymers so it gives the most stable hydrogen bonds and the main product of the reaction is a gel stabilized by random hydrogen bonds. In the case of PVPd, relatively weak H-bonds provide more chances to rearrangement of the macromolecular chains giving rise to well-ordered interpolymer complex. The precipitates consist of partially accrete submicron spherical particles (Figure 9) which agree with the competition between precipitating and jellification.

Figure 9. SEM images of the composite structures obtained by reaction of Poly-TEOS with PVI. Scale bar 1 μm.

Grafting of Polysiloxane Chains to Carbon-Chain Polymers

Ethoxysilane and silanol moieties are active in reactions with various functional groups: hydroxy, vinyloxy, oxirane, acetal, etc. These reactions are promising for synthesis of composite materials by grafting of short polysiloxane chains to polymers by hydrolysis of alkoxysilanes including TEOS in the presence of polymer with suitable functional groups. Indeed, alkoxysilanes had been hydrolyzed in the presence of a great number of polymers [2, 17, 23, 24, 67-69], but in the most of these works covalent grafting was not demonstrated and the other way to graft-copolymers is popular: copolymerization of vinyl monomer with unsaturated alkoxysilane derivative following with sol-gel reaction using this copolymer [70-72]. Here we have guarantee of the grafting but synthesis of special alkoxysilane-containing copolymer complicates and rises in price the whole process.

It seems that synthesis of soluble graft-copolymers using non-silicon polymers is possible in very narrow range of polymer structures and reaction conditions because of competition between hydrolysis/condensation of alkoxysilane and reaction with the polymer units. Attempts to facilitate grafting often result in acceleration of hydrolysis/condensation and the obtained product is a mixture of poly-alkoxysilane with non-silicon polymer. For example, we have failed trying to synthesize graft-copolymer by hydrolysis of TEOS in the presence of diethylcellulose (ETHOCEL polymers from The Dow Chemical Company). Apparently, the suitable polymer have to combine active groups with high flexibility of the chain in the solvents used in sol-gel chemistry because the hindered diffusion can decrease the polymer competitiveness in reaction with alkoxysilanes.

Recently [73] we have found the ability of such commercial polymer as poly(vinylbutyral) (PVB) to give graft-copolymers under acidic-catalyzed hydrolysis of TEOS in various solvents: ethanol, 1,2-dimethoxyethane (DME), acetone. PVB is a product of the reaction between butyraldehyde and poly(vinyl alcohol) which is obtained from poly(vinyl acetate), so macromolecules of PVB contain units of three types:

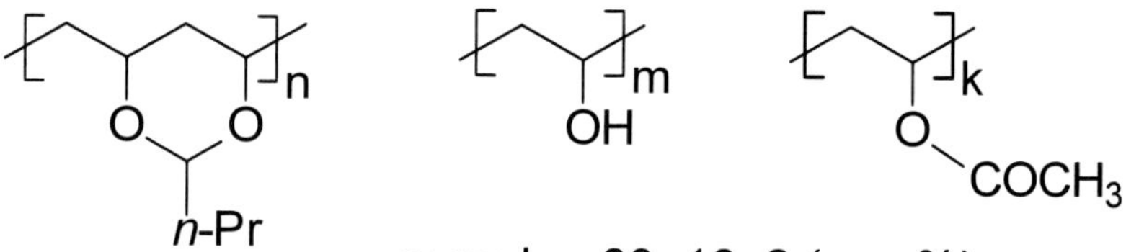

n: m: k = 80: 18: 2 (mas%)

Hydrolysis of TEOS in the presence of PVB has an anomalous feature: decrease of water content in the reaction mixture gives rise to less stability of the obtained solutions and at H_2O: TEOS < 2: 1 fast gelation of the mixture was observed (Figure 10). As usual, the increase of water content promote the gelation in TEOS-containing systems [1]. This phenomenon allowed us to suppose grafting of polysiloxane chains to PVB which can result in cross-linking at low water content and the increase of H_2O: TEOS ratio decrease the rate of cross-linking due to acceleration of parallel reaction of TEOS hydrolysis.

The formation of PVB-polysiloxane graft-copolymer is confirmed by turbidimetry titration and insolubility of the resulting solid product after evaporation of the solvent due to cross-linking. There are two possible routes for grafting of polysiloxane chains on PVB:

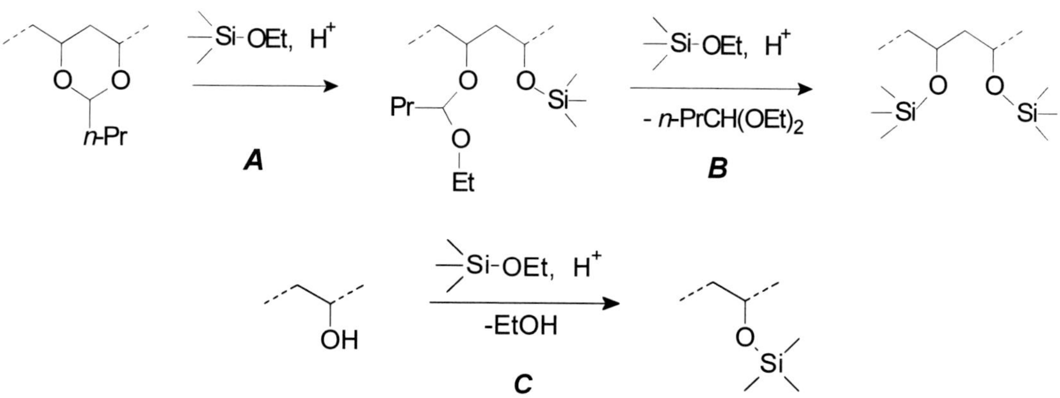

According to FTIR data, the reaction in DME results in considerable decrease of content of hydroxyl groups in the product comparing with the reaction in ethanol (route *C*). On the other hand, the formation of the graft copolymers in ethanol, in the excess of hydroxyl groups of solvent, points to significance of the *A* and *B* mechanisms in the interaction of TEOS and PVB. The eliminating of 1,1-diethoxybutane had been registered by gas-liquid chromatography and corresponds to reaction of 4.6 wt.-% of PVB units by mechanism *B* in anhydrous DME.

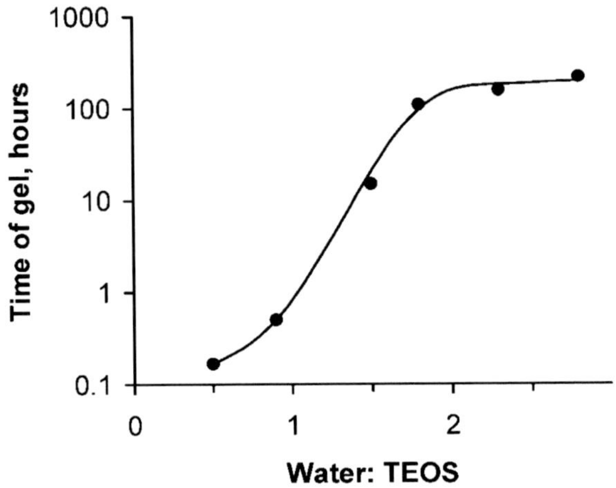

Figure 10. Dependence of jellification time on H_2O: TEOS ratio in acidic-catalyzed hydrolysis of TEOS in the presence of PVB. TEOS: PVB = 2: 1, ethanol, 6 wt.-% of PVB, boiling during 3 h, storage at room temperature.

Transparent flexible films can be obtained from the solutions of PVB-TEOS copolymers by pouring on polyethylene surface. SAXS data (Figure 11) show the presence of 1-2 nm silica particles in these composite films. Evidently, PVB macromolecules prevent polysiloxane chains from further condensation. Introduction of PEG capable of interaction with Si-OH results in larger particles because PEG compete with PVB in bonding with silica but the obtained hydrogen bonds are weaker than covalent ones and condensation of primary particles during drying becomes possible.

TGA and DSC data (Figure 12) show some differences in the thermal properties of PVB and PVB-TEOS copolymers. TGA curves have for PVB two regions of destruction at 330-370 and 480-540 °C which are visible in DSC data as exothermic peaks. Decomposition of the copolymers starts near 220°C and continues up to 350-360°C, this is an exothermic process too. The experiments have been done in air medium so we can suppose a thermal-oxidative destruction here. In the case of PVB-polysiloxane copolymers further propagation of the aforementioned reactions A, B, C and condensation of silanol with ethoxysilyl moieties are possible above 220°C. All these reactions result in volatile, easy-oxidizable products. The second exothermic peak for PVB and PVB-TEOS 1: 1 copolymer is connected with final burning-out of organic compounds but for the other copolymers we see exothermic peaks near 450°C without any remarkable decrease of the sample weight which is possible connected with stabilization of oxidized products in the matrix of silicon oxide.

The elaborated approach to synthesis of polysiloxane graft copolymers has been applied to other film-forming polymers: poly(glycidyl methacrylate) and copolymer of vinyl acetate and allyl glycidyl ether (AGE). Acid-catalyzed hydrolysis of TEOS in the presence of these polymers gives rise to stable insoluble films. Grafting of the polysiloxane chains proceeds by the oxirane moieties, these reactions results in decrease of oxirane bands in FTIR spectra (845 and 935 cm^{-1}).

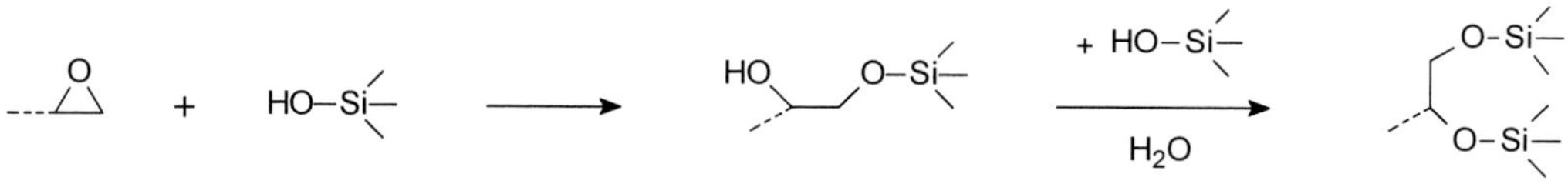

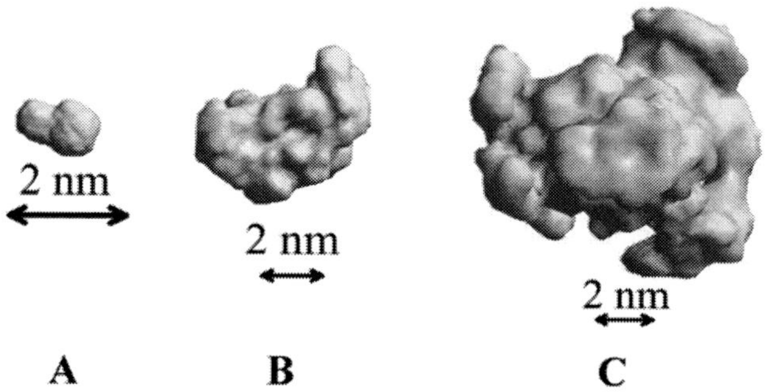

Figure 11. Typical shape of siloxane particles in hybrid organo-inorganic films restored with DAMMIN program [6] from SAXS data. PVB: TEOS = 1: 4 (**A**) and 1: 2 (**B** and **C**), TEOS: H_2O = 1: 2 (**A**), 1: 4.3 (**B**) and 1: 3.8 (**C**). PEG-300 had been added to **B** and **C** compositions (1: 1 to solid residue) before pouring of the film.

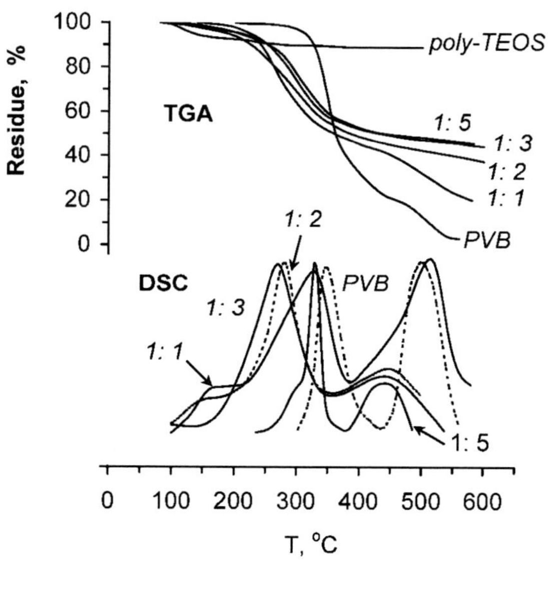

Figure 12. TGA and DCS data for PVB, poly-TEOS and PVB-TEOS graft-copolymers.

Application Areas of New Materials

Highly Functionalized Coatings

Design of solid-phase sensors (biological, chemical, electrochemical) requires highly-functionalized surfaces (glass, metals, organic polymers). For example, the present biochip technologies allow to produce inexpensive probes but the reading procedure is impossible without strong and expensive equipment [74-76]. The increase of analytical signal from solid-phase sensors is possible by the placement of active moieties not on the smooth surface only but above the surface using coatings with "hilly" relief and comb-like polymer structures. The objective of our work was to obtain amino- and oxirane-functionalized coatings with significantly increased surface density of the active groups. These moieties are the most used in technologies of DNA-biochips and they are also capable of binding with proteins in design of probes for immuno-analysis.

As mentioned above, PVB-TEOS copolymers are able to give films with silica nano-particles of various size. On the other hand, PVB itself and polysiloxane units have a lot of functional groups (-OH, ≡Si-OH, ≡Si-O-Et, acetal) which open various ways to modification of this material. We have synthesized a set of polymers with desired functions and "anchoring" groups for binding with PVB-TEOS copolymers. These polymers were obtained by radical polymerizations under the action of 2,2-azobis(isobutyronitrile) (AIBN):

a) Polymeric oxiranes: poly(allyl glycidyl ether) and poly(ethylene glycol vinyl glycidyl ether):

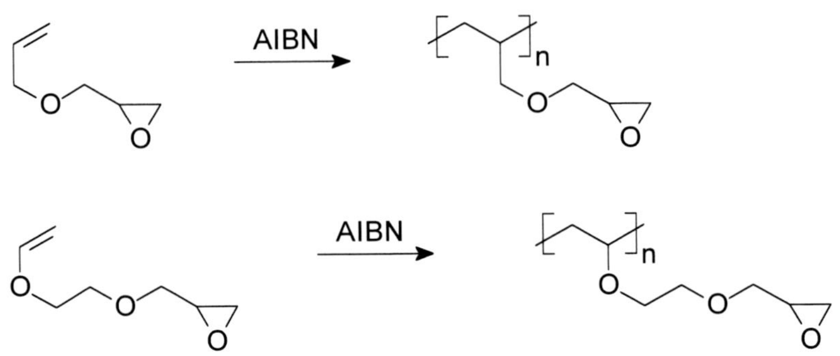

b) Copolymers of vinyl ether of ethanolamine (**VEEAm**) and divinyl ether of diethylene glycol:

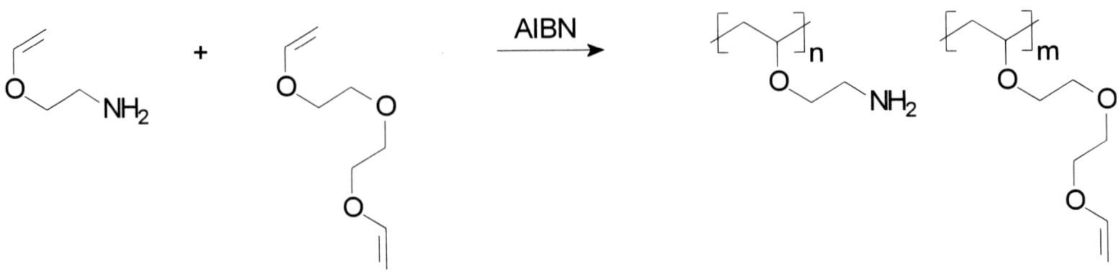

c) Copolymers on the basis of VEEAm and ethylene glycol vinyl glycidyl ether. Here the reaction between amino and oxirane groups was observed:

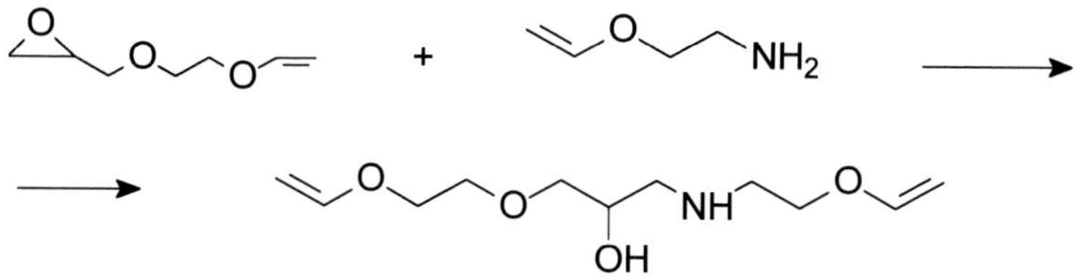

So, the resulting copolymer at excess of VEEAm has vinyloxy units of two types:

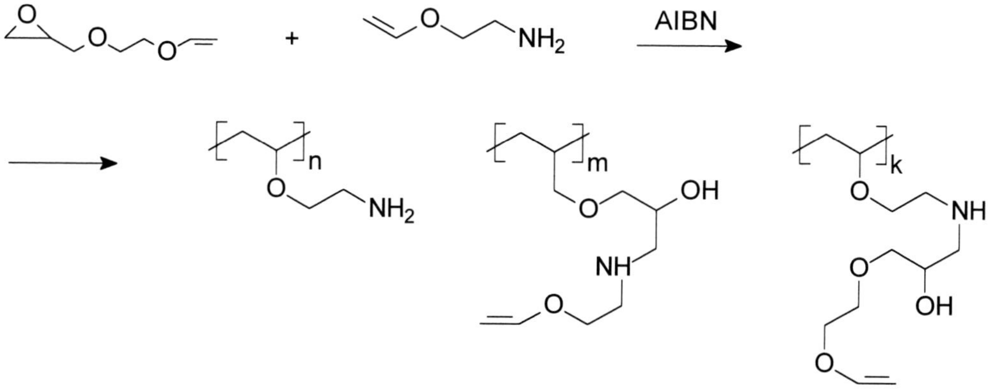

Structure of the copolymer is confirmed by FTIR spectra: bands at 965, 1200, 1320 and 1620 cm^{-1} of –O-CH=CH$_2$, absence of oxirane bands at 914 and 1254 cm^{-1}. Vinyloxy group is visible in ^{1}H NMR as quartet at 6.4-6.5 ppm (α-proton), the signals are doubled because of two types of vinyloxy units in the copolymer. According to NMR data, copolymers of VEEAm contain up to 20% of vinyloxy units retaining solubility.

Mixing of PVB-TEOS copolymers with amino- or oxirane-containing copolymers gives rise to transparent films on the glass or polystyrene surface. The coatings are insoluble in water and water-organic mixtures used in immobilization of DNA. Stabilization of the composite films occurs by the following interactions between polysiloxane chains and "anchoring" groups of the copolymers:

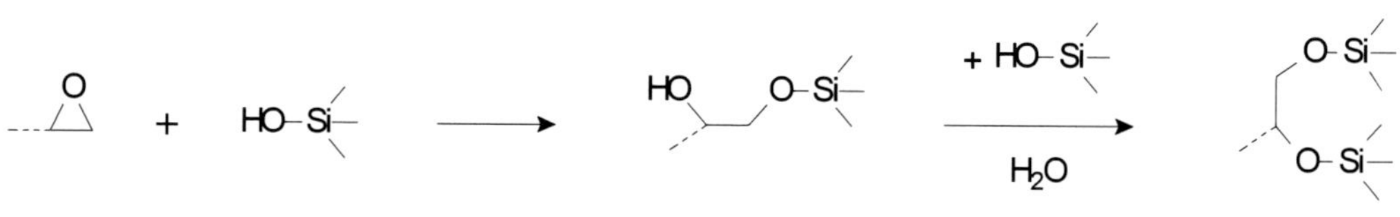

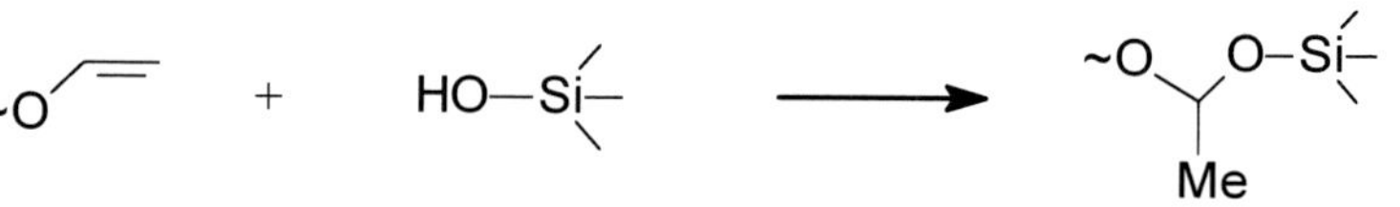

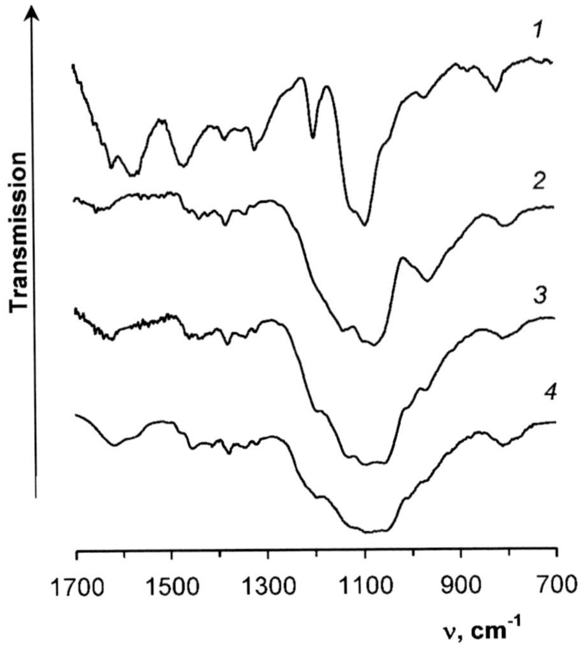

Figure 13. FTIR spectra of VEEAm and ethylene glycol vinyl glycidyl ether 95: 5 copolymer (**1**), PVB-TEOS 1: 2 graft-copolymer (**2**), functionalized coating from 1: 2 mixture of the copolymers **1** and **2** (**3**) and the coating heated at 95°C for 1 h (**4**).

FTIR spectra of the resulting films (Figure 13) show disappearance of Si-OH band near 960 cm^{-1} especially after heating at 95°C. Surface density of amino groups in the coatings have been determined by the reaction with picric acid and oxirane-containing films were treated with methylamine giving secondary amines which was detected by picric acid too. In the case of amino groups we have reached 700 units/nm^2 and for oxiranes – 100 units/nm^2. Certainly, these values are impossible for monolayer of the functional groups on smooth surface. The "hilly" structure of the coatings is clearly visible by SEM (Figure 14) and size of the "hills" (110±31 nm) is close to the size of composite from PVI and preliminary prepared poly(silicic acid) (Figures 5 and 6).

Membranes

The ability of PVB-TEOS compositions to give nano-structured coatings and films enables to test them as membranes. Indeed, permeability constant of thiourea in ethanol for PVB-TEOS (1: 2) film is $4.3 \cdot 10^{-3}$ cm/min at 20±1 μm thickness (the permeability have been determined according to [77]). The addition of PEG-300 to the film-forming composition before pouring increases permeability up to $8.2 \cdot 10^{-3}$ cm/min and after 3 h swelling in ethanol the permeability becomes $2.9 \cdot 10^{-2}$ cm/min.

The stability of PVB-TEOS films in organic solvents allows to try them as separators in lithium batteries. Recently [78-81] a mixture of PVB with alumina was proposed as battery separator but in spite of some positive results it seems inexpedient to use a mechanical mixture of polymer and alumina in organic medium containing DME and other solvents for

PVB. PVB-TEOS-PEG composition have been used for design of free-standing separators for lithium-sulfur batteries. The obtained results (Figure 15) show high efficiency of the composite separator comparing with standard polypropylene porous film (tonen). New separator is stable in the organic electrolyte in the presence of lithium salts and polysulfides in high concentration.

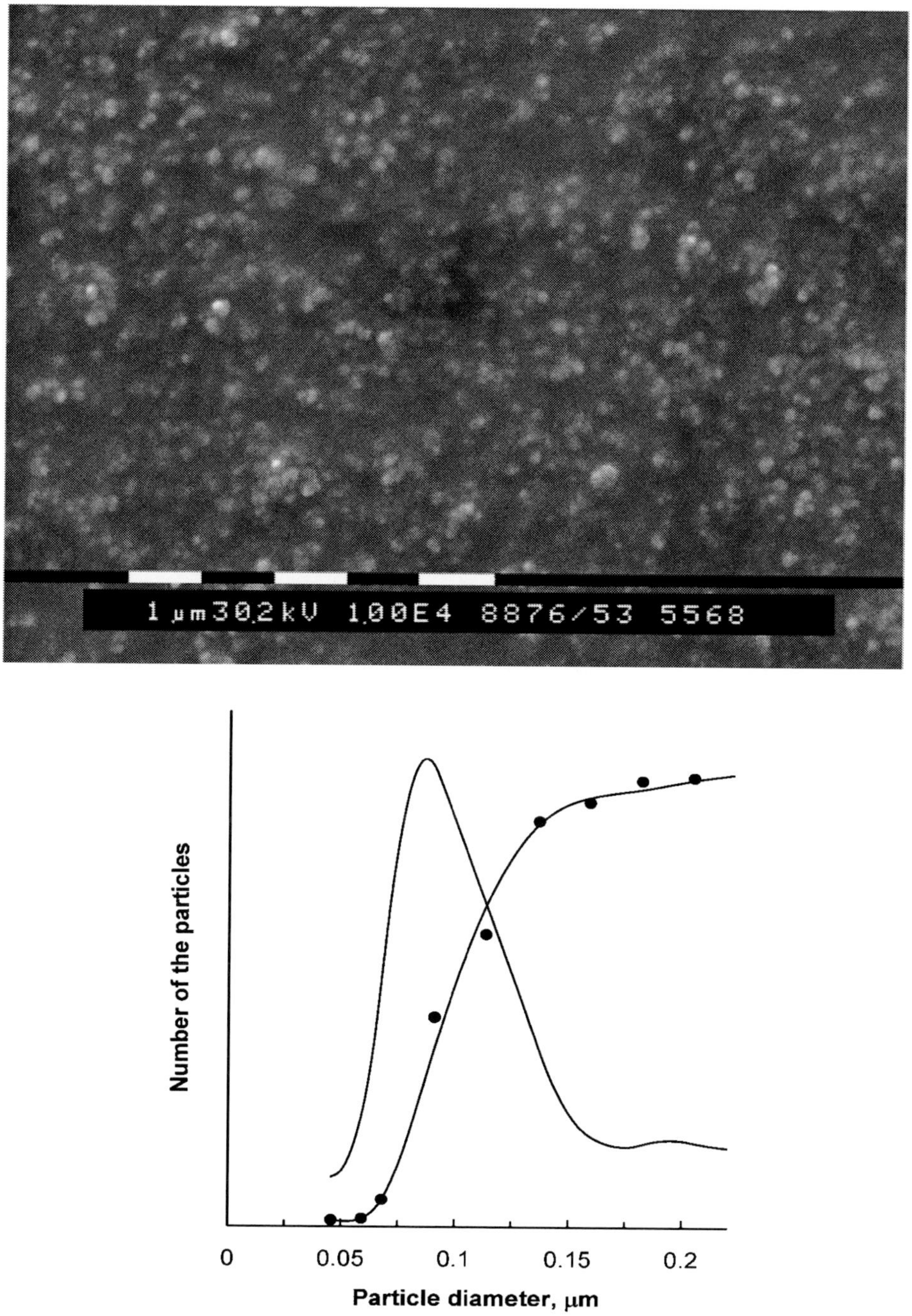

Figure 14. SEM data for functionalized coating (composition from Figure 13, **4**).

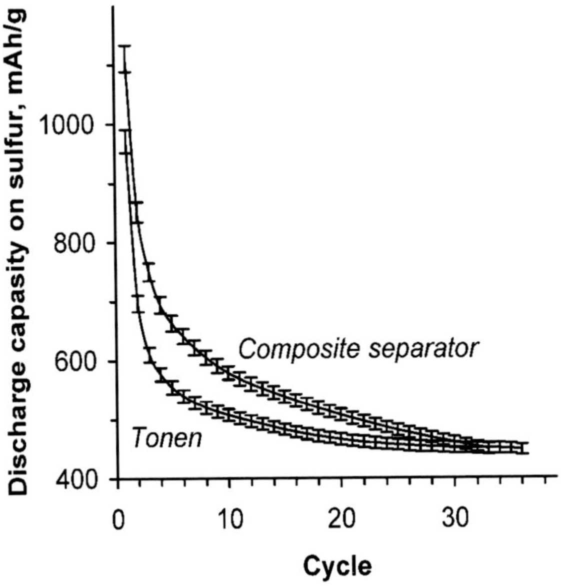

Figure 15. Dependence of discharge capacity on cycle for Li-sulfur rechargeable batteries with composite and tonen separators.

Metal Sorption

Sorption of heavy metals from water solutions by composite materials is progressing rapidly. Special companies [82] offer organo-inorganic sorbents for metal recovery in various branches:

- Catalyst Recovery: Remove precious metals from process streams containing homogeneous or residuals of "heterogeneous" catalysts systems.
- Power Plant: Allow recycle or disposal of boiler tube and other cleaning solutions containing metals or even nuclides.
- Agricultural: Removal of metals and toxic metalloids from irrigation and well water that persists in the natural cycle of irrigation, soil permeation and aquifer contamination.
- Metals Finishing: Extract valuable or polluting metals from plating baths or rinse flows.
- Rare Earths Speciation: Improve process separations efficiency of separating and refining rare earths.
- Microelectronics: Recover metals from water streams as well as in the recouping of metals from the recycle of electronics equipment and circuitry.
- Biotechnology: Protein and other macromolecule separations in columnar or chromatographic type affinity and ion exchange arrays.
- Process and Potable Water: Trace metals removal at point-of-use as required to meet high water quality standards.

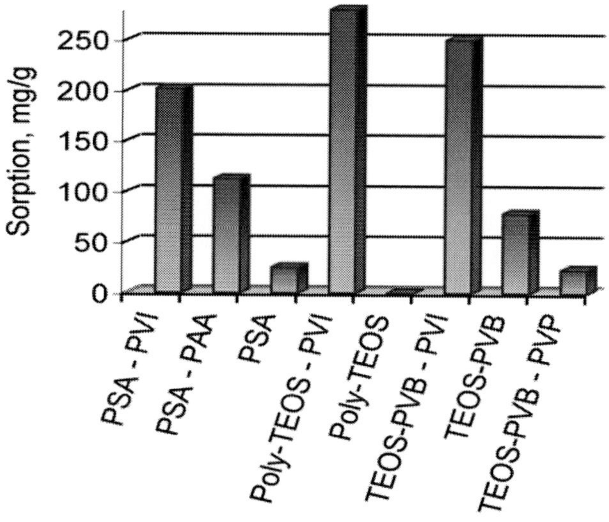

Figure 16. Maximum sorption capacity of composites towards Cu^{2+} ions.

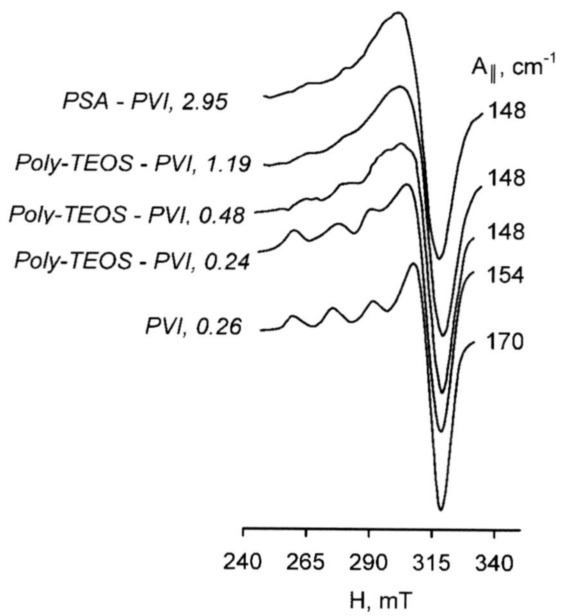

Figure 17. EPR spectra of Cu^{2+} complexes, copper loading (mmol/g) is presented after curves legends.

The combination of cheap, rigid and hydrophilic inorganic matrix with organic chelating groups defines advantages of organo/inorganic composites over traditional sorbents based on cross-linked organic polymers. The most of commercial composite sorbents are based on amino-modified silica but relatively high basicity of amines results in protonation of these groups at low pH which decrease their sorption capacity. Imidazole-containing systems (pK_{BH^+} = 5-6) are less sensitive to pH and can work in acidic media [83].

Sorption properties of new composites were studied using Cu^{2+} ions as an example (Figure 16). The sorption equilibrium attains in 20-60 min, sorption capacity is up to 280 mg/g. The most effective sorbents are composites on the basis of PVI, PVP decrease sorption capacity due to hydrophoby of this polymer.

The copper (II) complexes of the studied composites have EPR spectral patterns typical of Cu^{2+} tetragonal complexes with well-resolved hyper-fine structure in the parallel orientation (Figure 17). It should be noted that the spectra parameters ($A_{\parallel}$) are similar with each other irrespective of composite nature and copper loading but they differ from PVI-Cu^{2+} spectrum which is attributable to participation of oxygen atoms from silica in complexing.

Conclusion

The formation of poly(silicic acid) organo/inorganic composite with polymeric bases depends strongly on the conditions of silicic acid condensation. With the same polymer [poly(1-vinylimidazole)] various stoichiometry from 1: 2 to 2: 1 and various morphology from friable substance of 65±16 nm particles to spherical particles 2.9±1.1 μm in diameter can be obtained in aqueous medium. The similar effects were found with tetraethyl orthosilicate as precursor: hydrolysis/condensation in the presence of polymer matrix gives rise to more linear and soluble products but reaction of the polymeric bases with oligosiloxanes results in three-dimensional patterns (precipitate and gel). The use of poly(vinyl butyral) in sol-gel synthesis enables to fix primary oligosiloxane particles (1-2 nm) in polymeric matrix by covalent bonds. The obtained hybrid material gives stable transparent films and coatings. Membranes, highly functionalized surfaces for DNA and immuno-analysis, sorbents of heavy metals have been designed on the bases of new nanocomposites.

Acknowledgements

The authors thank Russian Academy of Sciences (Programs No. 10.3 and 25.5) and Siberian Branch of Russian Academy of Sciences (Program No. 126) for financial support. Appreciation is also extended to Prof. A. N. Ozerin for SAXS experiments.

References

[1] Brinker, C. J.; Scherer, G. W. *Sol-gel science: the physics and chemistry of sol-gel processing*; Academic Press: London, 1990.

[2] Yoshiki, C. In *Polymeric Materials Encyclopedia*; Salamone, J. P., Ed.; CRC Press: Boca Raton, New-York, London, Tokyo, 1996, pp 4793-4798.

[3] Bourgeat-Lami, E. *J. Nanosci. Nanotechnol.* 2002, 2, 1-24.

[4] Coradin, T.; Nassif, N.; Livage, *J. Appl. Microbiol. Biotechnol.* 2003, 61, 429-434.

[5] Moreau, J. J. E.;. Man, M. W. C. *Coordin. Chem. Rev.* 1998, 178-180, 1073-1084.

[6] Field, C.B.; Behrenfeld, M.J.; Randerson, J.T.; Falkowski, P. *Science* 1998, 281, 237-240.

[7] Vorobyova, S. S.; Pomazkina. G. V.; Baranova, E. Y.; Likhoshway, Y. V.; Sandgren, C. D. *J. Paleolimnol.* 1996, 15, 271-277.

[8] Likhoshway, E. V.; Pomazkina, G. V.; Nikiteeva, T. A. *Geol. Geofiz.* 1997, 38, 1445-1452.

[9] Scala, S.; Bowler, C. *Cell. Mol. Life Sci.* 2001, 58, 1666–1673.

[10] Poulsen, N.; Sumper, M.; Kroger, N. *PNAS* 2003, 100, 12075–12080.

[11] Kröger, N.; Lorenz, S.; Brunner, E.; Sumper, M. *Science.* 2002, 298, 584-586.

[12] Kröger, N.; Deutzmann, R.; Sumper, M. *J. Biol. Chem.* 2001, 276, 26066-26070.

[13] Patwarhan, S. V.; Clarson, S. J. *J. Inorg. Organomet. Polym.* 2003, 13, 49-52.Patwarhan, S. V.; Mukherjee, N.; Clarson, S. J. *J. Inorg. Organomet. Polym.* 2002, 11, 193-198.

[15] Patwarhan, S. V.; Clarson, S. J. *J. Inorg. Organomet. Polym.* 2003, 13, 109-115.

[16] Siddharth, V.; Patwardhan, S. V.; Mukherjee, N.; Steinitz-Kannan, M.; Clarson, S. J. *J. Chem. Commun.* 2003, 1122–1123.

[17] Eglin, D.; Ali, S. AM; Perry, C. C. *Polym. Int.* 2003, 52, 1807-1819.

[18] Perry, C. C.; Belton, D.; Shafran, K. L. In: *Silicon Biomineralization: Biology-Biochemistry-Molecular Biology-Biotechnology. Progress in Molecular and Subcellular Biology*; Müller, W. E. G. Ed; Springer: New York, 2003, 28 pp 269–299.

[19] Iler, R. K. *The Chemistry of Silica*; Plenum Press: New York 1979.

[20] Papisov, I. M.; Bolyachevskaya, K. I.; Litmanovich, A. A.; Matveenko, V. N.; Volchkova, I. L. *Europ. Polym. J.* 1999, 35, 2087-2094.

[21] Patwardhan, S. V.; Clarson, S. *J. Polym. Bull.* 2002, 48, 367-371.

[22] Patwardhan, S. V.; Mukherjee, N.; Clarson, S. *J. Silic. Chem.* 2002, 1, 47–55.

[23] Chan, C.-K.; Chu, I-M. *Polym.* 2001, 42, 6089-6093.

[24] Wilkes, C. L.; Wen, J. In *Polymeric Materials Encyclopedia*; Salamone J.P., Ed.; CRC Press: Boca Raton, New, York, London, Tokyo, 1996, pp 4782-4792.

[25] Habsuda, J.; Simon, G. P.; Cheng, Y.-B. *J. Polym. Sci., Polym. Chem.* 2001, 39, 1342-1352.

[26] Habsuda, J.; Simony, G. P.; Cheng, Y. B.; Hewitt, D. G., Diggins, D. R.; Cser H. T. F. *Polym.* 2002, 43, 4627-4638.

[27] Yermakova, L. N.; Frolov, Y. G.; Kasaikin, V. A.; Zezin, A. B.; Kabanov, V. A. *Vysokomolek. Soed.* 1981, 23A, 2328-2342.

[28] Yermakova, L. N.; Nuss, P. V.; Kasaikin, V. A.; Zezin, A. B.; Kabanov, V. A. *Vysokomolek. Soed.* 1983, 25A, 1391-1399.

[29] Yermakova, L. N.; Aleksandrova, T. A.; Nuss, P. V.; Kasaikin, V. A.; Zezin, A. B.; Kabanov, V. A. *Vysokomolek. Soed.* 1985, 27A, 1845-1851.

[30] Kasaikin, V. A.; Pavlova, N. V.; Yermakova, L. N.; Zezin, A. B.; Kabanov, V. A. *Colloid. J.* 1986, 48, 452-460.

[31] Kaluzhnaya, R. I.; Hulchaev, H. H.; Kasaikin, V. A.; Zezin, A. B.; Kabanov, V. A. *Vysokomolek. Soed.* 1994, 36A, 257-263.

[32] Newalkar, B. L.; Komarneni, S. *J. Sol-Gel Sci. Techn.* 2000, 18, 191-198.

[33] Adeogun, M.J.; Hay, J.N. *J. Sol-Gel Sci. Techn.* 2001, 20, 119-128.

[34] Philipp, B.; Dawydoff, W; Linow, K. *J. Z. Chem.* 1982, 22, 1-13.

[35] Petrick, R. A. *Stud. Phys. Theor. Chem.* 1982, 26, 594-622.

[36] Tsuchida, E.; Abe, K.; *Adv. Polym. Sci.* 1982, 45, 1-133.

[37] Papisov, I. M.; Litmanovich, A. A. In *Macromolecular Reactions*; Plate, N. A; Litmanovich, A. A.;.Noah, O. V., Eds.; J.Wiley & Sons: Chichester, 1995, Ch. 7, P. 283-344.
[38] Jiang, M; Li, M; Xiang, M; Zhou, H. *Adv. Polym. Sci.* 1999, 146, 121-196.
[39] Kotz, J. In *Polymeric Materials Encyclopedia*; Salamone, J. P., Ed.; CRC Press: Boca Raton, New-York, London, Tokyo, 1996; pp 5762-5791.
[40] Liu, K.-J.; Gregor, H. P. *J. Phys. Chem.* 1965, 69, 1252-1259.
[41] Liu, K.-J.; Gregor, H. P. *J. Phys. Chem.* 1965, 69, 1248-1251.
[42] Eskin, V. Y.; Magarik, S. Y.; Kuraev, U. B.; Rudkovskaya, S. Y. *Vysokomolek. Soed.* 1978, 20A, 2219-2223.
[43] Henrichs, P. M.; Whitlock, L. R.;. Sochor, A. R.; Tan, J. S. *Macromolecules* 1980, 13, 1375-1381.
[44] Sato, M.; Kondo, K.; Takemoto, K. J. *Macromol. Sci.-Chem.* 1980, 14A, 339-348.
[45] Lippert, J. L.; Robertson, J. A.; Havens, J. R.;. Tan, J. S. *Macromolecules* 1985, 18, 63-67.
[46] Kokorin, A. I.; Polynsky, A. S.; Pshegetsky, V. S.; Kuznetsova, N. P.; Ermakova, T. G.; Lopyrev, V. A.; Kabanov, V. A. *Vysokomolek. Soed.* 1985, 27A, 1834-1839.
[47] Skvortsova, G. G.; Frolova, V. L.; Skushnikova, A. I.; Kudaibergenov, S. E.; Domnina, E. S.; Bekturov, E. A. *Rus J. Appl. Chem.* 1986, 59, 1165-1168.
[48] Van de Grampel, H. T.; Tan, Y. Y.; Challa, G. *Macromolecules* 1991, 24, 3773-3778.
[49] Sakurai, M.; Imai, T.; Yamashita, F.; Nakamura, K.; Komatsu, T. *Polymer* 1994, 26, 658-664.
[50] Tan, Y. *Prog. Poly. Sci.* 1994, 19, 561-588.
[51] Kurimura, Y.; Abe, T.; Usul, Y.; Tsuchida, E.; Nishide, H. *J. Chem. Soc. Faraday Trans.* 1994, 90, 3563-3566.
[52] Skushnikova, A. I.; Svyatkina, L. I.; Domnina, E. S.; Merlina, T. V.; Frolova, V. L.; Bekturov, E. A. *Vysokomolek. Soed.*. 1995, 37B, 310-313.
[53] Kodama, H.; Miyajima, T.; Mori, T.; Takahashi, M.; Nishimura, H.; Ishiguro, S. *Colloid. Polym. Sci.* 1997, 275, 938-945.
[54] Luo, X.; Goh, S. H.; Lee, S. Y. *Macromol. Chem. Phys.* 1999, 200, 399–404.
[55] Pekel, N.; Guven, O. *Colloid. Polym. Sci.* 1999, 277, 570-573.
[56] Kodama, H.; Miyajima, T.; Tabuchi, H.; Ishiguro, S. *Colloid. Polym. Sci.* 2000, 278, 1-7.
[57] Pekel, N.; Guner, A.; Guven, O. *J. Appl. Polym. Sci.* 2002, 85, 376–384.
[58] Mazyar, N. L.; Annenkov, V. V.; Kruglova, V. A.; Anan'ev, S. M.; Danilovtseva, E. N.; Rohin, A. V.; Zinchenko, S. V. *Russ. Chem. Bull.* 2000, 12, 2047-2052.
[59] Annenkov, V. V.; Mazyar, N. L.; Kruglova, V. A.; Ananiev, S. M. *J. Mol. Liq.* 2001, 91, 109-114.
[60] Iliopoulos, I.; Audebert, R. *Macromolecules* 1991, 24, 2566-2575.
[61] Annenkov, V. V.; Mazyar, N. L.; Kruglova, V. A. *Polym. Sci.*. 2001, 43A, 807-812.
[62] Chan, C-K.; Peng, S-L.; Chu, I-M.; Ni, S-C. *Polymer* 2001, 42, 4189-4196.
[63] Papisov, I. M.; Litmanovich, A. A. *Adv. Polym. Sci.* 1989, 90, 140-178.
[64] Litmanovich, O. E.; Litmanovich, A. A.; Papisov, I. M. *Vysokomolek. Soed.* 1997, 39A,B, 1506-1510.
[65] Papisov, I. M.; Litmanovich, A. A. *Colloids and Surfaces A: Physicochem. Eng. Aspects* 1999, 151, 399-408.

[66] Hildebrand M. *Progress in Organic Coatings* 2003, 47, 256–266.
[67] Nishio, K.; Okubo, K.; Watanabe, Y.; Tsuchiya, T. *J. Sol-Gel Sci. Technol.* 2000, 19, 187–191.
[68] Liu, Y. L.; Lin, Y. L.; Chen, C. P.; Jeng, R. J. *J. Appl. Polym. Sci.* 2003, 90, 4047-4053.
[69] Messori, M.; Toselli, M.; Pilati, F.; Fabbri, E.; Fabbri, P., Busoli, S., Pasquali, L., Nannarone, S. *Polymer* 2003, 44, 4463-4470.
[70] Gunji, T.; Okonogi, H.; Sakan, T.; Takamura, N.; Arimitsu, K.; Abe, Y. *Appl. Organometal. Chem.* 2003, 17, 580-588.
[71] Wu, K. H.; Chang, T. C.; Wang, Y. T.; Chiu, Y. S. *J. Polym. Sci. A-Polym. Chem.* 1999, 37, 2275-2284.
[72] Chang, T. C.; Wang, Y. T.; Hong, Y. S. *Polym. Degrad. Stab.* 2000, 69, 317-322.
[73] Annenkov, V. V.; Danilovtseva, E. N.; Filina, E. A.; Mikhaleva, A. I; Skotheim, T. A.; Trofimov B. A. *Polym. Int.* 2004, 53, 772-776.
[74] Vo-Dinh , T.; Cullum, B. M.; Stokes, D. L. *Sensor. Actuat.* 2001, 74B, 2-11.
[75] Bier, F. F.; Kleinjung, F. Fresenius *J. Anal. Chem.* 2001, 371, 151-156.
[76] Lee, Y-S.; Mrksich, M. *Trends in Biotechnol.* 2002, 20, 14-18.
[77] Zhumadilova, G. T.; Gazizov, A. D.; Bimendina, L. A.; Kudaibergenov, S. E. *Polymer* 2001. 42. 2985-2989.
[78] Ying, Q.; Carlson, S. A.; Skotheim, T. A. *U.S. Patent 6,277,514, 2001.*
[79] Ying, Q.; Carlson, S. A.; Skotheim, T. A. *U.S. Patent 6,183,901, 2001.*
[80] Ying, Q.; Carlson, S. A.; Skotheim, T. A. U.*S. Patent WO 2000036671 A1, 2000.*
[81] Ying, Q.; Carlson, S. A.; Skotheim, T. A. *U.S. Patent WO 2000036670 A1, 2000.*
[82] Purity Systems Inc. Advertizing materials, available from: www.puritysystemsinc.com.
[83] Annenkov, V. V.; Danilovtseva, E. N.; Saraev, V.V.; Mikhaleva, A. I. *J. Polym. Sci. Part A: Polym. Chem.* 2003, 41, 2256-2263.

In: New Frontiers in Polymer Research
Editor: Robert K. Bregg, pp. 201-223
ISBN: 1-59454-956-7

Chapter 8

DIFFUSION CHARACTERISTIC OF POLYSTYRENE IN SOME THETA AND POOR SOLVENTS - DLS INVESTIGATIONS

Hanna Wilczura-Wachnik[**] ***and W. Alexander Van Hook***[**]
[*]The Faculty of Chemistry Warsaw University, 02-093Warsaw, Pasteura 1 Poland
[**]Chemistry Department University of Tennessee, Knoxville, Tennessee 37996

ABSTRACT

The parameters describing Dynamic Light Scattering (intensity (I), correlation radius (ζ) and diffusion coefficient (D)) are reported for some (polystyrene + solvent) systems. Measurements were performed for polystyrenes of molecular weight (Mw) 7.5k to 400k amu. dissolved in protiated or deuterated methylcyclohexane, cyclohexane and acetone. DLS measurements were performed under isothermal and/or isobaric conditions for semidilute polymer solutions at high enough concentrations to ensure precipitation (liquid-liquid demixing). The results describe measurements over broad ranges of temperature, pressure, solution concentration, and solvent isotope label (H/D). In this paper we concentrate on description of the diffusion of polymer chains in three different small molecule solvents. Two, cyclohexane and methylcyclohexane, are classified as θ solvents. The third, acetone, is a poor solvent.

Our investigations of the demixing of polystyrene-solvent systems beginning in the homogenous region show that DLS correlograms, initially monomodal, split during temperature and/or pressure quenching, and became much more complicated as precipitation is approached. There is an appreciable concentration dependence on the shape of the correlogram. As concentration increases a bimodal pattern develops. Changes in intensity and correlation radius of the diffusive mode as functions of temperature and pressure are discussed.

[*] corresponding author

INTRODUCTION

The scattering of electromagnetic radiation by liquids and solutions can be a useful source of structural information. Light scattering is consequent to refractive index differences (i.e. fluctuations) of adjacent volume elements caused by random thermal motion, and density and concentration fluctuations. For a scattering particle characterized by a velocity ν0, the frequency of the scattered (or Doppler displaced) wave, Δν , is Δν=ν0-νsc = v /c, where c is the speed of light and ν0 and νsc are source and scattered frequencies. The spectral distribution of the scattered light, Isc(ν), is broadened as compared to that of the incident wave, I0(ν). This effect (frequency broadening) caused by molecular motion is called Dynamic Light Scattering, DLS, (or inelastic Rayleigh scattering). DLS was discovered shortly after gas lasers were applied as light sources in optical scattering investigations. Lasers are high intensity sharply monochromatic light sources; as a consequence the spectral broadening of the low intensity Rayleigh scattered radiation is relatively easy to detect. The time-averaged intensity of scattered light at given angle, frequency, and averaging times, Isc(ν, θ, <t>) thus depends on the dynamics of decay of the refractive index fluctuations, and hence on relaxation rate (decay rate) and diffusion constant.

The implementation of DLS requires high-speed digital processing of stochastic optical signals and usually involves photon counting (detection) over regularly timed intervals of variable length. The statistical and stochastic concepts used in photon correlation techniques discussed in detail by Schätzel (1993).

In a DLS measurement one infers the time displaced photon intensity correlation function I(q,t) from the scattering data. Here t is of course time, and q is the momentum transfer wave number (sometimes called "the scattering vector"). Note that q is a function of scattering angle and incident light frequency, and varies with experimental condition (temperature, pressure, concentration, etc.).

$$q = \left(4n\pi\lambda^{-1}\right)\sin(\theta/2) \quad (1)$$

again, n is the refractive index of the medium, λ the wavelength in vacuum, and θ the scattering angle.

A method for calculation of the dynamic scattering function is clearly described by Akcasu (1993), but unfortunately numerically exact solutions are only possible for highly idealized cases. There is no tractable general theory that encompasses all regions of concentration and temperature. Even so, theoretical models do exist which describe the (n x n) dynamic scattering matrix, S(q,t), for limiting dilute or semi-dilute conditions. S(q,t), sometimes called the dynamic structure factor, can be represented by a superposition of exponential modes with different relaxation frequencies. The interpretation proceeds to resolve such modes into the so-called slow and fast (or short and long time) mobilities. The scattering lengths probed in DLS usually lie between $10^2 \div 10^4$ Å and are often expressed in terms of the inverse scattering vector q^{-1}. Recent advances in DLS techniques have made it possible to measure accurate autocorrelation functions covering 12 decades in delay time (i.e. from ~12.5ns to ~10^4 seconds) allowing practical application of DLS to the study of dilute and semidilute polymer solutions (Brown and Nicolai 1993).

In this chapter we focus on semidilute polymer solutions in theta solvents and poor solvents. Earlier DLS investigations on these kinds of polymer solutions have been reported by de Gennes (1976) and by Brochard (1983, 1979). Their calculation of the structure factor for semi-dilute solutions paralleled that of Tanaka et al., (1973) used for stabile gels. Later Adam and Delsanti (1985) and Chen and Berry (1990) derived expressions for S(q,t) for solutions of polymers in theta solvents. In summary, and following Brown and Nicolai (1993), one obtains for the dynamic structure factor

$$S(q,t) = A_1 e^{-\Gamma_1 \tau} + A_2 e^{-\Gamma_2 \tau} \tag{2}$$

where Γ_1 and Γ_2 are relaxation rates, and A_1 and A_2 their corresponding amplitudes. In semidilute polymer solutions the so-called "pseudogel" effect appears as the elastic modulus of the transient gel formed by the entangled polymer chains. Such entanglements which form the crosslinks of the transient gel are not permanent, but have a characteristic life time and viscoelascity. In the case where viscoelastic effects of the background "pseudogel" are negligible simplification of equation (2) to a form involving a single exponential is possible. One then obtains

$$\Gamma = D_c q^2 \tag{3}$$

where D_c is known as the cooperative diffusion coefficient of the system. A simplification of equation (2) occurs when, for example, the time required for the system to diffuse over a distance scale q^{-1} is much shorter than the characteristic disentanglement time of the polymer chain. In such situation the relaxation times and amplitudes are given by:

$$\Gamma_1 = D_g q^2 \text{ and } A_1 = D_c (D_g)^{-1} \tag{4}$$

$$\Gamma_2 = D_c (D_g T_r)^{-1} \text{ and } A_2 = 1 - D_c (D_g)^{-1} \tag{5}$$

where D_g is the cooperative diffusion constant of the pseudogel. Because the relaxation times, Γ_1 and Γ_2, satisfy the inequality $\Gamma_1 > \Gamma_2$ the exponentials are referred to as fast and slow modes respectively.

DLS studies of the dynamical behavior of polymer coils in semidilute solutions involve measurement of relaxation times which can be converted to correlation lengths only after the introduction of a set of assumptions relating time and distance. In the case of solutions of monodisperse small particles (of low enough molecular weight) DLS correlograms are well described in terms of a single exponential decay rate $g_i(t)=B\exp(-\Gamma t)$. In fact, $g_i(t)$ is the electric field correlation function which is directly measured in DLS technique. The constant B depends on the optical setup of the equipment (and reduces to a least-squares fitting constant), while Γ is the relaxation time which describes the exponential decay. The cooperative diffusion coefficient D_c calculated from equation (3) is $D_c = \Gamma / q^2$, and the assumption of monomodal exponential decay permits a calculation of the correlation length $(\xi)^0$ following the Stokes-Einstein equation

$$(\xi)^0 = \frac{k_B T}{6\Pi \eta_0 D_c} \tag{6}$$

In equation (6) k_B is the Boltzmann constant and η_0 the solvent viscosity. Values for Γ and thence D_c are obtained from the least-squares fit to the DLS monomodal correlogram.

At the low concentration and pressure limits the so-called hydrodynamic correlation length ξ_h is defined as the limiting value of ξ: $\lim_{c=0,p=0}(\xi)=\xi_h$ and, similarly, $\lim_{c=0,p=0}(D_c)=D_0$, defines the single particle diffusion coefficient. Both D_c and ξ_h, are important descriptors of polymer coil dynamics. They can be deduced from an appropriate DLS data set by the Laplace inversion computer program CONTIN (Prevencher 1982), or other procedures (such as CUMULANT analysis (Koppel 1972). Software packages for DLS data analysis including single and double exponential CUMULANT and CONTIN analysis are now standard tools in DLS analysis.

DLS correlograms obtained for polymer solutions in good solvents can usually be described with single exponential decay functions. Correlograms in θ solvents are more complex, and for those in poor solvents the complexity increases yet again. Correlograms in θ and poor semidilute polymer solutions, however, do show a certain regularity. As polymer concentration increases, the initially simple exponential form changes to a more complicated bimodal or polymodal pattern. The narrow peak found at a shorter time which is marked as a first component (so-called fast mode) , varies linearly with q^2 and is clearly diffusive in nature. Thus Szydlowski and Van Hook (1998) confirmed the diffusive nature of the fast mode experimentally in polystyrene/methylcyclohexane solution. The slow peak or peaks is(are) important at concentrations above entanglement. They broaden with increasing the polymer molecular weight and concentration, and as well their average position shifts to longer time. The slow modes are not diffusive and the hydrodynamic correlation lengths are determined from the parameters defining the short time decay (Brown and Nicolai 1993). The longer time relaxation process is only seen for polymer solution at sufficiently high viscosity and are observed only for solutions of reasonably high concentration and polymer molecular weight.

Although equation (6) is strictly applicable only at infinite dilution it is often applied to semidilute concentrations (Brown and Nicolai 1993; Gold et al. 1996; Sun 1994) even though, from a strictly phenomenological point of view, the concentration dependence of polymer solution viscosity should be accounted for. Van Hook et.al., (1999) proposed the correction of correlation radii (ζ DLS) obtained from equation (6) by using $\xi=\xi^0_{DLS}(\eta_c/\eta_0)$. Viscosities for semidilute solutions are generally significantly larger than solvent viscosities, $\eta/\eta_0 >> 1$. The correction is an empirical attempt to compensate for critical slowing down (Jlan 1996; Ikier 1995) and related effects.

In this chapter we describe DLS measurements of the dynamic properties of semidilute polymer solutions, focusing on a description of the diffusion of polystyrene chains in three small molecule solvents. Two, cyclohexane and methylcyclohexane are classified as θ solvents, the third, acetone, is a poor solvent for polystyrene. The parameters obtained in the present study: scattering intensity, correlation radius and diffusion coefficient for [polystyrene (7500 to 400,000amu) + solvent] are discussed as functions of temperature and pressure. It is worth noting that over the past twenty years dynamic properties of semidilute polystyrene solutions have been intensively studied by osmotic pressure (Noda et al., 1981), and light and neutron scattering (Daoud 1975; Cotton et. al., 1996; Amirzadeh et. al., 1982; Stepanek et. al., 1984; Wiltzius et. al., 1983). The results show that to first order both the osmotic pressure and the correlation length of the concentration fluctuation correlation function are independent of polymer molecular weight.

DLS measurements at elevated pressures have been previously reported by just a few groups. Fytas et al., (1982, 1984), Meier et al., (1984) studied acrylate polymers at near ambient temperature under pressure from 0.1 to 210MPa. More recently Smith, Freeman and Hall (1997) reported measurements on low molecular weight polypropylene oxide near the glass transition at 253K for pressures from 250 to 390MPa. Van Hook, Wilczura and Rebelo (1999) have reported DLS results on polystyrene in methylcyclohexane and in cyclohexane for pressures between 0.1 and 200MPa. The last report includes the first DLS measurements on polymer solutions in the vicinity of pressure induced demixing. Kiepen and Borchard (1988), Xiong and Kiran (1998) and Steinhoff et. al., (1997) have described cells used for static light scattering of polymer solutions during pressure induced precipitation from 0 to 80MPa. Lechner and Mattern (1991) reported static and dynamic light scattering measurements on three component system polywinyl alcohol + water + alcohol at high pressure and temperature but not in the vicinity of the demixing transition.

In this chapter we describe temperature and/or pressure quenches beginning under homogeneous conditions (absolutely transparent) and cooling(or warming) or pressurizing (or depressurizing) in order to induce liquid-liquid phase precipitation (demixing), the process being followed by DLS. Scattering intensities and correlation radii have been reported previously by Van Hook et al., (1999) and Van Hook, Wilczura and Rebelo (1999) and a detailed discussion of the relevant correlograms is available including the deconvolution and relative contribution of the fast and slow modes and their associated relaxation times. Consequently the shape of the autocorrelation function is not discussed in the present chapter.

Assuming that the cooperative diffusion coefficient is a measure of thermal motion of the polymer coils in the given medium, the Stokes-Einstein equation (6) can be written as $D_c = k_B T/f$. Here $f = 6\Pi\eta\xi$ is a friction factor for modes of radius ξ in a medium of viscosity η. The diffusion coefficient D depends strongly on temperature and pressure. In the usual case D increases with T and decreases with pressure, even though the viscosity of the medium normally increases with pressure. D also depends on molecular size. It is larger for small molecules because freedom of movement decreases with increasing collision cross-section.

Our attention is focused mainly on the semidilute polystyrene solutions. However, the concentrations which set the semidilute limit are not strongly fixed. From the theoretical point of view a polymer solution is considered semidilute when its concentration is above the so-called overlap concentration, C*. Ying and Chu (1987) discussed the definition of the overlap concentration of macromolecules in solution. The mostly used are given by the following relations: $3M(4\pi N_A R_g^3)^{-1}$, $M(N_A R_g^3)^{-1}$ or $(\eta)^{-1}$ where M is the molecular weight of the polymer chain, R_g is radius of gyration, N_A is Avogadro's number and η is the intrinsic viscosity.

DLS EXPERIMENTS; RESULTS

Scattering intensities, correlation lengths, and diffusion coefficients for polystyrene (7.5k ÷ 400k amu) were obtained in protio and deutero acetones, cyclohexanes, and methylcyclohexanes by DLS. Polystyrene samples of high-purity were purchased from Pressure Chemical Co. and have been specified in previous papers (Van Hook et al., 1999; Van Hook, Wilczura, Rebelo 1999; Szydlowski et al., 1998; Melnichenko et al., 1998; Szydlowski* et al., 1998). Solvents were purchased from Aldrich Chemical and used as received except for drying over molecular sieves and filtration through poly(tetrafluoroethylene) (0.2μm, Alltech or Whatman). DLS measurements were performed using a Brookhaven Instruments Corporation package which included a BI-200SM goniometer, BI-9025AT correlator and EMIB2FBK/RMI photomultiplier. A Spectrophysics model 127 50mw He-Ne laser, and a software package with single and double exponential CUMULANT and CONTIN data analysis were employed. The measurements were limited to the single scattering, angle $\theta=90^0$ due to constraints imposed by the high pressure cell design ($0 < P/MPa < 200$). Details are given in previous papers (Van Hook et al., 1999; Van Hook, Wilczura, Rebelo, 1999). In the data analysis of experiments at elevated temperature and pressure the solvent viscosities at given temperature and pressure were obtained from standard sources (Szydlowski et al., 1998; Yaws 1995; Tanaka et al., 1991; Bridgman 1958; Rabinowitch 1970) and smoothed by least-squares, see Table 1.

Table 1. Smoothing relation for solvents viscosities

Acetone	$\eta(0.1MPa)=2.2017-9.91564(10^{-3})T+1.199576(10^{-5})T^2$	(Ying et al., 1987)
MCH-d	$\eta(0.1MPa)=7.0063-0.0339\,T+4.2656(10^{-5})T^2$	(Ying et al., 1987)
CH	$\eta(0.1MPa)=12.9138-0.0671\,T+8.976\,(10^{-5})T^2$	(Szydlowski et al., 1998)

For protio: methylcyclohexane and cyclohexane according to Bridgman (1958) :

$$\log(\eta_p/\eta_{0.1MPa}) = [3.6367\times10^{-3} - 1.12\times10^{-5}(t-30)](p) + [-8.7347\times10^{-7} + 5.3910\times10^{-9}(t-30)](p)^2$$

$$\frac{\eta(d_{14})}{\eta(h_{12})} = 1.06$$ (Rabinowitch, 1970)

units: η [centipoises]; t [^{0}C]; T[K]; p[MPa]

Early in our studies we focused on demixing diagrams in the t=f(p) plane using pressure and/or temperature quenching (Van Hook et al., 1999; Van Hook, Wilczura, Rebelo 1999; Szydlowski et al., 1998; Melnichenko et al., 1998; Szydlowski* et al., 1998), with particular interest in scattering in the immediate vicinity of liquid-liquid demixing. Our principal interest was comparison of correlation radii and scattering intensities obtained from DLS with SANS (Small-Angle Neutron Scattering) measurements in the vicinity of critical demixing surface in the temperature, pressure, concentration projection (Szydlowski et al. 1998; Melnichenko et al. 1998; Szydlowski* et al., 1998). The SANS data permit a comparison of intra- and intermolecular correlations of polystyrene in poor and θ solvents. In the poor solvent (deutereted acetone) intermolecular correlations dominate over the intramolecular and generally their magnitudes approaches, or even falls below, single chain dimensions at the highest temperatures and pressures reached in the experiment. In the θ solvent (deutereted cyclohexane), on the other hand, the intermolecular correlations predominate only in a narrow region close to the critical line and become negligible above the θ point.

We have reported studies on the pressure dependence of the correlation function in theta solvents (Van Hook et al., 1999; Van Hook, Wilczura and Rebelo 1999). The experiments (Van Hook, Wilczura and Rebelo 1999) are of interest in distinguishing between interpretations correlogram complexity for semidilute polystyrene solutions at theta temperatures and below. Relatively few DLS studies on such solutions have been reported. Of these, Brown and Nicholai (1993) followed Adam's and Delsanti's theoretical analysis (1985) and interpreted the slow q-independent peak in terms of viscoelastic modes associated with polymer entanglements. Semenov (1990), Doi and Onuki (1992) and Genz (1994) offered further refinements of this approach. Contrarily, Sun and Wang (1994) maintain that polymer entanglement is not a prerequisite for the appearance of the viscoelastic modes, although entanglement is expected to enhance the slow mode in the DLS spectrum. Because the viscosity of liquids increases exponentially with pressure one expects the DLS spectrum to exhibit an appreciable pressure coefficient. In the context of the polymer entanglement interpretation of Brown and Nicholai (1993), we anticipated little or no pressure dependence for the DLS spectrum (Van Hook, Wilczura, and Rebelo 1999). We found that at high pressure the slow modes are characterized better by determining the q-dependences and comparing them with the longest chain relaxation time. We supported significant changes in the DLS patterns for polystyrene in cyclohexane and in methylcyclohexane solutions near critical concentrations with pressure increasing (pressure-induced slow modes under 90^0 scattering angle). For polystyrene 90k in cyclohexane slow modes appear prominent in solutions of near critical concentration at 40^0C at two pressures: at P=0.5MPa and at P=50MPa. Whereas for polystyrene with molecular weight significantly smaller than 90k amu. (around 30k amu.) slow mode contribution is vanishingly small at any temperature. Interesting appeared comparison of the relative contributions of the fast and slow modes to the CONTIN pattern deduced from DLS correlation functions for PS90k in methylcyclohexane at near critical concentrations (12.98 wt%) at 85^0C as a pressure function (Van Hook, Wilczura, Rebelo (1999). Presumably poorer solvent quality of methylcyclohexane comparing to cyclohexane is why slow mode pressure sensitivity in methylcyclohexane persists to higher temperature. An interesting complication occurs in the case of cyclohexane/PS-90k solutions due to the overlap of the cyclohexane liquid-solid phase transition (Van Hook, Wilczura, Rebelo 1999) under pressure with the liquid-liquid phase transition of the solutions.

In the material below we briefly summarize the principal results reported in this experimental program (Van Hook et al., 1999; Van Hook, Wilczura, Rebelo 1999; Tanaka et al., 1991; Bridgman 1958). Acetone is a poor solvent for polystyrene, and consequently the highest polymer molecular weight which could be investigated was well below the value at which DLS nondiffusive slow modes develop (Brown and Nicolai 1993; Van Hook et al., 1999). DLS data for (PS + protioacetone) and (PS+deuteroacetone) solutions were obtained at near-critical concentrations, ~ 20 wt% for polystyrenes between 7.5k and 25k amu. Tables 2 - 4 show DLS (isothermal) pressure quench and isobaric (temperature quench) data for polystyrene(PS)/protioacetone and PS/deuteroacetone solutions. Data-nets for PS-7.5k, 11.6k and 13k in deuteroacetone are shown in Figure 1, and for PS-13k and 25k in protioacetone in Figure 2. Equations of the correlation lines which describe demixing in the (T,P) plane are reported in Table 8. For these solutions the DLS correlograms are well described in terms of monomodal (simple) exponential decay $g_1(t)=B\exp(-\Gamma t)$ (CONTIN). Diffusion coefficients have been calculated from $\Gamma(q)$ data using equation $D=\Gamma/q^2$, and correlation radii from equation 6. The lowest and highest values of scattered light intensity, correlation radii and diffusion coefficients are collected in Tables 2 and 3 for PS/deuteroacetone, and Table 4 for PS/protioacetone systems. The experimental procedures used to investigate PS/methylcyclohexane and PS/cyclohexane solutions were the same as for PS/acetone systems. The relevant data are found in Tables 5÷7. Smoothing equations for the demixing lines are reported in Table 8. The data-nets are shown in Figures 3 and 4.

Table 2. The measurement parameters and DLS data for polystyrene 11.6k and 13k in deutero acetone

System: PS11.6k M_w/M_n= 1.03, 20.3wt% in Ac-d				
T / ^{0}C	**19.76**		**52.71**	
Pressure range / MPa	15.00	35.00	0.01	40.00
Relative intensity / arb. units	6310	453	5907	222
Correlation radius / nm	130	17.9	105	7.4
10^7diff.coeff. /cm^2s^{-1}	2.2	16.9	4.9	74.0
P / MPa	**0.10**		**20.00**	
Temperature range / ^{0}C	85.1	52.7	85.1	52.7
Relative intensity / arb. units	917	5907	177	308
Correlation radius / nm	20.26	105	8.6	10.6
10^7diff.coeff. /cm^2s^{-1}	40.4	4.9	95.4	48.1

System: PS13k M_w/M_n= 1.06, 20.3wt% in Ac-d						
T / ^{0}C	**20.4**		**5.4**		**54.1**	
Pressure range / MPa	14.10	65.00	4.90	40.00	0.90	55.00
Relative int. / arb. units	202	33.7	168	30.4	122	23.1
Correlation radius / nm	53	19.7	39.0	15.7	24.8	6.9
10^7diff.coeff. /cm^2s^{-1}	1.2	2.6	1.9	4.2	3.8	10.9

Table 3. The measurement parameters and DLS data for polystyrene 7.5k in deuteroacetone

System: PS7.5k M_w/M_n=1.06, 20.3wt% in Ac-d										
T / °C	**12.0**		**16.1**		**20.7**		**30.4**		**52.1**	
Pressure range / MPa	1.78	30.00	0.50	30.00	0.50	30.00	0.50	50.00	0.50	25
Relative int./arb. units	5789	298	4457	281	1549	252	569	209	331	164
Correlation radius/nm	115.0	11.4	43.4	10.6	28.6	9.8	12.2	6.4	8.7	5.8
10^7diff.coeff. /cm^2s^{-1}	0.22	2.27	0.46	2.64	1.06	3.10	2.92	5.59	11.1	16.7

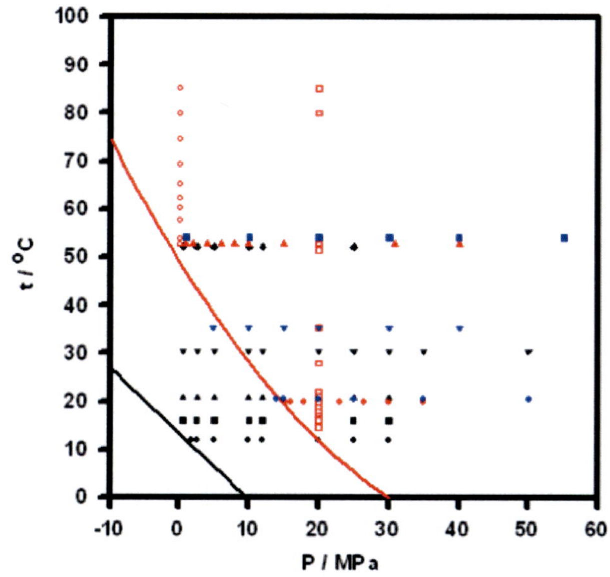

Figure 1. Location of DLS data on the LL demixing diagram (lines) of near critical polystyrene + deuteroacetone solutions. Black points and line mark data for PS7.5k; Red points and linemark data for PS11.6k; Blue points (blue line covers red one) marks data of PS13. Correlation equations are reported in Table 8. In each case the LL two phase region lies at the lower left.

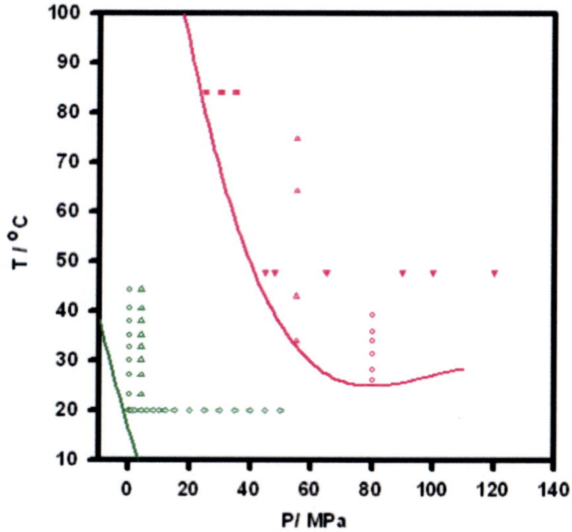

Figure 2. Location of DLS data on the LL demixing diagrams (lines) of near critical polystyrene + protioacetone solutions. Green points and line mark data of PS13k; Pink ponts and line mark data of PS25k (the solutions were prepared under pressure see ref. 17). Correlation equations are reported in Table 8.

Table 4. The measurement parameters and DLS data for polystyrene 13k and 25k in protio acetone

System: PS13k M_w/M_n= 1.06, 20.3wt% in Ac-h				
P / MPa	**0.10**		**4.00**	
Temperature range / ^{0}C	44.3	20.0	44.3	14.3
Relative intensity / arb. units	1010	6581	589	4272
Correlation radius / nm	18.7	67.6	13.3	50.1
10^7diff.coeff. /cm^{2}s^{-1}	24.3	4.5	34.1	5.3
T/ ^{0}C	**20.0**		______	
Pressure range / MPa	50.00	0.1	__	__
Relative intensity / arb. units	162	6581	__	__
Correlation radius / nm	6.9	67.6	__	__
10^7diff.coeff. /cm^{2}s^{-1}	43.6	4.5	__	__
System: PS25k M_w/M_n= 1.06, 13.8wt% in Ac-h				
T / ^{0}C	**47.8**		**84.0**	
Pressure range/ MPa	120.0	45.0	35.0	25.0
Relative intensity / arb. units	24.3	82.9	32.6	47.2
Correlation radius / nm	75.5	208.5	85.0	127.5
10^7diff.coeff. /cm^{2}s^{-1}	16.9	7.8	21.0	14.6
P / MPa	**55.0**		**80.0**	
Temperature range / ^{0}C	74.7	33.9	39.3	26.0
Relative intensity / arb. units	24.9	50.8	33.6	43.5
Correlation radius / nm	22.5	152.0	64.5	200.1
10^7diff.coeff. /cm^{2}s^{-1}	69.7	8.0	17.8	5.3

Table 5. The measurement parameters and DLS data for polystyrene 30k in deutero- methylcyclohexane

System: PS30k M_w/M_n= 1.06, 11.4wt% in MCH-d				
T / ^{0}C	**32.5**		**33.5**	
Pressure range / MPa	80.0	50.0	105	29.1
Relative intensity / arb. units	4097	4627	2533	4958
Correlation radius / nm	559	434	365	282
10^7diff.coeff. /cm^{2}s^{-1}	0.03	0.05	-	0.09
T / ^{0}C	**35.0**		**37.6**	
Pressure / MPa	136	18.9	175	7.9
Relative intensity / arb. units	3137	5158	4081	3920
Correlation radius / nm	457	299	704	181
10^7diff.coeff. /cm^{2}s^{-1}	0.02	0.11	0.01	0.19
T / ^{0}C	**40.7**		**52.0**	
Pressure / MPa	50	0.5	180	0.5
Relative intensity / arb. units	229	2171	128	205
Correlation radius / nm	22	113	35.8	27.7
10^7diff.coeff. /cm^{2}s^{-1}	1.09	0.35	0.34	1.68

Table 6. The measurement parameters and DLS data for polystyrene 30k, 90k and 400k in protiomethylcyclohexane

System: PS30k M_w/M_n= 1.06, 11.5wt% in MCH-h				
P / MPa	**0.1**		**40.0**	
Temperature range / ^{0}C	49.3	34.3	49.3	27.2
Relative intensity / arb. units	-	-	-	-
Correlation radius / nm	19.5	19.5	18.5	126
10^7diff.coeff. /cm^{2}s^{-1}	3.05	3.05	-	0.26
T/ ^{0}C	**30.1**		**34.3**	
Pressure / MPa	50.0	14.0	50.0	0.1
Relative intensity / arb. units	-	-	-	-
Correlation radius / nm	52.5	68.4	35.9	64.4
10^7diff.coeff. /cm^{2}s^{-1}	0.67	0.51	1.10	0.60
System: PS90k M_w/M_n= 1.04, 12.98wt% in MCH-h				
P / MPa	**0.1**		**11.6**	
Temperature range / ^{0}C	56.8	47.2	50.9	47.9
Relative intensity / arb. units	418	3356	72.3	93.3
Correlation radius / nm	29.8	275	23.5	34.6
10^7diff.coeff. /cm^{2}s^{-1}	-	0.18	1.89	1.23
P / MPa	**60.0**		**110.0**	
Temperature range / ^{0}C	47.9	36.2	38.0	37.0
Relative intensity / arb. units	48.5	500	187	418
Correlation radius / nm	23.5	218	114	264
10^7diff.coeff. /cm^{2}s^{-1}	1.24	0.11	0.15	0.06
T/ ^{0}C	**38.0**		**47.9**	
Pressure range / MPa	133.0	25.9	180.0	0.5
Relative intensity / arb. units	367	506	54	398
Correlation radius / nm	198	196	34	50.8
10^7diff.coeff. /cm^{2}s^{-1}	0.07	0.17	0.35	0.89
T/ ^{0}C	**73.2**		**85.0**	
Pressure range/ MPa	135.0	0.5	170	0.5
Relative intensity / arb. units	101	158	41	64
Correlation radius / nm	15.5	26.0	17.0	17.7
10^7diff.coeff. /cm^{2}s^{-1}	1.7	2.6	1.48	4.55
System: PS400k M_w/M_n= 1.06, 8.9wt% in MCH-h				
P / MPa	**4.0**		**30.0**	
Temperature range / ^{0}C	51.0	46.8	46.2	42.4
Relative intensity / arb. units	74.4	200	55.5	149
Correlation radius / nm	65	185	67.5	135
10^7diff.coeff. /cm^{2}s^{-1}	0.74	0.24	0.51	0.26
T/ ^{0}C	**53.1**		**-**	
Pressure range / MPa	50.1	3.7	-	-
Relative intensity / arb. units	27.7	64.7	-	-
Correlation radius / nm	33.8	47.0	-	-
10^7diff.coeff. /cm^{2}s^{-1}	1.01	1.05	-	-

Table 7. The measurement parameters and DLS data for polystyrene 90k in protio- cyclohexane

System: PS90k M_w/M_n= 1.04, 12.98wt% in CH-h				
P / MPa	**0.1**		**50.0**	
Temperature range / ^{0}C	34.8	20.1	67.5	32.4
Relative intensity / arb. units	53.0	45.3	17.0	41.0
Correlation radius / nm	25.0	129	9.5	60.0
10^7diff.coeff. /cm^2s^{-1}	1.2	0.17	3.31	0.32
T/ ^{0}C	**21.2**		**25.0**	
Pressure / MPa	26.6	0.5	31.2	0.5
Relative intensity / arb. units	787	268	157	153
Correlation radius / nm	382	181	164	71.0
10^7diff.coeff. /cm^2s^{-1}	0.05	0.13	0.11	0.35
T/ ^{0}C	**40.0**		**67.5**	
Pressure / MPa	60.0	0.5	124.0	0.5
Relative intensity / arb. units	30.0	25.0	19.0	23
Correlation radius / nm	20.5	14.5	9.0	8.5
10^7diff.coeff. /cm^2s^{-1}	0.94	2.75	1.9	6.60

Table 8. Correlation equations of LL demixing diagrams for polystyrene solutions

PS7.5k + AC-d (20.3wt%)	$T = 13.33 - 1.362*P$		(a)
PS11.6k + Ac-d (20.3wt%)	$T = 49.11 - 2.314(P) + 2.27*10^{-2}(P)^2$		(a)
PS13k+Ac-d (20.3wt%)	$T = 49.11 - 2.314(P) + 2.27*10^{-2}(P)^2$		(a)
PS13k + AC-h (20.3wt%)	$T = 16.33 - 2.02(P) + 1.80*10^{-2}(P)^2$		(a)
PS25k + Ac-h (20.3wt%)	$T = 168.1 - 4.676(P) + 4.97*10^{-2}(P)^2 - 1.704*10^{-4}(P)^3$		(a)
PS400k + MCH-h (8.9wt%)	$T = 49.3 - 0.46(P) + 9.76*10^{-3}(P)^2 - 7.2555*10^{-5}(P)^3$	$-40<P/MPa<50$	(b)
PS30k + MCH-d (11.4wt%)			(b)
	$T = 39.08 - 0.264(P) + 2.97*10^{-3}(P)^2 - 9.24*10^{-5}(P)^3$	$-40<P/MPa<108$	(b)
	$T = 33.63 + 0.058 (P-108)$	$108<P/MPa<200$	(b)
PS30k + MCH-h (11.5wt%)			(b)
	$T = [T(LLPS30k/MCH-d)] - 6.2$	$-40<P/MPa<1200$	(b)
PS90k + MCH-h (12.98wt%)			
	$T = 47.2499 - 0.4996(P) + 8.1457*10^{-3}(P)^2 - 5.7337*10^{-5}(P)^3 + 1.5342*10^{-7}(P)^4$	$-40<P/MPa<200$	(this chapter)
PS90k + CH-h (12.98wt%)			
	$T_{LL} = 19.8 + 0.045(P)$		(b)
	$P_{LS} = -12.4623 + 1.79268 (T) + 4.042 *10^{-3}(T)^2$		(c)

In all given above relations T is temperature in ^{0}C and P pressure in MPa.

(a) Van Hook et al., (1999)

(b) Van Hook, Wilczura and Rebelo, (1999)

(c) Matsum et al., (1984)

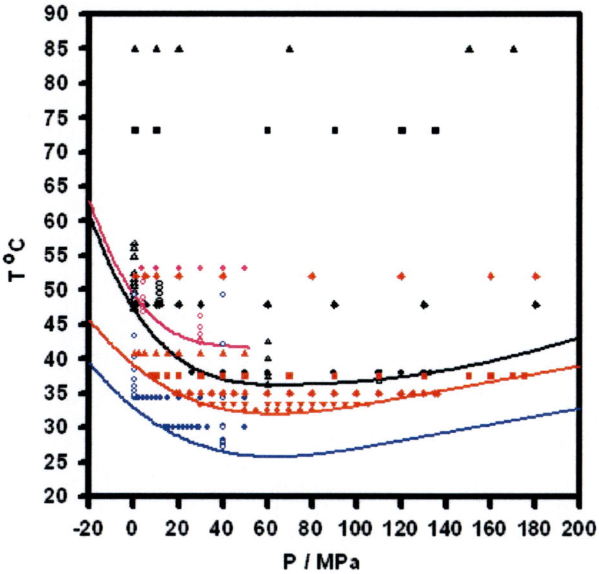

Figure 3. DLS data location on the LL demixing diagrams for polystyrene + methylcyclohexane and + deuteromethylcyclohexane solutions. Blue points and line mark PS30k in protio- methylcyclohexane solutions; Red poins and line mark PS30k in deuteromethylcyclohexane solutions; Black points and line mark PS90k in protio methylcyclohexane solutions; Pink points and line mark PS400k in protiomethylcyclohexane solutions. Correlation equation are reported in Table 8.

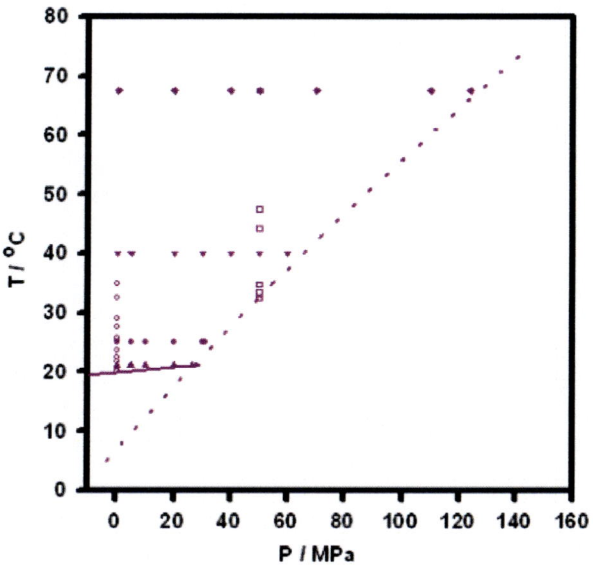

Figure 4. Location of DLS data (points) on the LL demixing diagram for polystyrene 90k in cyclohexane solutions (the graph reproduces part of the Figure 3a reported earlier by Van Hook et. al, [30]). Interrupted line is fit to low-pressure LS transitions and extrapolated [44]. The continuous line shows the LLE line for this system. Correlation equations are reported in Table 8.

The data reported in Tables 2 ÷ 7 establish that diffusion coefficients decrease markedly as the LL demixing locus is approached in θ solvents and poor solvents, probably due to increased polymer chains entanglement at lower temperature. Chu et. al., (1987) reported static (correlation length, mesh size) and dynamic (cooperative diffusion coefficient) properties of the polymer matrix for a quaternary polymer solution system containing of polystyrene, polymethyl methacrylate, toluene and α-chloronaphtalene. They found that even if the polymer molecules in semidilute solution form entanglements low molecular weight fractions still have the opportunity to move by translation diffusion. De Gennes (1976) described polymer entanglements using a reptation model. In a dilute solution each isolated chain has its own translation diffusion describing motion of the center mass as well as the internal modes. In the semidilute solution the situation is different. The motion of each chain is affected by entangled neighbouring polymer chains of comparable sizes. At sufficiently high concentrations, C>>C*, the polymer chains moves by reptation through a tube dynamically formed by entangled neighbouring polymer chains. Consequently the reptation motion should be corrected for the tube length fluctuation effect. Brown (1984) reported data on polymer dynamics in semidilute polyethylene oxide in water, and polystyrene (PS110k) in cyclopentane solutions. For cyclopentane which is a θ solvent they found at 20.4^{0}C (θ conditions) only a single-exponential relaxation in the concentration below 17%. Diffusion coefficients values were in the range from 5.91x10^{-7} cm^{2}s^{-1} till 3.11x10^{-7} cm^{2}s^{-1} for polystyrene concentrations 0.55% ÷ 16.9% suitable. These results are comparable to our values: D=4.55x10^{-7} cm^{2}s^{-1} for PS90k in MCH-h (12.98%) at 85^0 C under P= 0.5MPa, and D=2.75x 10^{-7} cm^{2}s^{-1} for PS90k in CH-h (12.98%) at 40^0 C under P=0.5MPa.

In the systems discussed in this chapter polymer concentrations barely exceed the overlap concentration C*, *i.e.* C>C* but not C>>C* . Under such conditions we can assume that polymer molecule is essentially surrounded by small solvent molecules. Phase separation follows spinodal decomposition/percolation. The numerical values of the cooperative diffusion coefficients indicate on relatively free diffusive motion (translation) of single polystyrene coils in the homogeneous regions. Our results are similar to those of Gold et. al., (1996) who studied the effect of solvent quality on the diffusion of polystyrene latex spheres in polymethyl methacrylate solutions. They had found that solvent quality is a crucial parameter in determining how the spheres behave in the context of the Stoke's-Einstein equation. Our data confirms that conclusion.

A brief discussion of the data presented in Figures 1÷4 is appropriate. In each case the deuterated isotopomer is a worse solvent than the protio. On Figure 3 the superposition of isopleths for PS30k, 90k and 400k in protio, and PS30k in deutero methylcyclohexane is shown. The one phase region in the deuterated solvent is smaller, *i.e.* deuterated methylcyclohexane is a worse solvent for polystyrene than is protio. Increasing molecular weight from 30k through 90k to 400k moves the liquid-liquid equilibrium line to higher T and P values. The same regularity was observed for polystyrene + acetone systems.

The DLS data-net location on the LL demixing diagrams for polystyrene in cyclohexane is qualitatively different (see Figure 4) than that for methylcyclohexane solutions. The DLS data are complicated by the fact they extended to pressures high enough to induce the liquid-solid solvent transition. At freezing point both scattered and transmitted light intensities drop discontinuously ending the experiment (see Figure 4) (Van Hook, Wilczura, Rebelo 1996).

In the other graphs reported in the following section scattered light intensities (I), correlation radii (ξ) and cooperative diffusion coefficients (D) are shown in three dimensional projections, I = I(T, P), ξ = ξ(T,P), and D = D(T,P). Figures 5 shows the isothermal data for PS7.5k+deuteroacetone solution (20.3wt%). The distinct rate of increase of scattered light intensity as pressure decreases is observed from 12^0C to 52.1^0 C (see Figure 5a). The cooperative diffusion coefficient as a function of correlation radius and pressure are shown in Figure 5b. In the homogeneous region at not too high a temperature the diffusion coefficient doesn't change much with pressure. However when demixing point is approached (during depressurising) the correlation radii which describe polystyrene chains entanglements increase, and the diffusion coefficients drop. The situation is different at higher temperature, where the solution is homogenous over the complete pressure range investigated (see Table 3, 5 and 6, and Figures 5b, 9b and 10b).

Scattered light intensity is a sensitive indicator of particle size. Comparing two isobars for PS11.6+deuteroacetone (20.3wt%) (Figure 6), one can see a noticeable increase in scattered light intensity at 70^0 C under 0.1MPa, while at 20MPa at the same temperature the polystyrene solution is still homogeneous (scattering intensity is low). Such regularity occurs also in all other polystyrene solutions under discussion.

The effect of molecular weight on particle size during the transition from the homogeneous region to critical demixing is shown in Figures 7 and 8 for polystyrene + deuteroacetone solutions. Such three dimensional diagrams clearly show the progressive growth of polymer chain entanglements during temperature/pressure quenching, and that their size rapidly increases in the vicinity of the LL demixing.

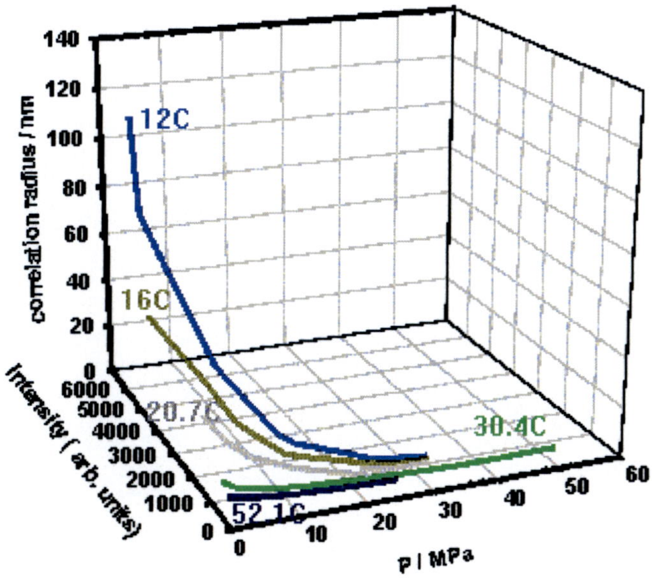

Figure 5a. 3D projection of correlation radii for PS7.5k in deuteroacetone as a function of scattered intensity and pressure - isothermal data.

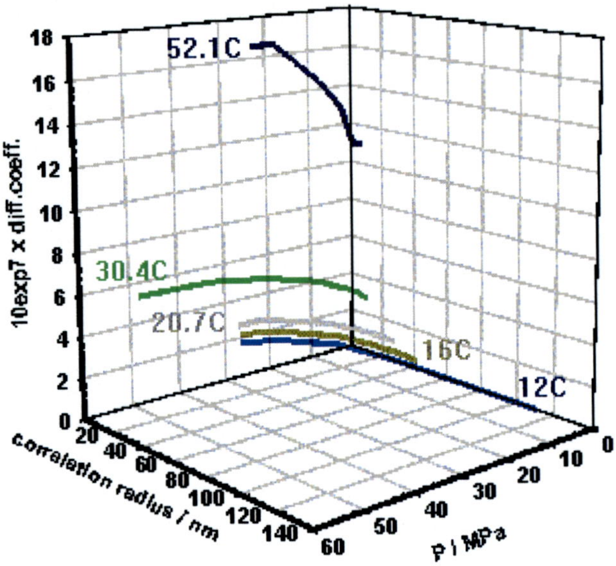

Figure 5b. 3D projection of diffusion coefficients for PS7.5k in deuteroacetone as a function of correlation radii and pressure - isothermal data.

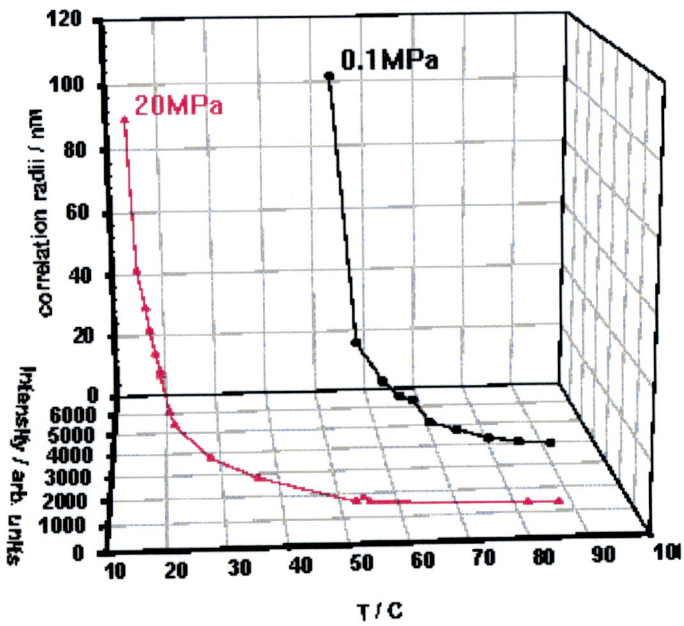

Figure 6. 3D projection of correlation radii for PS16.6k in deuteroacetone as a function of scattered intensity and temperature - isobaric data.

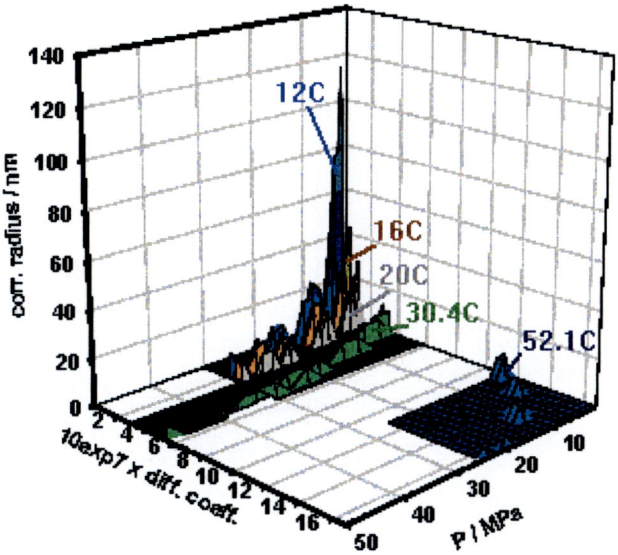

Figure 7. Isothermal 3D smoothed data of correlation radii for PS7.5k in deuteroacetone solutions.

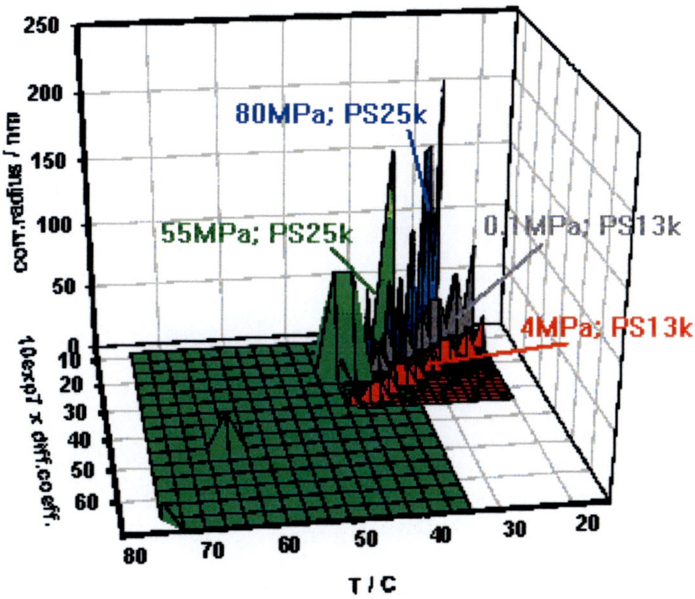

Figure 8. Comparison of isobaric 3D smoothed data of correlation radii for PS13k and 25k in deuteroacetone solutions.

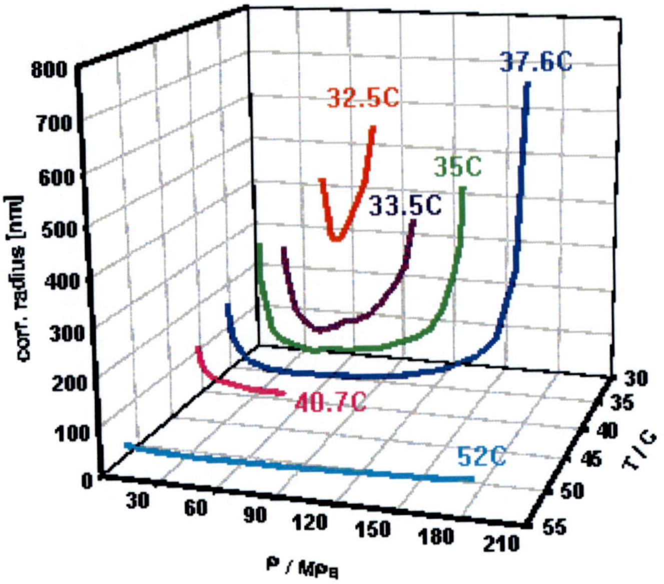

Figure 9a. 3D projection of correlation radii for PS30k in deuteromethylcyclohexane solutions as temperature and pressure functions - isothermal data.

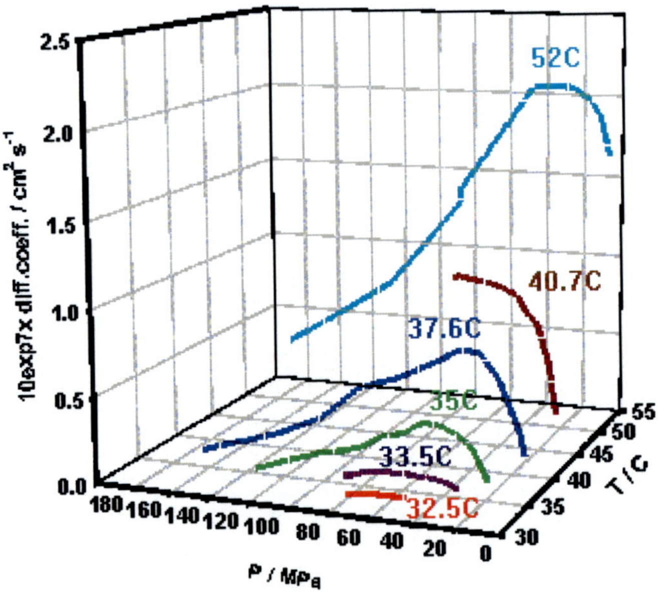

Figure 9b. 3D projection of diffusion coefficients for PS30k in deuteromethylcyclohexane solutions as temperature and pressure functions - isothermal data.

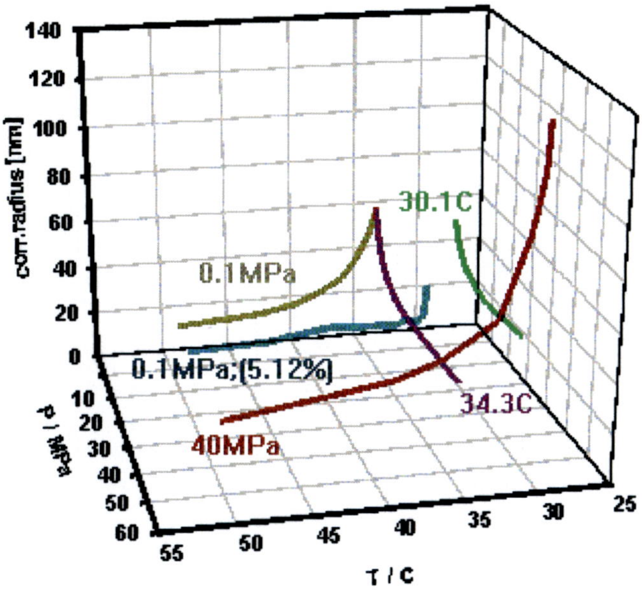

Figure 9c. 3D projection of correlation radii for PS30k in protiomethylcyclohexane solutions as a temperature and pressure functions - isotherms and isobars.

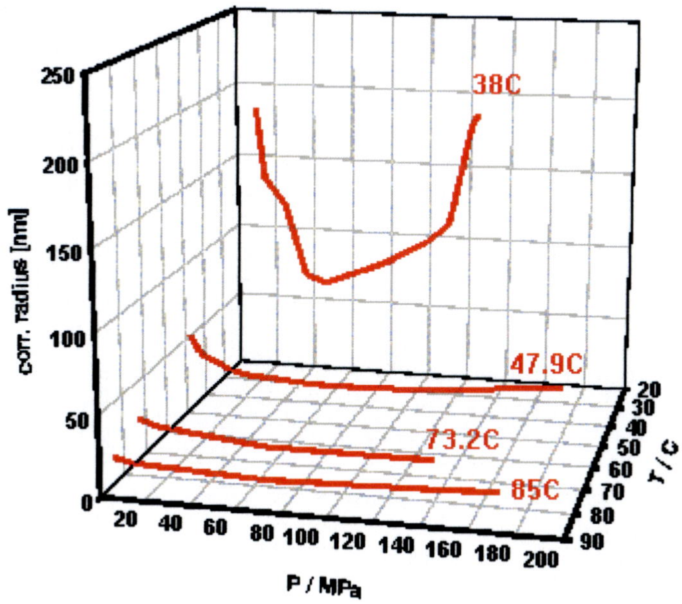

Figure 10a. 3D projection of correlation radii for PS90k in protiomethylcyclohexane solutions - isotherms.

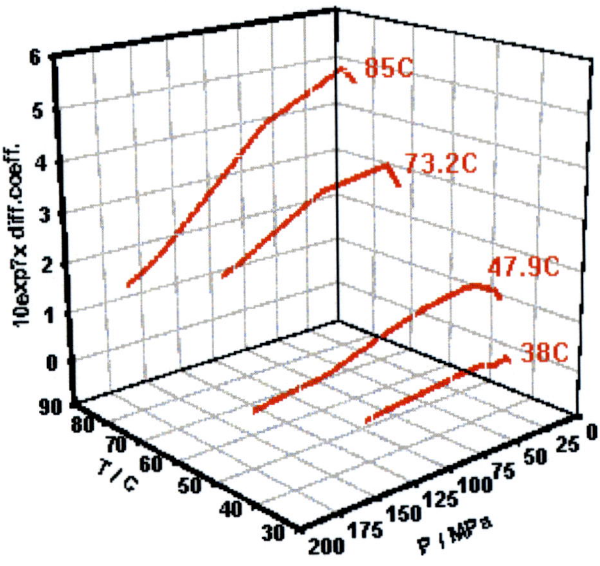

Figure 10b. 3D projection of diffusion coefficients for PS90k in protiomethylcyclohexane solutions - isotherms.

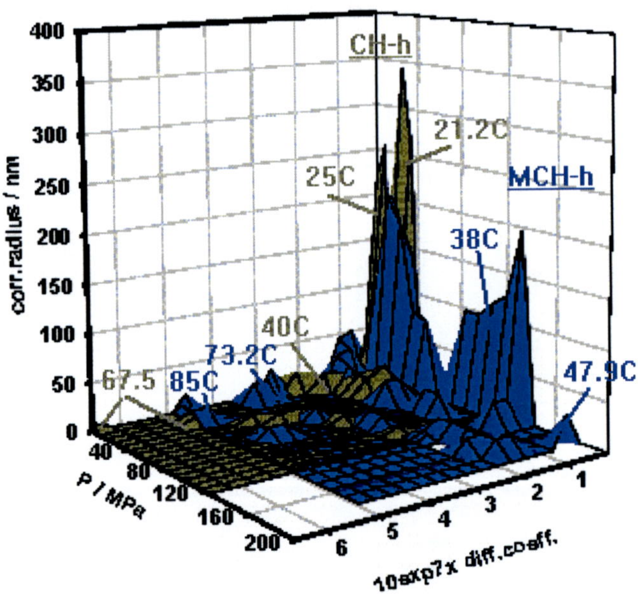

Figure 11. 3D smoothed data of correlation radii for PS90k in protiomethylcyclohexane (blue surfaces) and in protiocyclohexane (dark green surfaces) solutions - isotherms.

In Figures 9a and 9b the isothermal data for PS30k + deuteromethylcyclohexane and for PS30k + protiomethylcyclohexane are presented. The shape of isotherms which represent correlation radii as a function of pressure clearly supports the existence of a hypercritical (double critical) point which lies outside the present experimental pressure range. The hypothesis is consistent with Figure 8b.

Figures 10a and 10b contain isothermal data (ξ and D) as a temperature and pressure functions for PS90k in protiomethylcyclohexane (12.98wt%). The diffusion coefficient increases with pressure goes through a maximum, and then decreases rapidly when the phase transition is approached. The curves have similar shape to PS30k in deuteromethylcyclohexane solution.

The isothermal 3D smoothed data of correlation radii for PS90k in protiomethylcyclohexane and protiocyclohexane as functions of pressure and diffusion coefficient are compared on Figure 11. The particle sizes in cyclohexane at 67.5^0 C are smaller than their values in methylcyclohexane at 85^0 C. For polystyrene in cyclohexane solution was possible to detect LL demixing at 21.2^0 C when for methylcyclohexane solution the lower temperature was 38^0 C (the polymer concentration was equal in both solutions). This confirms that the cyclohexane is a better solvent for polystyrene than is methylcyclohexane.

CONCLUSIONS

This chapter discusses DLS data (scattered light intensity, correlation radius and diffusion coefficients) for polystyrene + solvent systems. The solutions contained polystyrene with molecular weights in the range 7.5k ÷ 400k amu. in acetone, cyclohexane, methylcyclohexane and their deuterated homologues. DLS data were recorded for semidilute polymer solutions under isothermal and/or isobaric conditions starting from homogeneous region. The LL phase transition was induced with pressure and temperature quenching. The polymer solutions concentrations were below the value at which DLS nondiffusive slow modes develops.

The scattered light intensity (I), correlation radius (ξ) and cooperative diffusion coefficient (D) of the diffusive mode as a functions of temperature and pressure are shown on the three coordinate diagrams. For chosen polystyrene solutions which were: PS25k+Ac-h; PS13k+Ac-h; PS7.5k+Ac-d; PS90k+MCH-h and PS90k+CH-h the correlation radii (geometrical parameter) as a function of pressure and diffusion coefficient (dynamic parameter) data are smoothed with 3D negative exponential. The DLS data clearly show the progressive growth of polymer chain entanglements in the vicinity of the LL demixing.

All solutions under discussion satisfied the criteria $c>c^*$ which indicate that polymer coils are surrounded by small solvent molecules. The correlation radii of polystyrene chains entanglements rapidly increase when demixing point is oncoming and as a consequence their diffusion coefficients diminish. In homogenous regions the opposite situation was observed. The numerical values of diffusion coefficient indicate on relatively free translation of single polymer chains in all solvents under consideration. Our results confirmed also the hypothesis that the solvent quality is the most important parameter in determining how polymer coils behave in the context of Stokes-Einstein equation.

REFERENCES

Adam, M.; Delsanti M. *Macromolecules* **1985**, 18, 1760-1770.

Akcasu.A.Z. In Book Brown, W. Nicolai T. *Dynamic Light Scattering: The Method and Some Application*"; Ed. Clarendon Press; Oxford University Press, **1993**; Chapter 1, pp 1-75.

Amirzadeh, J.; Mc Donnell, M.E. *Macromolecules* **1982**, 15, 927-933.

Bridgman, P.W. "*The Physics of High Pressure*" Ed. G.Bell and Sons; London, **1958**.

Brochard, F. *J.Phys.* **1983**, 44, 39-47.

Brochard, F.; de Gennes P.G. *Macromolecules* **1979**, 10, 1157-1161.

Brown W., Nicolai.T. In Book .Brown, W. Nicolai T. *Dynamic Light Scattering: The Method and Some Applications;* Ed. Clarendon Press; Oxford University Press, **1993**; Chapter 6, pp 272-318.

Brown, W. *Macromolecules* **1984**, 17, 66-72.

Chen, S.J.; Berry, G.C.; *Polymer* **1990**, 31, 793-801.

Chu, B.; Wu, D. *Macromolecules* **1987**, 20, 1606-1619.

Cotton, J.P.; Nierlich, M.; Boue, F.; Daoud, M.; Farnoux, B.; R.Duplesix, C.Picot *J.Chem.Phys*. **1976**, 65, 1101-1108.

Daoud, M.; Cotton, J.P.; Farnoux, B.; Jannink, G.; Sarma, G.; Benoit, H.; Duplesix, R.; Picot, C.; de Germes, P.G.; *Macromolecules* **1975**, 8, 804-818.

de Gennes *Macromolecules* P.G. **1976**, 9, 587-593.

de Gennes *Macromolecules* P.G. **1976**, 9, 594-598.

Doi, M.; A.Onuki, A. *J.Phys. II. Fr.* **1992**, 2, 1631-1639.

Fytas, G.; Patkowski, A.; Meier, G.; Dorfmuller, T. *Macromolecules* **1982**, 15, 870-874.

Fytas, G.; Patkowski, A.; Meier, G.; T.Dorfmuller, T. *J.Chem.Phys*. **1984**, 80, 2214-2220.

Genz, U. *Macromolecules* **1994**, 27, 3501-3512.

Gold, D.; C.Onyenemezu, C.; W.G. Miller, W.G. *Macromolecules* **1996**, 29, 5700-5709.

Hook, W.A. Wilczura, H.; Rebelo, L.P. *Macromolecules* **1999**, 32, 7299-7311.

Ikier, C.; Klein, H.; Woermann, D. *Macromolecules* **1995**, 28, 1003-1007.

Jlan, T.; Vlassopoulos, D.; Fytas, G.; Pakula, T.; Brown, W. *Colloid.Polym.Sci.* **1996**, 274, 1033-1039.

Kiepen, F.; Borchard, W. *Makromol.Chem.* **1988**, 189, 2595-2608.

Koppel, D.E.; *J.Chem. Phys.* **1972**, 57, 4814-4820

Lechner, M.D.; Mattern, R. *Macromol. Chem.* **1991**, 123/124, 45-54.

Matsum, S.; Van Hook, W.A. *J.Phys.Chem.* **1984**, 88, 1032-1041.

Meier, G.; Fytas, G.; Dorfmuller, T. *Macromolecules* **1984**, 17, 957-961.

Melnichenko, Y.; Wignall, G.D.; Van Hook, W.A.; Szydlowski, J.; Wilczura, H.; Rebelo, L.P. *Macromolecules* **1998**, 31, 8436-8438.

Noda, L.; Koto, N.; T.Kitano, T.; Nagasawa, M. *Macromolecules* **1981**, 14, 668-676.

Provencher, S.W.; *Comput.Phys.Commun.* **1982**, 27, 213-220.

Rabinowitch, I.B.; "*Influence of isotopy on the physicochemical properties of liquids*" Ed. Consultants Bureau, New York, **1970**.

Schätzel, K. In Book .Brown, W.; Nicolai T. "*Dynamic Light Scattering:The Method and Some Applications*"; Ed. Clarendon Press; Oxford University Press, **1993**; Chapter 2, pp 76-148.

Semenov, A.N.; *Physica A* **1990**, 166, 263-275.

Smith, S.W.; Freeman, B.D.; Hall, C.K. *Macromolecule* **1997**, 30, 2052-2057

Steinhoff, B.; Rullmann, M.; Kuhne, L.; Alig , I. *J.Chem.Phys* **1997**, 107, 5217-5226.

Stepanek, P.; Perzynski, R.; Delsanti, M.; Adam, M. *Macromolecules* **1984**, 17, 2340-2343.

Sun, S.F. "*Physical Chemistry of Macromolecules*" John Wiley and Sons; New York **1994**, Chapter 10.

Sun, Z.; Wang, C.H. *Macromolecules* **1994**, 27, 5667-5673.

Szydlowski*, J.; Rebelo, L.P.; Wilczura, H.; Van Hook,W.A.; Melnichenko, Y.; Wignall, G.D. *Fluid Phase Equilibria* **1998**, 150-151, 687-694.

Szydlowski, J.; Rebelo, L.P.; Wilczura, H.; Melnichenko, Y.; Wignall, G.D.; Van Hook, W.A. *Physica B* **1998**, 241-243, 1035-1037.

Szydlowski, J.; W.A.Van Hook, W.A. *Macromolecules* **1998**, 31, 3255-3265.

Szydlowski, J.; W.A.Van Hook, W.A. *Macromolecules* **1998**, 31, 3266-3274.

Tanaka, T.; Hocker, L.;O, Benedek, G.B. *J.Chem.Phys.* **1973**, 59, 5151 .

Tanaka, Y.; Husakawa, H.; Kubota, H.; Makita, T. *Int. J. Thermophys*. **1991**, 12, 245-252.

Van Hook, W.A.; Wilczura, H.; Imre, A.; Rebelo, L.P.; Melnichenko, Y. *Macromolecules* **1999**, 32, 7312-7318.

Wiltzius, P.; Haller, H.R.; Cannell, D.S.; Schaefer, D.W. *Phys.Rev.Lett*. **1983**, 51, 118.

Xiong, Y.; Kiran, E. *Rev. Sci. Instrum.* **1998**, 69, 1463-1472.

Yaws, C.L. Ed. *Handbook of Viscosity*, Gulf Publ.Co. Houston TX, **1995**.

Ying, Q.; Chu, B. *Macromolecules* **1987**, 20, 362-366.

INDEX

C

D

E

F

G

H

I

K

L

M

N

O

P

Q

R

S

T

U

V

W

X